Chiral Separation Techniques

Edited by
G. Subramanian

Chiral
Separation Techniques

A Practical Approach

Second, completely revised and
updated edition

Edited by
G. Subramanian

**WILEY-VCH**

Weinheim · Chichester · New York · Toronto · Brisbane · Singapore

Dr. Ganapathy Subramanian
60B Jubilee Road
Littlebourne
Canterbury
Kent CT3 1TP, UK

Library of Congress Card No. applied for

A catalogue record for this book is available from the British Library

Die Deutsche Bibliothek – CIP Cataloguing-in-Publication-Data
A catalogue record for this publication is available from Die Deutsche Bibliothek

© WILEY-VCH Verlag GmbH, D-69469 Weinheim (Federal Republic of Germany), 2001

ISBN 3-527-29875-4

Composition: TypoDesign Hecker GmbH, D-69181 Leimen
Printing: Strauss Offsetdruck, D-69509 Mörlenbach
Bookbinding: Osswald & Co., D-67433 Neustadt (Weinstraße)

Printed in the Federal Republic of Germany.

Preface

During the past two decades there has been intense interest in the development and application of chiral chromatographic methods, particularly in the pharmaceutical industries. This is driven both by desire to develop and exploit "good science" and by the increasing pressure by regulatory authorities over the past ten years against the marketing of racemic mixtures. The regulation of chiral drug provides a good demonstration of the mutual relationship between progress in scientific methodology and regulatory guidelines. It has also provided a common platform in establishing good understanding between international regulatory authorities and pharamceutical industries, leading to a consensus in recognition of the global nature of pharmaceutical development. This has provided a great challenge for the industries to seek techniques that are efficient, economical and easy to apply, in the manufacture of enantiopure products.

The versatility of chiral stationary phases and its effecitve application in both analytical and large-scale enantioseparation has been discussed in the earlier book 'A Practical Approach to Chiral Separation by Liquid Chromatography' (Ed. G. Subramanian, VCH 1994). This book aims to bring to the forefront the current development and sucessful application chiral separation techniques, thereby providing an insight to researchers, analytical and industrial chemists, allowing a choice of methodology from the entire spectrum of available techniques.

I am indebted to the leading international group of contributors, who have agreed to share their knowlegde and experience. Each chapter represents an overview of its chosen topic. Chapter 1 provider an overview of techniques in preparative chiral separation, while Chapter 2 provides an account on method development and optimisation of enantiomer separation using macrocyclic glycopeptide chiral stationary phase. Combinatorial approach and chirabase applications are discussed in Chapters 3 and 4. Chapter 5 details the development of membranes for chiral separation, while Chapter 6 gives an overview of implanting techniques for enantiopurification. Non chromatographic solid-phase purification of enantiomers is explained in Chapter 7, and Chapter 8 discusses modeling and simulation of SMB and its application in enantioseparation. A perspective on cGMP compliance for preparative chiral chromatography in discussed Chapter 9, and Chapter 10 provides an account of electrophoretically driven preparative chiral separation and sub- and supercritical fluid chromatography for enentioseparation is explained in Chapter 11. An insight into International Regulation of chiral drugs is provided in Chapter 12.

It is hoped that the book will be of value to chemists and chemical engineers who are engaged in the manufacture of enantiopure products, and that they will sucessfully apply some of the techniques described. In this way, an avenue will be provided for further progess to be made in this important field.

I wish to express my sincere thanks to Steffen Pauly and his colleagues for their enthusiasm and understanding in the production this book.

Canterbury, Kent, UK   *G. Subramanian*
April, 2000

Contents

List of Authors

Thomas E. Beesley
Advanced Separation Technologies, Inc.
37 Leslie Court
P. O. Box 297
Whippany, NJ 07981
USA

Jerald S. Bradshaw
Department of Chemistry and
Biochemistry
Brigham Young University
Provo, UT 84602
USA

Sarah K. Branch
Medicines Control Agency
Market Towers
1 Nine Elms Lane
London SW8 5NQ
UK

Y. L. Bruenning
IBC Advenced Technologies, Inc.
856 East Utah Valley Drive
P. O. Box 98
American Fork, UT 84003
USA

Jean M. J. Fréchet
Department of Chemistry
University of California
736 Latimer Hall
Berkeley, CA 94720-1460,
USA

Ingolf Heitmann
ENSSPICAM
University Aix-Marseille III
Avenue Escadrille Normandie-Niemen
13397 Marseille Cedex 20
France

Neil E. Izatt
IBC Advanced Technologies, Inc.
856 East Utah Valley Drive
P. O. Box 98
American Fork, UT 84003
USA

Reed M. Izatt
Department of Chemistry and
Biochemistry
Brigham Young University
Provo, UT 84602
USA

M. F. Kemmere
 Process Development Group
Department of Chemical Engineering
and Chemistry
Eindhoven University of Technology
P. O. Box 513
5600 MB Eindhoven
The Netherlands

Jos T.F. Keurentjes
Process Development Group
Department of Chemical Engineering
and Chemistry
Eindhoven University of Technology
P. O. Box 513
5600 MB Eindhoven
The Netherlands

K. E. Krakoviak
IBC Advanced Technologies, Inc.
856 East Utah Valley Drive
P. O. Box 98
American Fork, UT 84003
USA

J. T. Lee
Advanced Separation Technologies, Inc.
37 Leslie Court
P. O. Box 297
Whippany, New Jersey 07981
USA

Christina Minguillón
Laboratory Quimica Farmacia
Facultat de Farmacia
University of Barcelona
E-08028 Barcelona
Spain

Roger M. Nicoud
Novasep SAS
15, Rue du Bois de la Champelle
Parc Technologique de Brabois
B. P. 50
54502 Vandoeuvre-lès-Nancy Cedex,
France

Luís S. Pais
Laboratory of Separation and Reaction
Engineering
Faculty of Engineering
University of Porto
Rua dos Bragas
4050-123 Porto
Portugal

Scott R. Perrin
Novasep Inc.
480 S. Democrat Road
Gibbstown, NJ 08027-1297
USA

Karen W. Phinney
Analytical Chemistry Division
Chemical Science and Technology
Laboratory
National Institute of Standards and
Technology
100 Bureau Drive, Stop 8392
Gaithersburg, MD 20899-8392
USA

Johanna Pierrot-Sanders
ENSSPICAM
University Aix-Marseille III
Avenue Escadrille Normandie-Niemen
13397 Marseille Cedex 20
France

Patrick Piras
ENSSPICAM
University Aix-Marseille III
Avenue Escadrille Normandie-Niemen
13397 Marseille Cedex 20
France

Alírio E. Rodrigues
Laboratory of Separation and Reaction
Engineering
Faculty of Engineering
University of Porto
Rua dos Bragas
4050-123 Porto
Portugal

Christian Roussel
ENSSPICAM
University Aix-Marseille III
Avenue Escadrille Normandie-Niemen
13397 Marseille Cedex 20
France

Michael Schulte
Merck KGaA,
SLP Fo BS
Frankfurter Str. 250
D-64271 Darmstadt
Germany

Börje Sellergren
Department of Inorganic Chemistry and
Analytical Chemistry
Johannes Gutenberg University
Duesbergweg 10–14
55099 Mainz
Germany

Apryll M. Stalcup
Department of Chemistry
University of Cincinnati
P. O. Box 210172
Cincinnati, OH 45221-0172
USA

Ganapathy Subramanian
60 B Jubilee Road
Littlebourne
Kent CT3 1TP
UK

Frantisek Svec
Department of Chemistry
736 Latimer Hall
University of California
Berkeley, CA 94720-1460
USA

Andy X. Wang
Advanced Separation Technologies, Inc.
37 Leslie Court
P. O. Box 297
Whippany, NJ 07981
USA

Dirk Wulff
Department of Chemistry
University of California
736 Latimer Hall
Berkeley, CA 94720-1460
USA

Patrick Piras
ENSSPICAM
University Aix-Marseille III
Avenue Escadrille Normandie-Niemen
13397 Marseille Cedex 20
France

Alírio E. Rodrigues
Laboratory of Separation and Reaction
Engineering
Faculty of Engineering
University of Porto
Rua dos Bragas
4050-123 Porto
Portugal

Christian Roussel
ENSSPICAM
University Aix-Marseille III
Avenue Escadrille Normandie-Niemen
13397 Marseille Cedex 20
France

Michael Schulte
Merck KGaA,
SLP Fo BS
Frankfurter Str. 250
D-64271 Darmstadt
Germany

Börje Sellergren
Department of Inorganic Chemistry and
Analytical Chemistry
Johannes Gutenberg University
Duesbergweg 10–14
55099 Mainz
Germany

Apryll M. Stalcup
Department of Chemistry
University of Cincinnati
P. O. Box 210172
Cincinnati, OH 45221-0172
USA

Ganapathy Subramanian
60 B Jubilee Road
Littlebourne
Kent CT3 1TP
UK

Frantisek Svec
Department of Chemistry
736 Latimer Hall
University of California
Berkeley, CA 94720-1460
USA

Andy X. Wang
Advanced Separation Technologies, Inc.
37 Leslie Court
P. O. Box 297
Whippany, NJ 07981
USA

Dirk Wulff
Department of Chemistry
University of California
736 Latimer Hall
Berkeley, CA 94720-1460
USA

1 Techniques in Preparative Chiral Separations

Pilar Franco and Cristina Minguillón

1.1 Introduction

The recognition of differences in the pharmacological activity of enantiomeric molecules has created the need to administer them – and therefore to obtain them – as isolated enantiomers. However, nowadays this problem affects not only the pharmaceutical industry, but also the agrochemical industry and food additive producers, both of which are increasingly concerned by this subject.

When chiral, drugs and other molecules obtained from natural sources or by semisynthesis usually contain one of the possible enantiomeric forms. However, those obtained by total synthesis often consist of mixtures of both enantiomers. In order to develop commercially the isolated enantiomers, two alternative approaches can be considered: (i) enantioselective synthesis of the desired enantiomer; or (ii) separation of both isomers from a racemic mixture. The separation can be performed on the target molecule or on one of its chemical precursors obtained from conventional synthetic procedures. Both strategies have their advantages and drawbacks.

The separation of the enantiomers of a racemic mixture, when only one of them is required, implies an important reduction in yield during the production step of the target molecule. Techniques to racemize and recycle the unwanted enantiomer are used to reduce the extent of this problem. However, the same fact becomes an advantage in the development step of a drug, because it is the quickest way to have available both enantiomers in order to carry out the individual tests needed. In fact, even if the separation/racemization approach is considered to be "not elegant" by organic synthetic chemists, it is nowadays the most often used for the production of single enantiomers. The enantioselective synthetic approach has the main disadvantage of the cost and time that could take the development of a synthetic path leading to the desired enantiomer. Moreover, often the enantiomeric excesses obtained from an enantioselective procedure are not sufficient to fulfil the requirements of the regulatory authorities. In that case, an enrichment step must be added to the enantioselective process.

All separation techniques which allow the isolation of a certain amount of product can be qualified as being "preparative". In contrast, analytical techniques are devoted to detect the presence of substances in a sample and/or quantify them. How-

ever, not all preparative techniques are useful at the same scale; some are more easily adapted to the manipulation of large amounts of material, while others may only
be applicable to the isolation of few milligrams, or even less, though this may be
enough for given purposes.

In this chapter a number of the preparative techniques used in the resolution of
enantiomers is presented. Some of these techniques will be developed more fully in
following chapters.

1.2 Crystallization Techniques

Although crystallization is used routinely to separate solid compounds from impurities and by-products coming from secondary reactions in their synthesis, it may also
be applied in the isolation of individual enantiomers from a racemic or an enantiomerically enriched sample [1–3]. Indeed, until the development of chiral chromatographic techniques, crystallization was one of the few existing ways to resolve
enantiomers. Although crystallization is a very powerful technique for preparative
purposes, few industrial applications have been reported [3] for reasons of confidentiality. Moreover, the technique is far from being generally applicable, and thus
only those compounds which behave as conglomerates (different crystals for both
enantiomers) can be resolved from their equimolecular enantiomeric mixtures, either
by seeding their solutions with crystals of one enantiomer (*preferential crystallization*) [4–6], or by using a chiral environment to carry out the crystallization. The limitation of the preferential crystallization is, therefore, the availability of crystals of
the pure enantiomer.

A chiral environment can be produced by using a chiral solvent in the crystallization, but most of these are organic and therefore not useful for highly polar compounds. Therefore, a chiral co-solute is often used [7–9]. Applying this methodology, d,l-threonine was resolved into its enantiomers using small amounts of L-serine
or 4-(*R*)-hydroxy-L-proline [7]. Moreover, an inhibitory effect on the crystallization
of D-glutamic acid from its racemic mixture of some D-amino acids, such as lysine,
histidine or arginine has been described, while their L-counterparts inhibit crystallization of the L-enantiomer [8].

Unfortunately, the occurrence of conglomerates in nature is not common. On
occasion, the probability of obtaining a conglomerate can be increased by trans-
forming the considered compound into a salt, when possible. Racemic compounds
(both enantiomers in the same lattice) are more frequently encountered in nature.
Therefore, it is useful to know which is the behavior of the considered product, taking into account that it can change depending on the temperature of crystallization.
Several methods exist to determine such a point easily [3]. Racemic compounds may
be enriched by crystallization of a nonracemic mixture, in which case the success
and yield of the enrichment depends (among others) on the composition of the original mixture.

On that basis, crystallization is often used in combination with other enantiose-lective techniques, such as enantioselective synthesis, enzymatic kinetic resolution or simulated moving bed (SMB) chromatography [10, 11]. In general, when refer-ring to crystallization techniques, the aim is to obtain an enantiomeric enrichment in the crystallized solid. However, the possibility of producing an enrichment in the mother liquors [12, 13], even if this is not a general phenomenon [14], must be taken into account.

An additional strategy that is frequently used due to the reduced probability of the preceding situations is the separation of diastereomeric mixtures obtained from the reaction of the original enantiomeric mixture with chiral derivatizing agents [3, 15–18]. These should be easily cleaved from the target molecule with no racemiza-tion, and thus be readily available. Low cost and confirmed enantiomeric purity, in addition to their being recoverable and reusable, are also highly recommended prop-erties for a chiral derivatizing agent.

1.3 Chromatographic Techniques

1.3.1 Liquid Chromatography

Despite the fact that in former days liquid chromatography was reputed to be a very expensive and inefficient purification technique for preparative purposes, it is nowa-days one of the first choices to carry out a large-scale chiral separation. On the one hand, some technical developments, related to the equipment as well as to the pack-ing materials have improved the efficiency of the technique. On the other hand, applications such as the resolution of enantiomers, where the resulting products have a high added value, can partially balance the classically attributed high costs of li-quid chromatography. Furthermore, the relative short time necessary to develop a chromatographic method, and the availability of chromatographic systems, are inter-esting features that should be taken into account when the enantiomers under con-sideration need to be separated in the minimum time.

Analogously to crystallization techniques, the chromatographic separation pro-cess can be applied either to a mixture of enantiomers or to diasteromeric derivatives obtained by reaction with chiral derivatizing agents. In this case, it is a conventional chromatographic process which can be performed in achiral conditions, and the same drawbacks as with any other indirect method might be encountered. Thus, such indirect resolutions are strongly dependent on the enantiomeric purity of the deriva-tizing agent which must be cleaved without affecting the configuration of the stereo-genic elements in the target molecule.

Several chromatographic modes will be reviewed in this respect, and most will make use of a chiral support in order to bring about a separation, differing only in the technology employed. Only countercurrent chromatography is based on a li-quid–liquid separation.

1.3.1.1 High-Pressure/Medium-Pressure Liquid Chromatography (HPLC/MPLC)

HPLC separations are one of the most important fields in the preparative resolution of enantiomers. The instrumentation improvements and the increasing choice of commercially available chiral stationary phases (CSPs) are some of the main reasons for the present significance of chromatographic resolutions at large-scale by HPLC. Proof of this interest can be seen in several reviews, and many chapters have in the past few years dealt with preparative applications of HPLC in the resolution of chiral compounds [19–23]. However, liquid chromatography has the attribute of being a batch technique and therefore is not totally convenient for production-scale, where continuous techniques are preferred by far.

In order to carry out a direct preparative chromatographic resolution of enantiomers in batch elution mode, the same methods as are used in other nonchiral large-scale chromatographic separations are applied [24], such as multiple close injections, recycling or peak shaving. All of these are addressed to reduce cost and increase yield, while saving solvent and making full use of the stationary phase. Thus, the injections are sometimes performed repeatedly (*multiple close injection*), in such a way that most of the chromatographic support is involved in the separation at any moment. When the resolution is not sufficient, it can be improved by *recycling* of the partially overlapped peaks [23]. This has almost the same effect as is obtained when using a longer column, but clearly a broadening of peaks occurs. However, after several cycles partially overlapped peaks can be completely resolved. This method is often combined with the so-called *peak shaving*, which allows the recovery of the part of the peaks corresponding to pure enantiomers while the overlapped region is recycled. In fact, to date this method is the most often used.

The chiral environment needed for enantiomeric separations is furnished by the chiral support into the column. Scale-up of a chiral separation can be made having as a reference an analytical resolution, but optimization of the preparative process is critically dependent upon the nature of the CSP. In order to increase the throughput, the column is usually used in overloading conditions. The loading capacity of a chiral stationary phase depends not only on the chiral selector density, but also on the type of selector. Therefore, some types of CSPs are more suitable than others for preparative purposes [21, 25]. Not all the commercially available CSPs used for analytical purposes are appropiated for large-scale resolutions (Table 1-1). CSPs with a large application domain, such as those derived from proteins with a vast applicability for analytical purposes, have a very low loadability and, therefore, they are not well adapted to preparative separations [26, 27]. This is also the case of molecular imprinted polymers (MIPs) [28–30]. The limited number of recognition sites restricts their loading capacity and thus also their use in large-scale chromatography.

Some ligand-exchange CSPs have been used at preparative level [31, 32]. In this case it must be taken into account that an extraction process, to remove the copper salts added to the mobile phase, must be performed following the chromatographic process [33]. Teicoplanin, in contrast, resolves all ordinary α and β-amino acids with mobile phases consisting of alcohol/water mixtures. No buffer is needed in the mobile phase [34]; hence evaporation of the solution derived from the preparative separation leaves a pure product. Cyclodextrin-based CSPs [35, 36], antibiotics [34, 37–39] such as the above-mentioned teicoplanin, and certain types of Pirkle phases have been utilized with preparative purposes. In those cases, although the loading capacity is not high, other advantages, such as a broad application domain for antibiotics, or the ease of preparation and the chemical stability, for multiple interaction supports, can balance this limitation. Polyacrylamides [40, 41] have also been used extensively, often in nonreported separations. Nevertheless, polysaccharide-derived CSPs are the most commonly applied for preparative chromatography. This is due both to their substantial loading capacity and broad enantiodiscrimination scope. They are the best adapted supports for this purpose either in HPLC, or in medium-pressure liquid chromatography (MPLC) [19].

Several types of polysaccharide-derived CSPs can be considered (Table 1-1). The support with the highest number of applications described, either at high or medium pressure, is microcrystalline cellulose triacetate (CTA). Other polysaccharide derivatives, mainly some benzoates, are used in their pure form, as beads which are directly packed [42, 43]. Coated CSPs, consisting of a polysaccharide derivative, benzoate or aryl carbamate of cellulose or amylose, on a matrix of aminopropylsilanized silica gel, are also among the most extensively utilized CSPs [44–49]. However, these supports have the limitation of the choice in the mobile phase composition. CSPs whose chiral selectors are bonded to the chromatographic matrix (usually silica gel) can perform in a number of different conditions and compositions in the mobile phase. This is the case of the already mentioned Pirkle phases, cyclodextrin, antibiotic or polyacrylamide-derived CSPs. Unfortunately, polysaccharide derivatives in coated CSPs often swell, or even dissolve, in a number of solvents. Thus, the compatible mobile phases with these supports are mixtures of a hydrocarbon (hexane or heptane) with an alcohol (ethanol or isopropanol), though many compounds have a reduced solubility in these mixtures. This feature, which is not a problem when these CSPs are used for analytical purposes, can be a major disadvantage at preparative scale. The low solubility limits the amount of product that can be injected in a single run and, therefore, the maximum loadability of the column cannot be attained [45]. This limitation can be overcome when the chiral selector is bonded to the chromatographic matrix [50–57]. In this case, a broader choice of solvents can be considered as mobile phase or simply to dissolve the racemate to be separated [58, 59]. It must be taken into account that, changes in selectivity as well as in the loading capacity of the CSP occur when solvents are changed [58].

In this context, the enantiomeric pair containing the eutomer of cyclothiazide can be resolved by HPLC on cellulose-derived coated CSPs. Nevertheless, the poor solubility of this compound in solvents compatible with this type of support makes this separation difficult at preparative scale. This operation was achieved with a cellulose carbamate fixed on allylsilica gel using a mixture of toluene/acetone as a mobile phase [59].

On occasion, the broad choice of existing phases is not enough to resolve a particular problem successfully. Derivatization with achiral reagents can be useful to introduce additional interacting groups in a poorly functionalized substrate, or to adapt it to be resolved in a particular CSP. On these occasions, derivatization can increase the chances of success in a given resolution [60].

Table 1-1. Preparative chiral separations.

	Packing name	Chiral selector	Examples of (semi)preparative applications	Supplier or reference of the CSP
	CTA	Crystalline cellulose triacetate	[19,42,61–62]	Daicel, Merck
	TBC	Cellulose tribenzoate beads	[19,42]	[42]
	MMBC	Cellulose tris(3-methylbenzoate) beads	[19,42]	[43]
	PMBC	Cellulose tris(4-methylbenzoate) beads	[19,42]	[43]
	Chiralcel OD	Cellulose tris(3,5-dimethylphenylcarbamate) coated on silica gel	[11,19,67,68]	Daicel, [67,68]
	Chiralcel OC	Cellulose tris(phenylcarbamate) coated on silica gel	[19]	Daicel
	Chiralcel OJ	Cellulose tris(4-methylbenzoate) coated on silica gel	[19]	Daicel
Polysaccharides	Chiralcel OB	Cellulose tribenzoate coated on silica gel	[19]	Daicel
	Chiralpak AD	Amylose tris(3,5-dimethylphenylcarbamate) coated on silica gel	[19,69]	Daicel
	Chiralpak AS	Amylose tris[1-(S)-phenylethylcarbamate] coated on silica gel	[70]	Daicel
	–	Mixed cellulose 10-undecenoate/tris(3,5-dimethyl-phenylcarbamate) bonded on silica gel	[59,71]	[51]
	–	Mixed amylose 10-undecenoate/tris(3,5-dimethyl-phenylcarbamate) bonded on silica gel	[71]	[56]
Cyclodextrins	Cyclobond I	Cyclodextrin immobilized on silica gel	[19,22,72]	Astec
	Hyd-β-CD	Hydroxypropyl-β-cyclodextrin	[23]	Merck,[23]
	DNBPG-co	3,5-Dinitrobenzoylphenylglycine covalently bonded on silica gel	[19,21,73–78]	Regis
	ChyRoSine-A	3,5-Dinitrobenzoyltyrosine butylamide	[79]	Sedere
	Pirkle-1J	3,5-Dinitrobenzoyl-β-lactam derivative	[80]	Regis
Pirkle type	α-Burke 2	Dimethyl N-3,5-dinitrobenzoyl-α-amino-2,2-dimethyl-4-pentenyl phosphonate bonded to silica	[80]	Regis
	ULMO	N-dinitrobenzoyl-N'-undecenoyl-diphenylethanediamine	[81]	Regis
	–	Cis-3-(1,1-dimethylethyl)-4-phenyl-2-azetidinone	[82]	[82]
		Quaternary ammonium derivative of 3,5-Dinitro-benzoyl-L-leucine on α-zirconium phosphate	[83]	[83]
	–	(S)-N-undecenoylproline 3,5-dimethoxyanilide bonded on silica gel	[80]	[80]
	Poly-PEA	Poly[(S)-N-acryloylphenylethylamine ethyl ester]	[21,84,85]	[84]
Polyacrylamides	PolyCHMA	Poly[(S)-N-methacryloyl-2-cyclohexylethylamine]	[84]	[84]
	D-ChiraSpher	Poly[(S)-N-acryloylphenylalanine ethyl ester]	[11,23,86]	Merck
	Polystyrene-Prol	L or D-proline bonded to polystyrene	[87]	[87]
LEC	Chirosolve-pro	L or D-proline bonded to polyacrylamide	[88]	UPS Chimie
	NucleosilChiral-1	L-hydroxyproline Cu^{2+} complexes bonded on silica gel	[33]	Macherey-Nagel
Antibiotics	Teicoplanin	Chirobiotic T	[89]	Astec
	Vancomicin	Chirobiotic V	[89]	Astec

1.3.1.2 Flash Chromatography

Flash chromatography is widely employed for the purification of crude products obtained by synthesis at a research laboratory scale (several grams) or isolated as extracts from natural products or fermentations. The solid support is based on silica gel, and the mobile phase is usually a mixture of a hydrocarbon, such as hexane or heptane, with an organic modifier, e.g. ethyl acetate, driven by low pressure air. (Recently the comparison of flash chromatography with countercurrent chromatography (CCC), a technique particularly adapted to preparative purposes, has been studied for the separation of nonchiral compounds [90].)

With regard to the resolution of enantiomers, some applications can be found with modified silica gel supports. Thus, a Pirkle-type CSP was used for the separation of 200 mg of a racemic benzodiazepinone [75]. Also tris-(3,5-dimethylphenyl)carbamate of cellulose coated on silica C_{18} [91, 92] was applied successfully to the resolution of the enantiomers of 2-phenoxypropionic acid and to oxprenolol, alprenolol, propranolol among other basic drugs. However, the low efficiency of this technique and the relative high price of the CSPs limits its use to the resolution of milligram range of sample.

1.3.1.3 Simulated Moving Bed (SMB)

The simulated moving bed (SMB) technology was patented in the early 1960s as a binary continuous separation technique. It consists of a series of several columns connected to each other head-to-tail, simulating an annular column. The eluent source, the feed of mixture to process and the two collecting positions move along this circle in such a way that mimics a relative countercurrent movement between the mobile and the stationary phases. This makes compatible the continuous injection of mixture to be purified, and the recovery of two different fractions with the chromatographic process [93]. The feature of being continuous was considered an advantage in order to be included in a production chain, when related to other existing separation techniques that act mainly in a batch basis. Although the ability to obtain two fractions from a mixture might be seen as a limitation, SMB found very important applications in the petro-chemical and sugar industries [94]. However, it was not until some decades later that such a binary technique was realized to be advantageous and especially suited to the separation of enantiomeric mixtures.

Since the first separation of enantiomers by SMB chromatography, described in 1992 [95], the technique has been shown to be a perfect alternative for preparative chiral resolutions [10, 21, 96, 97]. Although the initial investment in the instrumentation is quite high – and often prohibitive for small companies – the savings in solvent consumption and human power, as well as the increase in productivity, result in reduced production costs [21, 94, 98]. Therefore, the technique would be specially suitable when large-scale productions (≥100 g) of pure enantiomers are needed. Despite the fact that SMB can produce enantiomers at very high enantiomeric excesses, it is sometimes convenient to couple it with another separation technique, often crystallization [11, 94], in order to increase the global productivity [10].

The type of CSPs used have to fulfil the same requirements (resistance, loadability) as do classical chiral HPLC separations at preparative level [99], although different particle size silica supports are sometimes needed [10]. Again, to date the polysaccharide-derived CSPs have been the most studied in SMB systems, and a large number of racemic compounds have been successfully resolved in this way [95–98, 100–108]. Nevertheless, some applications can also be found with CSPs derived from polyacrylamides [11], Pirkle-type chiral selectors [10] and cyclodextrin derivatives [109]. A system to evaporate the collected fractions and to recover and recycle solvent is sometimes coupled to the SMB. In this context the application of the technique to gas can be advantageous in some cases because this part of the process can be omitted [109].

Enantiomeric drugs or intermediates in their synthesis are the compounds most often purified with this technology and reported in the literature, although many resolutions performed in the industry have not been published for reasons of confidentiality. Some of the most recent examples in the field are summarized in Fig. 1-1.

1.3.1.4 Closed-loop Recycling with Periodic Intra-profile Injection (CLRPIPI)

An intermediate approach between HPLC and SMB chromatography, called "closed-loop recycling with periodic intra-profile injection" (CLRPIPI) has been described recently [110]. This is a new binary preparative separation technique whose concept implies the combination of recycling with peak shaving and SMB. Thus, once the pure fractions of the peaks are collected, the partially resolved fraction is recycled into the column. A new injection of fresh sample is then produced just between the two partially resolved peaks. The new mixture passes through the column, at the end of which pure fractions are collected while the partially resolved fraction is recycled again, and the process is repeated. This is similar to SMB as it is a binary technique, but it is not continuous. The capital cost of this system is substantially lower than that of SMB devices but a high productivity is maintained. It can be a good alternative when the amount of enantiomers to purify is not high enough to justify the investment of a SMB instrument. Some examples of the use of this technique in the purification of enantiomers, either by derivatization and separation, on a nonchiral column [111], or by direct resolution on a CSP (Chiralpak AS) [112] can be found in the literature.

1.3.1.5 Countercurrent Chromatography (CCC/CPC)

Countercurrent chromatography (CCC) refers to a chromatographic technique which allows the separation of solutes in a two-phase solvent system subjected to a gravitational field. Two immiscible liquid phases, constituted by one or more solvents or solutions, are submitted to successive equilibria, where the solutes to be separated

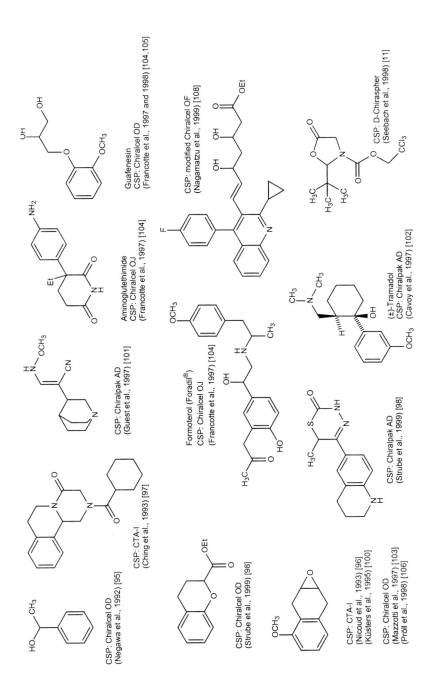

Fig. 1-1. Some of the structures of the racemates resolved recently by SMB.

are partitioned on the basis of their different affinity for one or the other phase. The chromatographic process occurs between them without any solid support. The CCC instruments maintain one of the liquid phases as stationary by means of the centrifugal force, while the other is pumped through as mobile phase [113–115]. Interest in the technique has favored the development of improved devices based on the same principle, namely the retention of the liquid phases by means of a centrifugal field, but with slight technical modifications. Thus, classical CCC devices use a variable-gravity field produced by a two-axis gyration mechanism, while centrifugal partition chromatography (CPC) devices are based on the use of a constant-gravity field produced by a single-axis rotation mechanism [113–115]. Both CCC and CPC preparative-scale instruments are available commercially [116].

The technique has some advantages relating to the traditional liquid–solid separation methods. The most important of these is that all the stationary phase takes part in the separation process, whereas the activity of a solid phase is mainly concentrated in the surface of the support, an important part of this being completely inert. This fact increases the loading capacity of the phase, and this is the reason why CCC is especially suited for preparative purposes. Therefore, modern CCC overcomes the disadvantages of direct preparative chromatography by HPLC with regard to the high cost of the chiral solid stationary phase and its relatively limited loadability.

From the pioneering studies of Ito et al. [117], CCC has been mainly used for the separation and purification of natural products, where it has found a large number of applications [114, 116, 118, 119]. Moreover, the potential of this technique for preparative purposes can be also applied to chiral separations. The resolution of enantiomers can be simply envisaged by addition of a chiral selector to the stationary liquid phase. The mixture of enantiomers would come into contact with this liquid CSP, and enantiodiscrimination might be achieved. However, as yet few examples have been described in the literature.

The first partial chiral resolution reported in CCC dates from 1982 [120]. The separation of the two enantiomers of norephedrine was partially achieved, in almost 4 days, using (*R,R*)-di-5-nonyltartrate as a chiral selector in the organic stationary phase. In 1984, the complete resolution of d,l-isoleucine was described, with *N*-dodecyl-L-proline as a selector in a two-phase buffered *n*-butanol/water system containing a copper (II) salt, in approximately 2 days [121]. A few partial resolutions of amino acids and drug enantiomers with proteic selectors were also published [122, 123].

However, it was not until the beginning of 1994 that a rapid (<1.5 h) total resolution of two pairs of racemic amino acid derivatives with a CPC device was published [124]. The chiral selector was *N*-dodecanoyl-L-proline-3,5-dimethylanilide (1) and the system of solvents used was constituted by a mixture of heptane/ethyl acetate/methanol/water (3:1:3:1). Although the amounts of sample resolved were small (2 ml of a 10 mM solution of the amino acid derivatives), this separation demonstrated the feasibility and the potential of the technique for chiral separations. Thus, a number of publications appeared subsequently. Firstly, the same chiral selector was utilized for the resolution of 1 g of (±)-*N*-(3,5-dinitrobenzoyl)leucine with a modified system of solvents, where the substitution of water by an acidified solution ensured the total retention of the chiral selector in the stationary phase [125]. The separation of 2 g of the same leucine derivative employing the pH-zone refining technique with the same instrument was later described [127]. (The elution pattern of pH-zone-refining CCC bears a remarkable resemblance to that observed in displacement chromatography and allows the displacement of ionizable molecules through the CCC column by means of a pH gradient [116, 126].)

Recently, two examples of the separation of enantiomers using CCC have been published (Fig. 1-2). The complete enantiomeric separation of commercial d,l-kynurenine (**2**) with bovine serum albumin (BSA) as a chiral selector in an aqueous–aqueous polymer phase system was achieved within 3.5 h [128]. Moreover, the chiral resolution of 100 mg of an estrogen receptor partial agonist (7-DMO, **3**) was performed using a sulfated β-cyclodextrin [129, 130], while previous attempts with unsubstituted cyclodextrin were not successful [124]. The same authors described the partial resolution of a glucose-6-phosphatase inhibitor (**4**) with a Whelk-O derivative as chiral selector (**5**) [129].

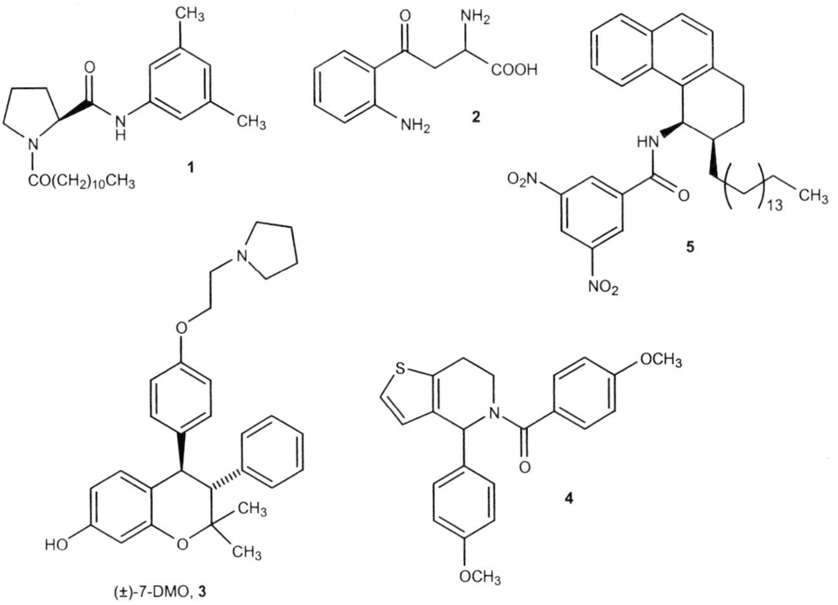

Fig. 1-2. Several racemates resolved by CCC (**2, 3, 4**) and some of the chiral selectors used (**1, 5**) (see text).

The CCC instruments have even been used as enzymatic reactors to carry out enantioselective processes. Thus, the hydrolysis of 2-cyanocyclopropyl-1,1-dicarboxylic acid dimethylester including a bacterial esterase in the stationary phase was reported [131]. After 8 h, the procedure yielded the desired product automatically, without any extraction and with an 80 % e.e.

1.3.2 Subcritical and Supercritical Fluid Chromatography

Supercritical fluid chromatography (SFC) refers to the use of mobile phases at temperatures and pressures above the critical point (supercritical) or just below (subcritical). SFC shows several features that can be advantageous for its application to large-scale separations [132–135]. One of the most interesting properties of this technique is the low viscosity of the solvents used that, combined with high diffusion coefficients for solutes, leads to a higher efficiency and a shorter analysis time than in HPLC.

As a matter of fact, the main advantage in comparison with HPLC is the reduction of solvent consumption, which is limited to the organic modifiers, and that will be nonexistent when no modifier is used. Usually, one of the drawbacks of HPLC applied at large scale is that the product must be recovered from dilute solution and the solvent recycled in order to make the process less expensive. In that sense, SFC can be advantageous because it requires fewer manipulations of the sample after the chromatographic process. This facilitates recovery of the products after the separation. Although SFC is usually superior to HPLC with respect to enantioselectivity, efficiency and time of analysis [136], its use is limited to compounds which are soluble in nonpolar solvents (carbon dioxide, CO_2). This represents a major drawback, as many of the chemical and pharmaceutical products of interest are relatively polar.

Although some applications for preparative-scale separations have already been reported [132] and the first commercial systems are being developed [137, 138], examples in the field of the resolution of enantiomers are still rare. The first preparative chiral separation published was performed with a CSP derived from (*S*)-*N*-(3,5-dinitrobenzoyl)tyrosine covalently bonded to γ-mercaptopropyl silica gel [21]. A productivity of 510 mg/h with an enantiomeric excess higher than 95 % was achieved for **6** (Fig. 1-3).

Examples with other Pirkle-type CSPs have also been described [139, 140]. In relation to polysaccharides coated onto silica gel, they have shown long-term stability in this operation mode [141, 142], and thus are also potentially good chiral selectors for preparative SFC [21]. In that context, the separation of racemic glibenclamide analogues (**7**, Fig. 1-3) on cellulose- and amylose-derived CSPs was described [143].

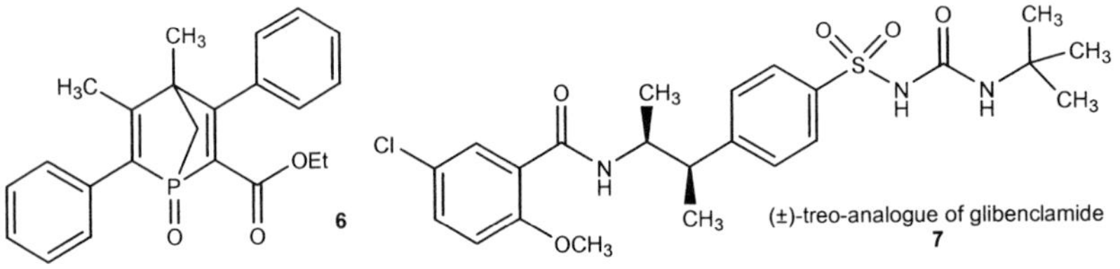

Fig. 1.3. Chemical structures of racemic compounds resolved by SFC.

ᵤoses, but is
pheromones or
ᵤd. Nevertheless, in
ᵤ as an economical alter-
ᵤmers were performed on
ᵤily on very few occasions were

ᵤne of the latest resolutions of the anesthetic enflurane (**8**) has been performed by preparative GC on a γ-cyclodextrin CSP, the process later being scaled-up via SMB [109] (Fig. 1-4). This is the first GC-SMB separation described.

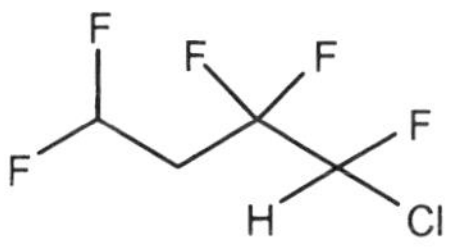

CSP: Octakis(3-O-butanoyl-2,6-di-O-*n*-pentyl)-γ-cyclodextrin
Temperature: 50°C
Productivity: 10 mg of enantiomers/g CSP/day

(±)-enflurane

Fig. 1-4. Resolution of enflurane by GC.

1.4 Enantioselective Membranes

Membrane-based separation techniques constitute nowadays well-established process methods for industrial treatments of fluids. Like SMB, membrane-based separations can be performed in continuous mode. In the field of preparative-scale enantiodiscrimination, much effort has been invested in this subject due its high potential [154, 155]. (Chapter 5 of this book is devoted to the subject, and further discusses the advantages and applications of membrane technologies.)

The first successful chiral resolutions through enantioselective membranes have been published recently, but few cases are applicable to the preparative scale, mainly due to mechanical and technical limitations. Low flow rates, saturation of the chiral selectors and loss of enantioselectivity with time are some of the common problems encountered and that should be solved in the near future.

Enantioselective transport processes can be achieved either with solid or liquid membranes (Fig. 1-5). In this latter case, the liquid membrane can be supported by a porous rigid structure, or it can simply be an immiscible liquid phase between two solutions with the same character (aqueous or nonaqueous), origin and destination

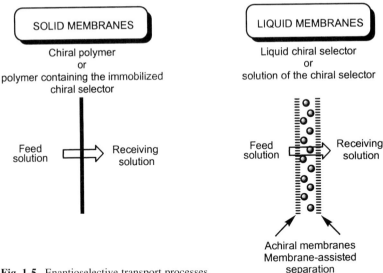

Fig. 1-5. Enantioselective transport processes.

of the compound to be transported [154]. The membrane is then simply a technical tool which permits a type of liquid–liquid extraction to be performed. In all cases the membrane should contain the chiral selector to carry out the separation of enantiomers.

The nature of enantioselective solid membranes can be very diverse. Chiral synthetic and semisynthetic polymers have been applied directly for this purpose, but other chiral molecules have also proved to be useful after immobilization on a nonchiral porous membrane. Polysaccharide derivatives, especially cellulose carbamates [156–159], acrylic polymers, poly(α-amino acids) [160–162] and polyacetylene-derived polymers are some of the polymeric selectors that have been successful in the resolution of racemic mixtures by this method. The high loadability of these compounds, already demonstrated in HPLC and other classical applications, makes them very attractive in continuous processes. Moreover, the filmogenic properties of some of them, such as the polysaccharide derivatives, are interesting characteristics when the formation of a membrane is envisaged. More recently, the introduction of molecular imprinted polymers (MIPs) to membrane technologies has been described as a promising alternative [163–166]. Among the chiral molecules immobilized on a nonchiral rigid support membrane to perform an enantioselective separation are amino acids and proteins, such as BSA [167–169]. The main limitation in the case of solid membranes is the silting that occurs when all recognition sites have been occupied and there is no real transport through the membrane. An ingenious system has been described [159] to take advantage of this phenomenon for the separation of enantiomers.

Liquid membranes can be constituted by liquid chiral selectors used directly [170] or by solutions of the chiral molecules in polar or apolar solvents. This later possibility can also be an advantage since it allows the modulation of the separation con-

re.
the solve...
gradient in concentration of pH ......
back common to all these systems, it should be mentioned ....
enantiomer usually decreases when the enantiomer ratio in the permeate ...
Nevertheless, this can be overcome by designing a system where two opposite selectors are used to transport the two enantiomers of a racemic solution simultaneously,
as it was already applied in W-tube experiments [171].

Most of the chiral membrane-assisted applications can be considered as a modal-
ity of liquid–liquid extraction, and will be discussed in the next section. However, it
is worth mentioning here a device developed by Keurentjes et al., in which two miscible chiral liquids with opposing enantiomers of the chiral selector flow countercurrently through a column, separated by a nonmiscible liquid membrane [179]. In
this case the selector molecules are located out of the liquid membrane and both
enantiomers are needed. The system allows recovery of the two enantiomers of the
racemic mixture to be separated. Thus, using dihexyltartrate and poly(lactic acid),
the authors described the resolution of different drugs, such as norephedrine, salbutamol, terbutaline, ibuprofen or propranolol.

1.5 Other Methods

1.5.1 Chiral Extractions

Liquid-liquid extraction is a basic process already applied as a large-scale method.
Usually, it does not require highly sophisticated devices, being very attractive for the
preparative-scale separation of enantiomers. In this case, a chiral selector must be
added to one of the liquid phases. This principle is common to some of the separation techniques described previously, such as CCC, CPC or supported-liquid membranes. In all of these, partition of the enantiomers of a mixture takes place thanks
to their different affinity for the chiral additive in a given system of solvents.

The instrumentation which until now has been used in chiral extraction experiments is very diverse, ranging from the simple extraction funnel [123, 180], the U-
or W-tubes [171, 181], to more sophisticated devices, such as hollow-fiber extraction apparatus [175] or other membrane-assisted systems. Most of these experiments

have been brought about at a reduced scale, though the potential of the extraction techniques is very promising. In principle, the type of chiral additives used can be the same as the selectors applied to supported-liquid membranes or CCC. Nevertheless, as all the chiral recognition process occurs in solution, and an aqueous phase is often involved, the solvation of selector and racemate molecules competes with the chiral interactions selector-enantiomers, especially those implying hydrogen bonds. Therefore, it is very often the case that chiral selectors with very high chiral recognition abilities are needed [171, 182, 183]. The main disadvantage of liquid–liquid extraction in the separation of enantiomers is the need for an additional treatment to separate the chiral selector from the phase containing one of the enantiomers of the resolved racemate.

Early examples of enantioselective extractions are the resolution of α-aminoalcohol salts, such as norephedrine, with lipophilic anions (hexafluorophosphate ion) [184–186] by partition between aqueous and lipophilic phases containing esters of tartaric acid [184–188]. Alkyl derivatives of proline and hydroxyproline with cupric ions showed chiral discrimination abilities for the resolution of neutral amino acid enantiomers in *n*-butanol/water systems [121, 178, 189–192]. On the other hand, chiral crown ethers are classical selectors utilized for enantioseparations, due to their interesting recognition abilities [171, 178]. However, the large number of steps often required for their synthesis [182] and, consequently, their cost as well as their limited loadability makes them not very suitable for preparative purposes. Examples of ligand-exchange [193] or anion-exchange selectors [183] able to discriminate amino acid derivatives have also been described.

Proteins (BSA or ovomucoid, OVM) have also been successful in the preparative resolution of enantiomers by liquid–liquid extraction, either between aqueous and lipophilic phases [181] or in aqueous two-phase systems (ATPS) [123, 180]. The resolution of d,l-kynurenine [180] and ofloxacin and carvedilol [123] were performed using a countercurrent extraction process with eight separatory funnels. The significant number of stages needed for these complete resolutions in the mentioned references and others [123, 180, 189], can be overcome with more efficient techniques. Thus, the resolution of d,l-kynurenine performed by Sellergren et al. in 1988 by extraction experiments was improved with CCC technologies 10 years later [128].

It is worth noting that the extractive process can be performed continuously. Thus, the separation of (±)-mandelic acid into its enantiomers was achieved with a liquid particle extractor described by Abe et al. [190–192] using *N*-docecyl-L-proline as chiral selector.

1.5.2 Preparative Gel Electrophoresis and Thin-Layer Chromatography

Recently, the separation of some milligram quantities of terbutaline by classical gel electrophoresis has been reported [194]. A sulfated cyclodextrin impregnated on the agarose gel was used as a chiral selector and the complete resolution was achieved in 5 h. Analogously, small amounts of enantiomers can be isolated using thin-layer chromatographic supports impregnated with chiral selectors [195–197]. However, these techniques seem, at present, far from being applicable to the resolution of important amounts of racemates.

1.5.3 Enantioselective Distillation and Foam Flotation

Among the existing separation techniques, some – due to their intrinsic characteristics – are more adapted than others to processing large amounts of material. Such processes, which already exist at industrial level, can be considered in order to perform an enantioselective separation. This is the case for techniques such as distillation and foam flotation, both of which constitute well-known techniques that can be adapted to the separation of enantiomers. The involvement of a chiral selector can be the clue which changes a nonstereoselective process into an enantioselective one. Clearly, this selector must be adapted to the characteristics and limitations of the process itself.

Several chiral selectors have been used in the separation of enantiomers by distillation [198]. Among them, the bisalcohol **8** (Fig. 1-6) has permitted obtainment of the ketone (+)-**9** with an enantiomeric excess of 95 %. This example shows the feasibility of the process even though, in this particular case, the price of the chiral selector might prohibit scale-up of the separation.

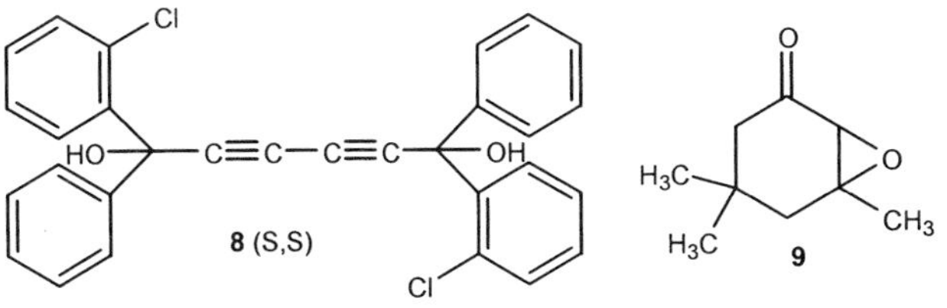

Fig. 1-6. Chemical structures of the chiral selector (**8**) used in the resolution of **9** by distillation.

In another example of enantioselective distillation, it was the enantiomeric mixture to resolve itself which contributed to create a chiral environment. Thus, non-racemic mixtures of α-phenylethylamine were enantiomerically enriched by submitting to distillation different salts of this amine with achiral acids [199].

The main advantage of distillation over other separation techniques is the absence of solvent involved. This feature can contribute to a reduction in the price of an enantioselective separation. In the search of other economical process-scale enantiomeric separations, foam flotation or froth flotation can be considered. To our knowledge, only one example has been described [200] regarding the application of foam flotation to the separation of enantiomers that shows the method to be feasible. Several derivatives of L-hydroxyproline, β-cyclodextrin derivatives, such as permethylated β-cyclodextrin, vancomicin and digitonin were used as chiral foaming agents for the separation of racemic amino acid derivatives and drugs such as warfarin, and with different results. The best separation described is the obtainment of N-*tert*-butoxycarbonyl-D-phenylalanine (**10**) with a 76 % e.e. when using permethyl-β-cyclodextrin in a foaming column of 40 cm length (Fig. 1-7).

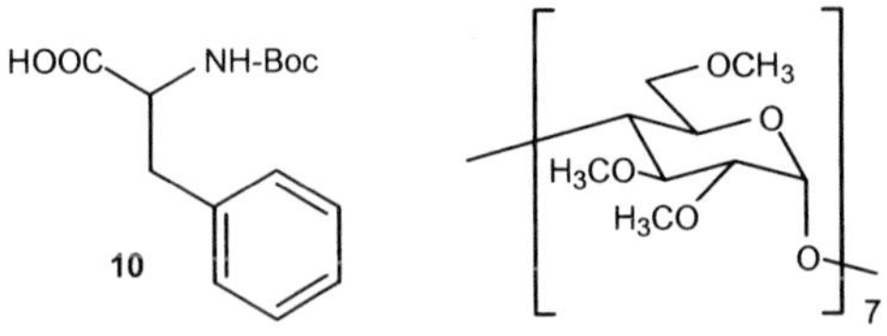

Fig. 1-7. Racemic aminoacid derivatives resolved and chiral selector used as foaming agent.

permethyl-ß-cyclodextrin

1.6 Global Considerations

All enantioselective separation techniques are based on submitting the enantiomeric mixture to be resolved to a chiral environment. This environment is usually created by the presence of a chiral selector able to interact with both enantiomers of the mixture, albeit with different affinities. These differences in the enantiomer–selector association will finally result in the separation that is sought.

Ideal chiral selectors to be used in preparative separations should fulfil certain properties. In general, high loadability is one of the most interesting features for large-scale purposes, but high enantioselectivity, high chemical stability, low cost and broad applicability are also very important issues. None of these properties can be considered independently.

Very high values of enantioselectivity can be attained with specific selectors for particular enantiomers [182, 201–204]. Nevertheless, the application domain is reduced consequently. Values of chromatographic enantioselectivity over 40 have been reported either in normal phase conditions [201, 202] or in aqueous mobile phases [204]. These high enantioselectivities can represent an increase in the loading capacity that could cause a reduction of the global cost of the separation. However, the long elution time of the second enantiomer would not be convenient for practical purposes when using chromatography. In contrast, other separation techniques would take profit of this characteristic. Thus, liquid–liquid extraction of aromatic amino acids was successfully achieved with a highly enantioselective synthetic receptor by Mendoza et al. [182]. In this case, however, the scale-up of this separation would be hardly feasible due to the large number of steps needed for the synthesis of the receptor.

A compromise among all the properties mentioned herein should be established, depending on the technique used and on the particular application. Preparative separation of enantiomers is still an open subject which requires further investigation in the search of new chiral selectors and techniques well adapted to large scale processes.

References

[1] J. Jacques, A. Collet, S. H. Wilen, *Enantiomers, racemates and resolutions*, Wiley Interscience, New York (1981).

[2] C. R. Bayley, N. A. Vaidya, in *Chirality and Industry*, A. N. Collins, G. N. Sheldrake, J. Crosby (Eds.), Wiley, Chichester (1992) pp. 69–77.

[3] V. M. L. Wood, "Crystal science techniques in the manufacture of chiral compounds" in *Chirality and Industry II. Developments in the Manufacture and applications of optically active compounds*, A. N. Collins, G. N. Sheldrake, J. Crosby (Eds.), John Wiley & Sons, New York (1997) Chapter 7.

[4] T. Shiraiwa, H. Miyazaki, M. Ohkubo, A. Ohta, A. Yoshioka, T. Yamame, H. Kurokawa, Chirality, 8 (1996) 197.

[5] T. Shiraiwa, H. Miyazaki, A. Ohta, K. Motonaka, E. Kobayashi, M. Kubo, H. Kurokawa, Chirality, 9 (1997) 656.

[6] T. Shiraiwa, H. Miyazaki, T. Watanabe, H. Kurokawa, Chirality, 9 (1997) 48.

[7] T. Shiraiwa, H. Miyazaki, H. Kurokawa, Chirality, 6 (1994) 654.

[8] T. Buhse, D. K. Kondepudi, B. Hoskins, Chirality, 11 (1999) 343.

[9] D. K. Kondepudi, M. Culha, Chirality, 10 (1998) 238.

[10] J. Blehaut and R. M. Nicoud, Analusis, 26 (1998) M60.

[11] D. Seebach, M. Hoffmann, A. R. Sting, J. N. Kinkel, M. Schulte, E. Küsers, J. Chromatogr. A, 796 (1998) 299.

[12] R. Tamura, T. Ushio, H. Takahashi, K. Nakamura, N. Azuma, F. Toda, K. Endo, Chirality, 9 (1997) 220.

[13] R. Tamura, H. Takahashi, T. Ushio, Y. Nakajima, K. Hirotsu, F. Toda, Enantiomer, 3 (1998) 149.

[14] Takahashi, R. Tamura, T. Ushio, Y. Nakajima, K. Hirotsu, Chirality, 10 (1998) 705.

[15] T. Vries, H. Wynberg, E. van Echten, J. Koek, W. ten Hoeve, R. M. Kellogg, Q. B. Broxterman, A. Minnaard, B. Kaptein, S. van der Sluis, L. Hulshof, J. Kooistra, Angew. Chem. Intl. Ed. Engl., 37 (1998) 2349.

[16] T. Toda, K. Tanaka, M. Watanabe, T. Abe, N. Harada, Tetrahedron Asymm., 6 (1995) 1495.

[17] Q.-S. Hu, D. Vitharana, L. Pu, Tetrahedron Asymm., 6 (1995) 2123.

[18] R. Yoshioka, O. Ohtsuki, T. Da-Te, K. Okamura, M. Senuma, Bull. Chem. Soc. Jpn., 67 (1994) 3012.

[19] E. Francotte, J. Chromatogr. A, 666 (1994) 565.

[20] E. Francotte, Chimia, 51 (1997) 717.

[21] E. Francotte, "Chromatography as a separation tool for the preparative resolution of racemic compounds" in *Chiral separations, applications and technology*, S. Ahuja (Ed.), American Chemical Society, Washington (1997) Chapter 10.

[22] W. H. Pirkle and B. C. Hamper, "The direct preparative resolution of enantiomers by liquid chromatography on chiral stationary phases" in *Preparative Liquid Chromatography*, B. A. Bidlingmeyer (Ed.), Journal Chromatography Library Vol. 38, 3rd Edition, Elsevier Science Publishers B. V., Amsterdam (1991) Chapter 7.

[23] J. Dingenen, J. N. Kinkel, J. Chromatogr. A, 666 (1994) 627.

[24] A. Siedel-Morgenstern, Analusis, 26 (1998) M46.

[25] G. B. Cox, Analusis, 26 (1998) M70.

[26] S. G. Allenmark, *Chromatographic Enantioseparations: methods and applications*, Chichester, Ellis Horwood (1991) 2nd Ed.

[27] S. G. Allenmark, "Separation of enantiomers by protein-based chiral phases" in *A practical approach to chiral separations by liquid chromatography*, G. Subramanian, VCH, Weinheim (1994) Chapter 7.

[28] B. Sellergren, "Enantiomer separation using tailor-made phases prepared by molecular imprinting" in *A practical approach to chiral separations by liquid chromatography*, G. Subramanian, VCH, Weinheim (1994) Chapter 4.

[29] V. T. Remcho, Z. J. Tan, Anal. Chem., 71 (1999) 248A.

[30] O. Ramström, R. J. Ansell, Chirality, 10 (1998) 195.

[31] V. A. Davankov, "Ligand-exchange phases" in *Chiral Separations by HPLC*, A. M. Krstulovic, Ellis Horwood Ltd., Chichester (1989) Chapter 15.

[32] V. A. Davankov, J. Chromatogr. A, 666 (1994) 55.
[33] S. B. Thomas, B. W. Surber, J. Chromatogr., 586 (1991) 265.
[34] A. Berthod, Y. Liu, C. Bagwill, D. W. Armstrong, J. Chromatogr. A, 731 (1996) 123.
[35] T. J. Ward, D. W. Armstrong, "Cyclodextrin-stationary phases" in *Chromatographic chiral separations*, M. Zief, L. J. Crane (Eds.), Chromatographic Science Series, Vol. 40, Marcel Dekker, New York (1988) Chapter 5.
[36] A. M. Stalcup, "Cyclodextrin bonded chiral stationary phases in enantiomer separations" in *A practical approach to chiral separations by liquid chromatography*, G. Subramanian, VCH, Weinheim (1994) Chapter 5.
[37] D. W. Armstrong, Y. Tang, S. Chen, Y. Zhou, C. Bagwill, J. R. Chen, Anal. Chem., 66 (1994) 1473.
[38] D. W. Armstrong, Y. Liu, K. H. Ekborg-Ott, Chirality, 7 (1995) 474.
[39] A. Berthod, U. B. Nair, C. Bagwill, D. W. Armstrong, Talanta, 43 (1996) 1767.
[40] G. Blaschke, "Substituted polyacrylamides as chiral phases for the resolution of drugs" in *Chromatographic chiral separations*, M. Zief, L. J. Crane (Eds.), Chromatographic Science Series, Vol. 40, Marcel Dekker, New York (1988) Chapter 7.
[41] J. N. Kinkel, "Optically active polyacrylamide/silica composites and related packings and their application to chiral separations" in *A practical approach to chiral separations by liquid chromatography*, G. Subramanian, VCH, Weinheim (1994) Chapter 8.
[42] E. Francotte, R. M.Wolf, Chirality, 3 (1991) 43.
[43] E. Francotte, R. M. Wolf, J. Chromatogr., 595 (1992) 63.
[44] J. Dingenen, "Polysaccharide phases in enantioseparations" in *A practical approach to chiral separations by liquid chromatography*, G. Subramanian, VCH, Weinheim (1994) Chapter 6.
[45] E. Yashima, Y. Okamoto, Bull. Chem. Soc. Jpn., 68 (1995) 3289.
[46] E. Yashima, C. Yamamoto, Y. Okamoto, Synlett, 4 (1998) 344.
[47] Y. Okamoto, E. Yashima, Angew. Chem. Intl. Ed. Engl., 37 (1998) 1020.
[48] K. M. Kirkland, J. Chromatogr. A., 718 (1995) 9.
[49] K. Oguni, H. Oda, A. Ichida, J. Chromatogr. A, 694 (1995) 91-100.
[50] L. Oliveros, P. López, C. Minguillón, P. Franco, J. Liq. Chromatogr., 18 (1995) 1521.
[51] C. Minguillón, P. Franco, L. Oliveros, P. López, J. Chromatogr. A, 728 (1996) 407.
[52] C. Minguillón, P. Franco, L. Oliveros, J. Chromatogr. A, 728 (1996) 415.
[53] L. Oliveros, P. Franco, A. Senso, C. Minguillón, Am. Biotech. Lab., 14 (1996) 34.
[54] C. Minguillón, A. Senso, L. Oliveros, Chirality, 9 (1997) 145.
[55] L. Oliveros, A. Senso, P. Franco, C. Minguillón, Chirality, 10 (1998) 283.
[56] P. Franco, A. Senso, C. Minguillón, L. Oliveros, J. Chromatgr. A, 796 (1998) 265.
[57] A. Senso, L. Oliveros, C. Minguillón, J. Chromatgr. A, 839 (1999) 15.
[58] P. Franco, C. Minguillón, L. Oliveros, J. Chromatogr. A, 793 (1998) 239.
[59] L. Oliveros, C. Minguillón, B. Serkiz, F. Meunier, J. P. Volland, A. Cordi, J. Chromatogr. A, 729 (1996) 29.
[60] E. Francotte, Chirality, 10 (1999) 492.
[61] E. Francotte, D. Lohmann, Helv. Chim. Acta, 70 (1987) 1569.
[62] E. Francotte, R. M.Wolf, Chirality, 2 (1990) 16.
[63] E. Francotte, R. W. Lang, T. Winkler, Chirality, 3 (1991) 177.
[64] J. M. Janssen, S. Copinga, G. Gruppen, R. Isaksson, D. T. Witte, C. J. Grol, Chirality, 6 (1994) 596.
[65] R. Oehrlein, R. Jeschke, B. Ernst, D. Bellus, Tetrahedron Letters, 30 (1989) 3517.
[66] R. M.Wolf, E. Francotte, J. Hainmüller, Chirality, 5 (1993) 538.
[67] J. Wagner, H. J. Hamann, W. Dopke, A. Kunath, E. Hoft, Chirality, 7 (1995) 243.
[68] M. E. Tiritan, Q. B. Cass, A. del Alamo, S. A. Matlin, S. J. Grieb, Chirality, 7 (1995) 573.
[69] E. Küsters, J. Nozulak, Chromatographia, 47 (1998) 440.
[70] L. Miller, C. Orihuela, R. Fronek, D. Honda, O. Dapremont, J. Chromatogr. A, 849 (1999) 309.
[71] C. Cativiela, M. D. Díaz-de-Villegas, A. I. Jiménez, P. López, M. Marraud, L. Oliveros, Chirality, 11 (1999) 583.
[72] C. J. Shaw, P. J. Sanfilippo, J. J. McNally, S. A. Park, J. B. Press, J. Chromatogr., 631 (1993) 173.
[73] W. H. Pirkle, J. M. Finn, J. Org. Chem., 47 (1982) 4037.
[74] W. H. Pirkle, A. Tsipouras, J. Chromatogr., 291 (1984) 291.
[75] W. H. Pirkle, A. Tsipouras, T. J. Sowin, J. Chromatogr., 319 (1985) 392.
[76] W. H. Pirkle, T. C. Pochapsky, G. S. Mahler, D. E. Corey, D. S. Reno, D. M. Alessi, J. Org. Chem., 51 (1986) 4991.

[77] W. H. Pirkle, T. J. Sowin, J. Chromatogr., 396 (1987) 83.

[78] A. Tambuté, P. Gareil, M. Caude, R. Rosset, J. Chromatogr., 363 (1986) 81.

[79] M. Caude, A. Tambuté, L. Siret, J. Chromatogr., 550 (1991) 357.

[80] W. H. Pirkle, M. E. Koscho, J. Chromatogr. A, 840 (1999) 151.

[81] N. M. Maier, G. Uray, J. Chromatogr. A., 732 (1996) 215.

[82] W. H. Pirkle, J. E. McCune, J. Chromatogr., 441 (1988) 311.

[83] M. E. Garcia, J. L. Naffin, N. Deng, T. E. Mallouk, Chem. Mater., 7 (1995) 1968.

[84] G. Blaschke, J. Liq. Chromatogr., 9 (1986) 341.

[85] G. Blaschke, J. Maibaum, J. Chromatogr., 366 (1986) 329.

[86] D. Seebach, S. G. Müller, U. Gysel, J. Zimmermann, Helv. Chim. Acta, 71 (1988) 1303.

[87] V. A. Davankov, Y. A. Zolotarev, A. A. Kurganov, J. Liq. Chromatogr., 2 (1979) 1191.

[88] G. Jeanneret-Gris, C. Soerensen, H. Su, J. Porret, Chromatographia, 28 (1989) 337.

[89] Chirobiotic™ Handbook, Guide to using macrocyclic glycopeptide bonded phases for chiral LC separations, Advanced Separation Technologies Inc. (ASTEC), 2nd Ed. Whippany, New York (1997).

[90] W. D. Conway, E. L. Bachert, A. M. Sarlo, C. W. Chan, J. Liq. Chrom. & Rel. Technol., 21 (1998) 53.

[91] S. A. Matlin, S. J. Grieb, A. M. Belenguer, J. Chem. Soc., Chem. Commun., (1995) 301.

[92] S. J. Grieb, S. A. Matlin, A. M. Belenguer, J. Chromatogr. A, 728 (1996) 195.

[93] R. M. Nicoud, LC-GC Int., 5 (1992) 43.

[94] B. Pynnonen, J. Chromatogr. A, 827 (1998) 143.

[95] N. Negawa, F. Shoji, J. Chromatogr., 590 (1992) 113.

[96] R. M. Nicoud, G. Fuchs, P. Adam, M. Bailly, E. Küsters, F. D. Antia, R. Reuille, E. Schmid, Chirality, 5 (1993) 267.

[97] C. B. Ching, B. G. Lim, E. J. D. Lee and S. C. Ng, J. Chromatogr., 634 (1993) 215.

[98] J. Strube, A. Jupke, A. Epping, H. Schmidt-Traub, M. Schulte and R. Devant, Chirality 11 (1999) 440.

[99] M. Schulte, R. Ditz, R. M. Devant, J. N. Kinkel, F. Charton, J. Chromatogr. A, 769 (1997) 93.

[100] E. Küsters, G. Gerber, F. D. Antia, Chromatographia, 40 (1995) 387.

[101] D. W. Guest, J. Chromatogr. A, 760 (1997) 159.

[102] E. Cavoy, M. F. Deltent, S. Lehoucq, D. Miggiano, J. Chromatogr. A, 769 (1997) 49.

[103] M. Mazzotti, G. Storti, M. Morbidelli, J. Chromatogr. A, 769 (1997) 3.

[104] E. R. Francotte, P. Richert, J. Chromatogr. A, 769 (1997) 101.

[105] E. Francotte, P. Richert, M. Mazzotti, M. Morbidelli, J. Chromatogr. A, 796 (1998) 239.

[106] T. Pröll, E. Küsters, J. Chromatogr. A, 800 (1998) 135.

[107] L. S. Pais, J. M. Loureiro, A. E. Rodrigues, J. Chromatogr. A, 827 (1998) 215.

[108] S. Nagamatzu, K. Murazumi, S. Makino, J. Chromatogr. A, 832 (1999) 55.

[109] M. Juza, O. di Giovanni, G. Biressi, V. Schurig, M. Mazzotti, M. Morbidelli, J. Chromatogr. A, 813 (1998) 333.

[110] C. M. Grill, J. Chromatogr. A, 796 (1998) 101.

[111] G. Nadler, C. Dartoois, D. S. Eggleston, R. C. Haltiwanger, M. Martin, Tetrahedron Asymm., 9 (1998) 4267.

[112] C. M. Grill, L. Miller, J. Chromatogr. A, 827 (1998) 359.

[113] Y. Ito, B. Mandava (Eds.), *Countercurrent chromatography: theory and practice*, Chromatographic science Series, Vol. 44, Marcel Dekker, New York (1988).

[114] W. D. Conway (Ed.), *Countercurrent Chromatography: apparatus, theory and applications*, VCH Publ., New York (1990).

[115] A. P. Foucault (Ed.), *Centrifugal Partition Chromatography*, Chromatographic Science Series, Vol. 68, Marcel Dekker, New York (1995)

[116] A. P. Foucault, L. Chevolot, J. Chromatogr. A, 808 (1998) 3.

[117] Y. Ito, M. Weinstein, I. Aoki, R. Harada, E. Kimura, K. Nonugaki, Nature, 212 (1966) 985.

[118] A. Marston, K. Hostettmann, J. Chromatogr. A, 658 (1994) 315.

[119] I. A. Sutherland, L. Brown, S. Forbes, G. Games, D. Hawes, K. Hostettmann, E. H. McKerrell, A. Marston, D. Wheatley, P. Wood, J. Liq. Chrom. & Rel. Technol., 21 (1998) 279.

[120] B. Domon, K. Hostettmann, K. Kovacevik, V. Prelog, J. Chromatogr., 250 (1982) 149.

[121] T. Takeuchi, R. Horikawa, T. Tanimura, J. Chromatogr., 284 (1984) 285.

[122] B. Ekberg, B. Sellergen, P.-A. Albertsson, J. Chromatogr., 333 (1984) 211.

[123] T. Arai, H. Kuroda, Chromatographia, 32 (1991) 56.
[124] L. Oliveros, P. Franco-Puértolas, C. Minguillón, E. Camacho-Frias, A. Foucault, F. Le Goffic, J. Liq. Chromatogr., 17 (1994) 2301.
[125] Y. Ma, Y. Ito, A. P. Foucault, J. Chromatogr. A, 704 (1995) 75.
[126] Y. Ito, J. Chromatogr. A, 753 (1996) 1.
[127] Y. Ma, Y. Ito, Anal. Chem., 67 (1995) 3069.
[128] K. Shinomiya, Y. Kabasawa, Y. Ito, J. Liq. Chrom. & Rel. Technol., 21 (1998) 135.
[129] J. Breinholt, A. Varming, Poster presented at the 11th International Symposium on Chiral Discrimination, Chicago, July 1999.
[130] J. Breinholt, S. V. Lehmann, A. Varming, Chirality (1999) 11 (1999) 768.
[131] O. R. Bousquet, J. Braun, F. Le Goffic, Tetrahedron Letters, 36 (1995) 8195.
[132] P. Jusforgues, M. Shaimi, Analusis, 26 (1998) M55.
[133] K. L. Williams, L. C. Sander, S.A.Wise, J. Pharm. Biomed. Anal., 15 (1997) 1789.
[134] K. L. Williams, L. C. Sander, J. Chromatogr. A, 785 (1997) 149.
[135] M. Perrut, J. Chromatogr., 658 (1994) 293.
[136] C. Wolf, W. H. Pirkle, J. Chromatogr. A, 799 (1998) 177.
[137] K. D. Bartle, C. D. Bevan, A. A. Clifford, S. A. Jafar, N. Malak, M. S. Verrall, J. Chromatogr. A, 697 (1995) 579.
[138] G. Fuchs, L. Doguet, D. Barth, M. Perrut, J. Chromatogr., 623 (1992) 329.
[139] P. Macaudière, M. Lienne, M. Caude, R. Rosset, A. Tambuté, J. Chromatogr., 467 (1989) 357.
[140] A. M. Blum, K. G. Lynam, E. C. Nicolas, Chirality, 6 (1994) 302.
[141] K. G. Lynam, E. C. Nicolas, J. Pharmacol. Biochem. Anal., 11 (1993) 1197.
[142] K. W. Phinney, L. C. Sander, S. A. Wise, Anal. Chem., 70 (1998) 2331.
[143] J. Whatley, J. Chromatogr. A, 697 (1995) 251.
[144] M. Lindström, N. Torbjörn, J. Roeraade, J. Chromatogr., 513 (1990) 315.
[145] V. Schurig, H. Grosenick, B. S. Green, Angew. Chem. Int. Ed. Engl., 32 (1993) 1662.
[146] I. Hardt, W. A. König, J. Chromatogr. A, 666 (1994) 611.
[147] G. Fuchs, M. Perrut, J. Chromatogr. A, 658 (1994) 437.
[148] D. U. Staerk, A. Shitangkoon, G. Vigh, J. Chromatogr. A, 663 (1994) 79.
[149] D. U. Staerk, A. Shitangkoon, G. Vigh, J. Chromatogr. A, 677 (1994) 133.
[150] D. U. Staerk, A. Shitangkoon, G. Vigh, J. Chromatogr. A, 702 (1995) 251.
[151] V. Schurig, H. Grosenick, J. Chromatogr. A, 666 (1994) 617.
[152] M. Juza, E. Braun, V. Schurig, J. Chromatogr. A, 769 (1997) 119.
[153] V. Schurig, U. Leyer, Tetrahedron Asymm., 1 (1990) 865.
[154] J. T. F. Keurentjes, F. J. M. Voermans, "Membrane separations in the production of optically pure compounds" in *Chirality and Industry II. Developments in the Manufacture and applications of optically active compounds*, A. N. Collins, G. N. Sheldrake, J. Crosby (Eds.), John Wiley & Sons, New York (1997) Chapter 8.
[155] L. J. Brice, W. H. Pirkle, "Enantioselective transport through liquid membranes" in *Chiral separations, applications and technology*, S. Ahuja (Ed.), American Chemical Society, Washington (1997) Chapter 11.
[156] E. Yashima, J. Noguchi, Y. Okamoto, Chem. Letters, (1992) 1959.
[157] E. Yashima, J. Noguchi, Y. Okamoto, J. Appl. Polym. Sci., 54 (1994) 1087.
[158] E. Yashima, J. Noguchi, Y. Okamoto, Macromolecules, 28 (1995) 1889.
[159] E. Yashima, J. Noguchi, Y. Okamoto, Tetrahedron Asymm., 6 (1995) 8368.
[160] N. Ogata, Macromol. Symp., 77 (1994) 167.
[161] T. Aoki, S. Tomizawa, E. Oikawa, J. Membr. Sci., 99 (1995) 117.
[162] A. Maruyama, N. Adachi, T. Takatsuki, M. Torii, K. Sanui, N. Ogata, Macromolecules, 23 (1990) 2748.
[163] S. A. Piletsky, T. L. Panasyuk, E. V. Piletskaya, I. A. Nicholls, M. Ulbricht, J. Membr. Sci., 157 (1999) 263 and references therein.
[164] A. Dzgoev, K. Haupt, Chirality, 11 (1999) 465.
[165] M. Yoshikawa, J. Izumi, T. Kitao, S. Koya, S. Sakamoto, J. Membr. Sci., 108 (1995) 171.
[166] M. Yoshikawa, J.-I. Izumi, T. Ooi, T. Kitao, M. D. Guiver, G. P. Robertson, Polym. Bull., 40 (1998) 517.
[167] S. Kiyohara, M. Nakamura, K. Saito, K. Sugita, T. Sugo, J. Membr. Sci., 152 (1999) 143.
[168] M. Nakamura, S. Kiyohara, K. Saito, K. Sugita, T. Sugo, Anal. Chem., 71 (1999) 1323.

[169] A. Higuchi, M. Hara, T. Horiuchi, T. Nakagawa, J. Membr. Sci., 93 (1994) 157.

[170] M. Bryjak, J. Kozlowski, P. Wieczorek, P. Kafarski, J. Membr. Sci., 85 (1993) 221.

[171] M. Newcomb, J. L. Toner, R. C. Helgeson, D. J. Cram, J. Am. Chem. Soc., 101 (1979) 4941.

[172] T. Yamaguchi, K. Nishimura, T. Shinbo, M. Sugiura, Chem. Letters, (1985) 1549.

[173] M. Pietraszkiewicz, M. Kozbial, O. Pietraszkiewicz, Enantiomer, 2 (1997) 319.

[174] D. W. Armstrong, H. L. Jin, Anal. Chem., 59 (1987) 2237.

[175] W. H. Pirkle, E. M. Doherty, J. Am. Chem. Soc., 111 (1989) 4113.

[176] W. H. Pirkle, W. E. Bowen, Tetrahedron Asymm., 5 (1994) 773.

[177] C. M. Heard, J. Hadgraft, K. R. Brain, Bioseparation, 4 (1994) 111.

[178] P. J. Pickering, J. B. Chaudhuri, Chirality, 9 (1997) 261.

[179] J. T. F. Keurentjes, L. J. W. M. Nabuurs, E. A. Vegter, J. Membr. Sci., 113 (1996) 351.

[180] B. Sellergren, B. Ekberg, P.-A. Albertsson, K. Mosbach, J. Chromatogr., 450 (1988) 277.

[181] H. Nakagawa, K. Shimizu, K.-I. Yamada, Chirality, 11 (1999) 516.

[182] A. Galán, D. Andreu, A. M. Echevarren, P. Prados, J. de Mendoza, J. Am. Chem. Soc., 114 (1992) 1511.

[183] K.-H. Kellner, A. Blasch, H. Chmiel, M. Lämmerhofer, W. Lindner, Chirality, 9 (1997) 268.

[184] V. Prelog, Z. Stojanac, K. Kovacevic, Helv. Chim. Acta, 65 (1982) 377.

[185] V. Prelog, S. Mutak, K. Kovacevic, Helv. Chim. Acta, 66 (1983) 2279.

[186] V. Prelog, M. Dumic, Helv. Chim. Acta, 69 (1986) 5.

[187] Y. Abe, T. Shoji, M. Kobayashi, W. Qing, N. Asai, H. Nishizawa, Chem. Pharm. Bull., 43 (1995) 262.

[188] Y. Abe, T. Shoji, S. Fukui, M. Sasamoto, H. Nishizawa, Chem. Pharm. Bull., 44 (1996) 1521.

[189] T. Takeuchi, R. Horikawa, T. Tanimura, Sep. Sci. Technol., 25 (1990) 941.

[190] H. Nishizawa, K. Tahara, A. Hayashida, Y. Abe, Anal. Chem., 9 (1993) 611.

[191] H. Nishizawa, Y. Watanabe, S. Okimura, Y. Abe, Anal. Chem., 5 (1989) 339.

[192] H. Nishizawa, Y. Watanabe, S. Okimura, Y. Abe, Anal. Chem., 5 (1989) 345.

[193] H. Tsukube, J. Uenishi, T. Kanatani, H. Itoh, O. Yonemitsu, Chem. Commun. (1996) 477.

[194] A. M. Stalcup, K. H. Gahm, S. R. Gratz, R. M. C. Sutton, Anal. Chem., 70 (1998) 144.

[195] L. Lepri, J. Planar Chromatogr., 10 (1997) 320.

[196] R. Suedee, C. M. Heard, Chirality, 9 (1997) 139.

[197] R. Bhushan, G. Thuku Thiongo, Biomed. Chromatogr., 13 (1999) 276.

[198] G. Kaupp, Angew. Chem. Intl. Ed. Engl., 33 (1994) 728.

[199] D. Kozma, Z. Madarász, M. Acs, E. Fogassy, Chirality, 7 (1995) 381.

[200] D. W. Armstrong, E. Y. Zhou, S. Chen, K. Le, Y. Tang, Anal. Chem., 66 (1994) 4278.

[201] W. H. Pirkle, P. G. Murray, J. Chromatogr., 641 (1993) 11.

[202] F. Gasparrini, D. Missiti, C. Villani, A. Borchardt, M. D. Burger, W. C. Still, J. Org. Chem., 60 (1995) 4314.

[203] M. Martin, C. Raposo, M. Almaraz, M. Crego, C. Caballero, M. Grande, J. R. Moran, Angew. Chem. Intl. Ed. Engl., 35 (1996) 3286.

[204] N. M. Maier, W. Oberleitner, M. Lämmerhofer, W. Lindner, Oral presentation at the 11th International Symposium on Chiral Discrimination, Chicago, July 1999.

2 Method Development and Optimization of Enantiomeric Separations Using Macrocyclic Glycopeptide Chiral Stationary Phases

Thomas E. Beesley, J. T. Lee and Andy X. Wang

2.1 Introduction

Enantiomeric separations have become increasingly important, especially in the pharmaceutical and agricultural industries as optical isomers often possess different biological properties. The analysis and preparation of a pure enantiomer usually involves its resolution from the antipode. Among all the chiral separation techniques, HPLC has proven to be the most convenient, reproducible and widely applicable method. Most of the HPLC methods employ a chiral selector as the chiral stationary phase (CSP).

Currently, several hundred CSPs have appeared in publications, and over 110 of them are available commercially [1]. These CSPs are made by using either a polymeric structure or a small ligand (MW < 3000) as the chiral selector. The polymeric CSPs include synthetic chiral polymers [2] and naturally occurring chiral polymers [3–5]. The most commonly used natural polymers include proteins and carbohydrates (cellulose and amylose). The chiral recognition mechanisms for these polymeric CSPs are relatively complicated. A protein, for example, is often complex enough to contain several chiral binding sites, in which case the major (high-affinity) site may differ for any given pair of enantiomers [6]. The other type of CSPs, with a small molecule as the chiral selector, include ligand exchange CSPs [7], π–complex (Pirkle-type) CSPs [8, 9], crown ether CSPs [10], cyclodextrin CSPs [11–15] and macrocyclic glycopeptide CSPs [16–19]. Compared to the polymeric CSPs, the separation mechanisms on these CSPs are better characterized and understood. Macrocyclic glycopeptides, which were introduced by Armstrong in 1994 [16], are the newest class of CSPs. Three macrocyclic glycopeptides – vancomycin, teicoplanin and ristocetin A – are now available commercially [20]. Much research effort has been made on the characterization and application of these CSPs, and on a wide variety of chiral compounds.

2.2 Characteristics of Macrocyclic Glycopeptide CSPs

2.2.1 Chiral Recognition Mechanisms

Vancomycin, ristocetin A and teicoplanin are produced as fermentation products of *Streptomyces orientalis, Nocardia lurida* and *Actinoplanes teichomyceticus,* respectively. All three of these related compounds consist of an aglycone "basket" made up of fused macrocyclic rings and pendant carbohydrate moieties (Fig. 2-1). The macrocycles contain both ether and peptide linkages. The aglycones of vancomycin and teicoplanin contain two chloro-substituted aromatic rings, while the analogous portion of ristocetin A contains no chloro substituents.

Vancomycin is the smallest of the three molecules, consisting of three macro-cyclic rings and an attached disaccharide comprising D-glucose and vancosamine. The other two glycopeptides are somewhat larger and have four fused macrocyclic rings and different types of pedant sugar moieties. Teicoplanin has three attached monosaccharides, two of which are D-glucosamine and one of which is D-mannose. Ristocetin A has a pendant tetrasaccharide and two monosaccharide moieties. These saccharides include D-arabinose, D-mannose, D-glucose, and D-rhamnose. Teicoplanin has one unique characteristic: namely, it has a hydrophobic acyl side chain ("hydrophobic tail") attached to a 2-amino-2-deoxy-β-D-glucopyranosyl moiety. The structural characteristics of the three macrocycles are outlined in Table 2-1.

Table 2-1. Structural characteristics of macrocyclic glycopeptides.

	Vancomycin	Teicoplanin	Ristocetin A
Molecular weight	1,449	1,877	2,066
Stereogenic centers	18	23	38
Macrocycles	3	4	4
Sugar moieties	2	3	6
Hydroxyl groups	9	15	21
Amine groups	2	1	2
Carboxyl groups	1	1	0
Amido groups	7	7	6
Aromatic groups	5	7	7
Methyl esters	0	0	1
Hydrophobic tail	0	1	0
pI values	7.2	4–6.5	7.5

All three glycopeptides have analogous ionizable groups which control their charge and are thought to play a major role in their association with analytes and chiral recognition. For example, there is an amine on the aglycone portion of each compound. Vancomycin has a secondary amine, while the other two macrocycles have primary amine groups. All three compounds also have amino saccharide moieties. However, teicoplanin is unique in that it has two amino saccharides, both of which are N-acylated. There is a carboxylic acid moiety on the aglycone of both vancomycin and teicoplanin, while the equivalent group on ristocetin A is esterified. The

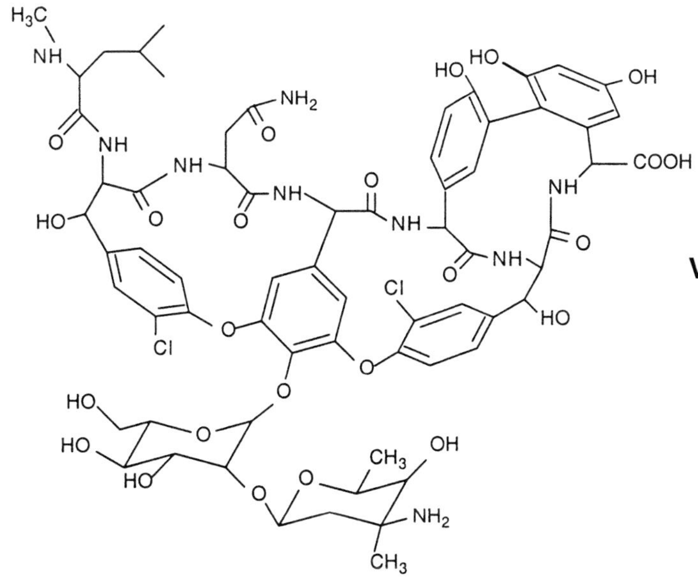

Vancomycin

1a)

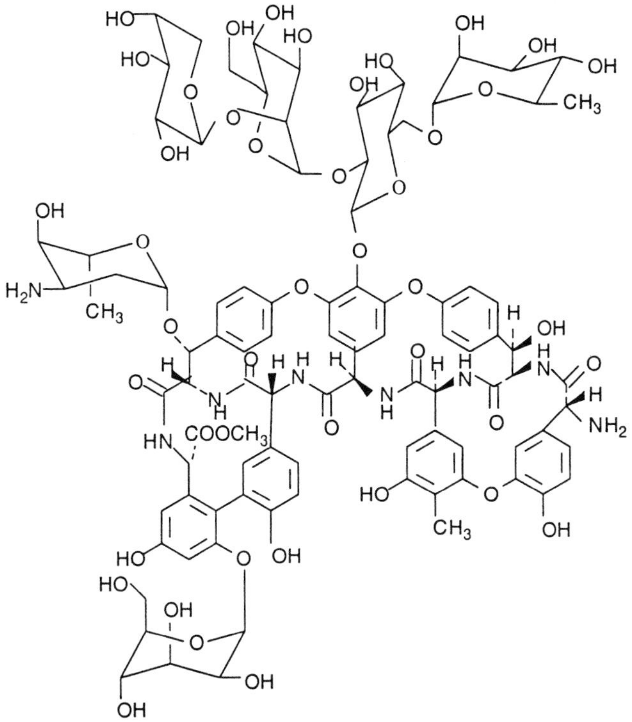

Ristocetin A

1b)

1c)

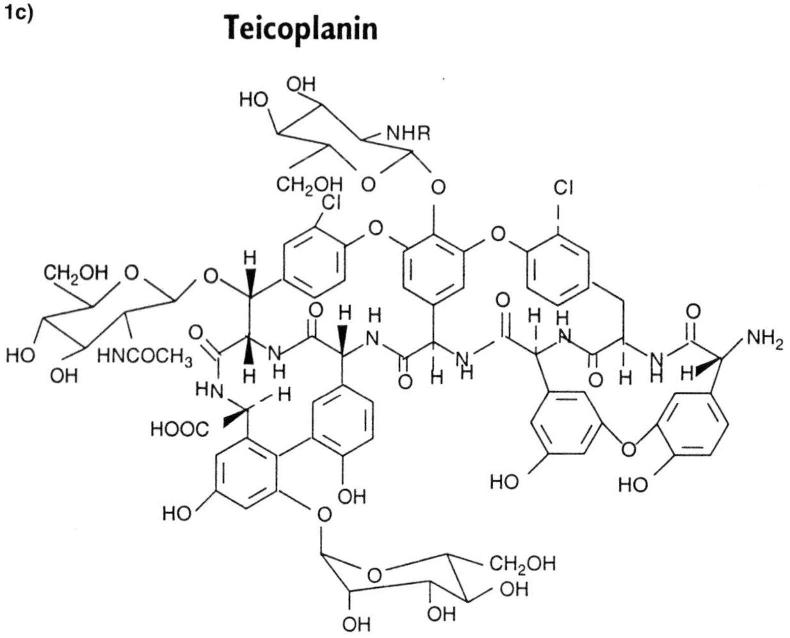

Teicoplanin

Fig. 2-1 Proposed structures of three macrocyclic glycopeptides. On teicoplanin, R = 8-methyl-nonanoic acid.

only other ionizable groups on these structures are the phenolic moieties. At pH values 4–7, they are generally protonated and probably serve mainly as nonchiral hydrogen bonding sites. The unique structure and functionalities on the macrocyclic glycopeptides provides a variety of possible interactions for chiral recognition. A summary of the possible interactions is listed in Table 2-2.

Table 2-2. The relative strength of potential interactions between glycopeptide CSPs and chiral analytes.

π-π Complexation	Very strong
Hydrogen bonding	Very strong
Inclusion	Weak
Dipole stacking	Medium strong
Steric interactions	Weak
Anionic or cationic binding	Strong

2.2.2 Multi-modal CSPs

The macrocyclic glycopeptides CSPs are capable of operating in three different mobile phase systems: reversed phase, normal phase, and the new polar organic mode. The new polar organic mode refers to the approach when methanol is used as the mobile phase with small amounts of acid and/or base as the modifier to control selectivity. Since the macrocycles are covalently bonded to silica gel through multiple linkages, there is no detrimental effect when a column switches from one mobile phase system to another.

The enantioselectivity of the macrocyclic CSPs are different in each of the operating modes, probably because of different separation mechanisms functioning in the different solvent modes. The possible chiral recognition mechanisms for three mobile phase compositions on glycopeptide phases are listed in Table 2-3 in descending order of strength.

Statistically, of the compounds enantioresolved by macrocyclic glycopeptide CSPs, new polar organic mode accounts for more than 40 %, balanced by reversed-phase mode, while typical normal-phase operation resulted in approximately 5 % of separations. Some categories of racemic compounds that are resolved on the glycopeptide CSPs at different operating modes are listed in Table 2-4.

Table 2-3. Possible separation mechanisms for three mobile phase compositions on glycopeptide CSPs.

New polar organic mode	Ionic interaction Hydrogen bonding Steric interaction
Reversed phase	Ionic interaction Hydrophobic inclusion Hydrogen bonding Steric interaction
Normal phase	Hydrogen bonding π-π interaction Dipole stacking Steric interaction

Table 2-4. Typical categories of racemic compounds resolved on glycopeptide CSPs in three mobile phase modes.

	Vancomycin	Teicoplanin	Ristocetin A
New polar organic mode	(Cyclic) Amines Amino alcohols	Amino alcohols N-blocked amino acids	(α-Hydroxyl/halogen) acids Substituted aliphatic acids Profens N-blocked amino acids
Reversed phase	Amines Imides Acids Profens Amides	(α-Hydroxyl) acids Oxazolidinones Native amino acids Small peptides N-blocked amino acids	(α-Hydroxyl) acids Substituted aliphatic acids Profens N-blocked amino acids Amino esters Hydantoins Small peptides
Normal phase	Hydantoins Barbiturates Imides Oxazolidinones	Hydantoins Imides	Imides Hydantoins N-blocked amino acids

2.2.3 Predictability of Enantioselectivity

One of the characteristics of glycopeptide CSPs is the predictability of chiral separation for racemates with the same stereogenic environment. If a compound is separated on a glycopeptide column under certain mobile phase conditions, it is very
likely that the molecules with similar stereogenic configuration will be separated
under the same conditions. For example, when the compound albuterol was separated in the new polar organic mode on a teicoplanin column, a number of other
β-blockers were tested and found to be baseline resolved as well by using the same
mobile phase on this stationary phase (Fig. 2-2). This indicates that further separation is likely to be achieved for analogous compounds with an aromatic moiety, a
hydroxyl group on the chiral center and a secondary amine group in the β-position
to the chiral center. This predictability is extremely beneficial to the combinatory
chemists who synthesize huge number of analogous chiral compounds. The predictability is further shown with enantiomeric separation of profens on vancomycin
in reversed-phase (Fig. 2-3) and α-hydroxyl/halogen carboxylic acids on ristocetin
A in the new polar organic mode (Fig. 2-4).

2.2.4 Complementary Separations

One thing that makes the macrocyclic glycopeptide CSPs unique and different from
other CSPs is the "principle of complementary separations" [17, 19]. This refers to
an empirical observation that an increase in selectivity is obtained when one glycopeptide phase is switched to another under the same or similar mobile phase. Warfarin, for example, is only partially separated on a teicoplanin column under optimized reversed-phase conditions. When the same mobile phase is used with a vancomycin column, a baseline resolution of warfarin enantiomers is achieved with
shorter analysis time. Some other examples are given in Fig. 2-5 to show the phenomenon of complementary separations. One advantage of the complementary separation is that one can switch from one column to another to achieve better selectivity without changing the mobile phase. Another advantage is that two or three different glycopeptide columns can be coupled together in order to achieve broader
selectivity and to screen a larger number of analytes. The column coupling technique
will be discussed later in this chapter.

2a)

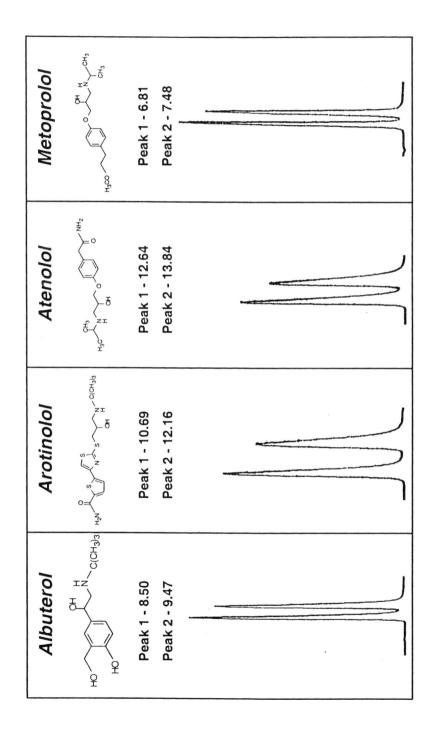

2b)

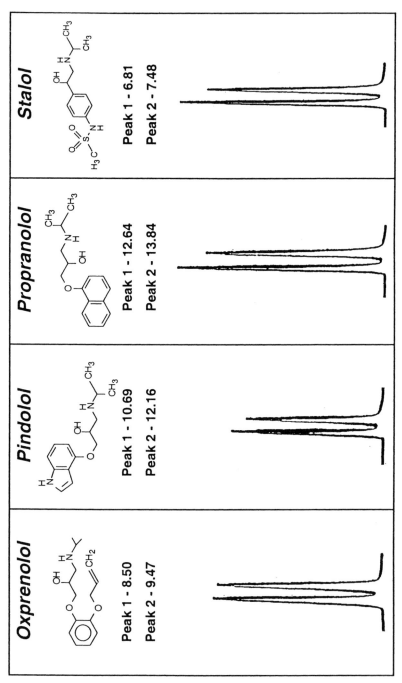

Fig. 2-2. The enantiomeric separation of β-blockers on teicoplanin CSP (250 × 4.6 mm) with the same mobile phase composition: methanol with 0.1 % acetic acid and 0.1 % triethylamine (v/v). The flow rate was 1.0 mL min⁻¹ at ambient temperature (23 °C).

3a)

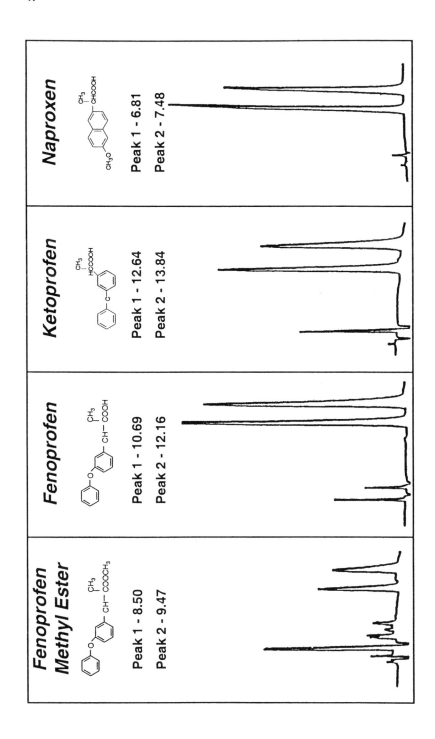

3b)

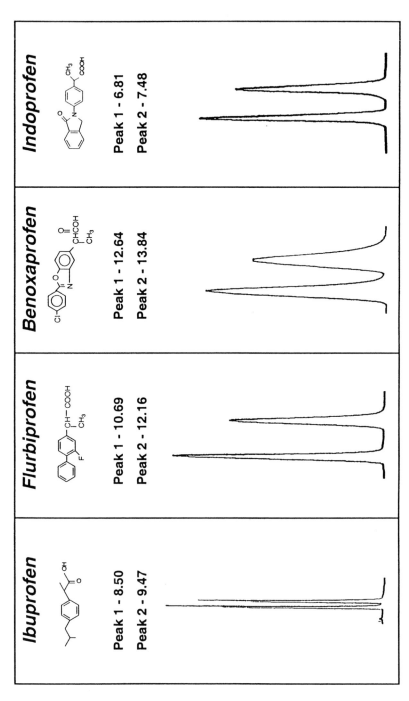

Fig. 2-3. The enantiomeric separation of profens on vancomycin CSP (250 × 4.6 mm) with the same mobile phase composition: tetrahydrofuran: 20 mM sodium citrate (10/90 v/v) pH 6.3. The flow rate was 1.0 mL min⁻¹ at ambient temperature (23 °C).

4a)

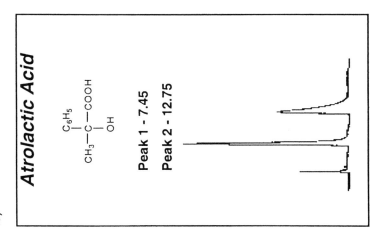

Atrolactic Acid

C₆H₅ | CH₃—C—COOH | OH

Peak 1 - 7.45

Peak 2 - 12.75

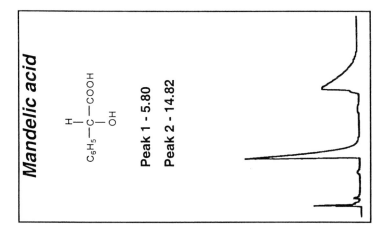

Mandelic acid

H | C₆H₅—C—COOH | OH

Peak 1 - 5.80

Peak 2 - 14.82

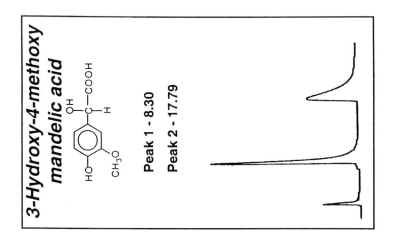

3-Hydroxy-4-methoxy mandelic acid

Peak 1 - 8.30

Peak 2 - 17.79

4b)

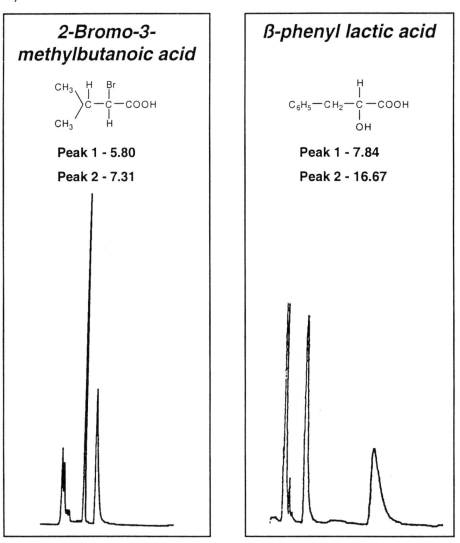

Fig. 2-4. The enantiomeric separation of α-hydroxy/halogen acids on ristocetin A CSP (250 × 4.6 mm) with the same mobile phase composition: methanol with 0.02 % acetic acid and 0.01 % triethylamine (v/v). The flow rate was 1.0 mL min^{-1} at ambient temperature (23 °C).

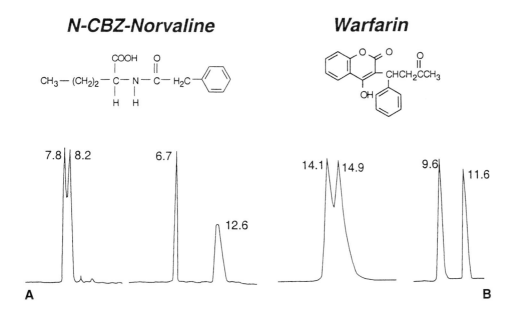

A   B

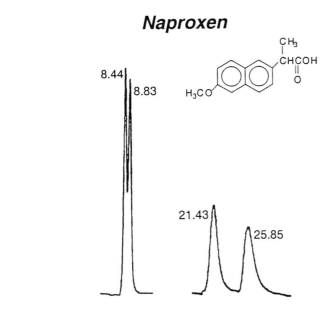

C

Fig. 2-5. Examples showing the complementary separations on glycopeptide CSPs. (A) Separation of N-CBZ-norvaline on vancomycin (left) and teicoplanin (right). The mobile phase was methanol: 1 % triethylammonium acetate (20/80 v/v) pH 4.1. (B) Separation of warfarin on teicoplanin (left) and vancomycin (right) CSPs. The mobile phase was acetonitrile: 1 % triethylammonium acetate (10/90 v/v) pH 4.1. (C) Separation of naproxen on teicoplanin (left) and ristocetin A (right). The mobile phase was methanol: 0.1 % triethylammonium acetate (30/70 v/v) pH 4.1. All columns were 250×4.6 mm i.d. The flow rate for all the separations was 1 mL min^{-1} at ambient temperature (23 °C).

2.3   Method Development with Glycopeptide CSPs

2.3.1   Method Development Protocols

The glycopeptide CSPs are multi-modal phases and can switch from one mobile phase to another without deleterious effects. The new polar organic mobile phase offers the advantages of broad selectivity, high efficiency, low back-pressure, short analysis time, high capacity and excellent prospects for preparative-scale separation. Whenever a racemic compound is targeted for separation, its structure can give a hint as to which mobile phase should be investigated. If the compound has two or more functional groups which are capable of interacting with the CSPs, and at least one of these functional groups is near or on the stereogenic center, the new polar organic phase is recommended to be tested first. In this context, functional groups include: hydroxyl groups, halogens (I, Br, Cl, and F), nitrogen in any form (primary, secondary and tertiary), carbonyl and carboxyl groups as well as oxidized forms of sulfur and phosphorus.

In the new polar organic mode, the ratio of acid/base in the mobile phase affects the selectivity and the concentration of acid and base controls the retention. It is suggested to start the method development with a medium concentration (0.1 %) for both acid and base. If retention is too long or too short, the concentration can be increased to 1 % or reduced to 0.01 %. If no selectivity is observed in this mode, reversed phase is recommended as the next step in the protocols.

In reversed phase, the selectivity and retention is affected by such parameters as the type and percentage of organic modifier, the type of aqueous buffer, pH and concentration of buffer. Fortunately, there is an empirically optimal mobile phase composition for each glycopeptide phase. Vancomycin has shown its best performance for most compounds when tetrahydrofuran (THF) is used as the organic modifier at a concentration of 10 % with ammonium nitrate at a concentration of 20 mM and pH 5.5 (no pH adjustment is needed). Methanol seems to be the preferred organic modifier for teicoplanin and ristocetin A CSPs, and 20 % is usually a good starting composition. Triethylammonium acetate (TEAA) at the concentration of 0.1 % is a suitable buffer for both of these latter columns, although teicoplanin usually shows better selectivity at pH 4.1 while ristocetin A has better selectivity at pH 6.8.

When analytes lack the selectivity in the new polar organic mode or reversed-phase mode, typical normal phase (hexane with ethanol or isopropanol) can also be tested. Normally, 20 % ethanol will give a reasonable retention time for most analytes on vancomycin and teicoplanin, while 40 % ethanol is more appropriate for ristocetin A CSP. The hexane/alcohol composition is favored on many occasions (preparative scale, for example) and offers better selectivity for some less polar compounds. Those compounds with a carbonyl group in the α or β position to the chiral center have an excellent chance to be resolved in this mode. The simplified method development protocols are illustrated in Fig. 2-6. The optimization will be discussed in detail later in this chapter.

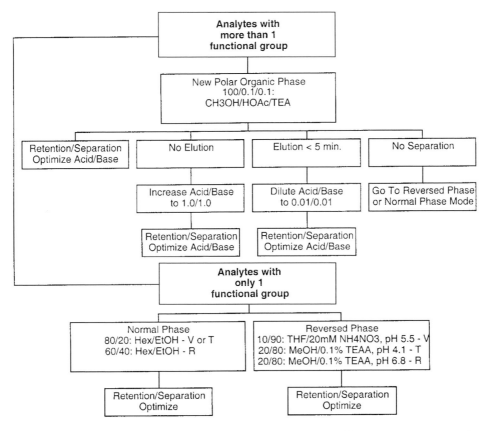

Fig. 2-6. Method development protocols on 250 × 4.6 mm glycopeptide columns. V: vancomycin, T: teicoplanin, R: ristocetin A.

2.3.2 Column Coupling Technique

The direct coupling of two columns through a low dead volume connector is currently a common and useful practice. The coupled columns are used with the same mobile phase conditions. Various kinds of combinations have been made to directly couple columns of similar or distinctive selectivities. A standard reversed-phase (C_{18}) column, for example, has been coupled with: a cyano column [21], molecularly imprinted polymer [22], phenyl [23], C_8 [24], and silica [25] columns in order to achieve broader selectivity. For the analysis of a mixture containing one or more chiral analytes, the column coupling becomes more important. While a chiral column is designed to resolve optical isomers, its performance was enhanced and lifetime increased by coupling it with an achiral column [26]. A variety of chiral pharmaceuticals were enantiolresolved and separated from impurities or their metabolites in human plasma by coupling achiral columns with polymeric chiral stationary phases

[27, 28]. Two chiral columns were also coupled to resolve three pairs of enantiomers simultaneously [29].

Each glycopeptide CSP has unique selectivity as well as complementary characteristics, and a considerable number of racemates have been resolved on all three of them. Interestingly, most of the resolved enantiomers have the same retention order on these macrocyclic CSPs. When they are mixed or coupled with each other, the selectivity on one CSP will not be canceled by another. Even if some compounds may not have the same retention order, the complementary effects will result in an identifiable selectivity. Therefore, the coupled chiral columns can be used as a screening tool and save chromatographers substantial time in method development.

In order to compare the retention and selectivity of coupled phases with individual CSPs, the parameters such as mobile phase composition, flow rate, detection wavelength and operating temperature were maintained constant during the screening process. In the new polar organic mode, the mobile phase was methanol with 0.02 % glacial acetic acid and 0.01 % anhydrous triethylamine (v/v) and the flow rate 2 mL min^{-1}. In reversed phase, the mobile phase was 25 % methanol and 75 % triethylammonium acetate buffer (0.1 %, pH 6) (v/v) and the flow rate 1 mL min^{-1}. In normal phase, the mobile phase was 40 % hexane and 60 % ethanol (v/v) and the flow rate 1.5 mL min^{-1}. The minimum column length was determined to be 10 cm to achieve efficient selectivity. The appropriate sequence of columns was important to the coupling practice. It proved to be beneficial that ristocetin A column, which usually has the lowest retaining ability, is put at the first position, whereas teicoplanin – which usually has the highest retaining ability – is at the last position.

In the new polar organic mode, it was found that the capacity factors of the analytes on the coupled columns, i.e., ristocetin A plus teicoplanin (R+T), ristocetin A plus vancomycin (R+V), and ristocetin A plus vancomycin plus teicoplanin (R+V+T), is approximately the average of those on the individual columns [30]. Although there is not enough evidence to show a linear relationship between retention and the composition of a stationary phase, it is likely that each stationary phase contributes equally to the retention. The contribution of each stationary phase to the selectivity and resolution is more complicated. In some cases, where vancomycin (V) and teicoplanin (T) had slightly different selectivity, the coupled columns (V+T) showed selectivity as good or better than the two individual columns. For these compounds, the coupled 10 cm columns (V+T) worked as well as a single 25 cm analytical column. However, for most compounds the coupled columns showed a medium selectivity compared to the individual columns. The individual column with the lower selectivity acts as a diluting factor. However, in the new polar organic mode, the diluting effect from the column with lower selectivity is not very strong (Fig. 2-7). Therefore, it is always beneficial to couple three columns in this mobile phase in order to achieve the broadest selectivity in the shortest time.

One potential problem associated with column coupling in reversed phase is relatively high back-pressure ($\sim$ 2600 psi at 1 mL min^{-1}). This will place a limit on the flow rate, which in turn limits the further reduction of analysis time. Also, compared to the new polar organic mode, the retention in reversed phase on coupled columns is deviated more from the average retention on the individual stationary phases,

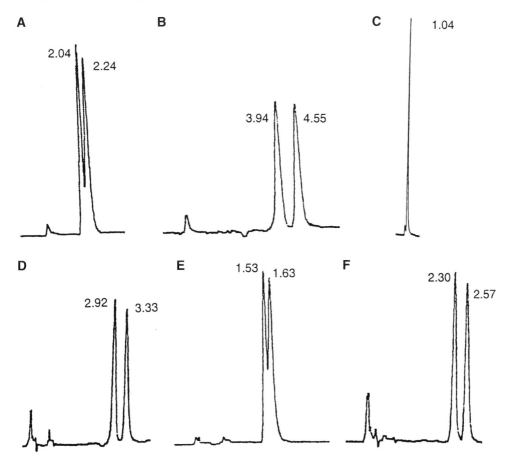

Fig. 2-7. Chromatograms of albuterol in the new polar organic phase on: vancomycin (A), teicoplanin (B), ristocetin A (C), vancomycin + teicoplanin (D), ristocetin A + vancomycin (E), and ristocetin A + vancomycin + teicoplanin (F). All columns were 100×4.6 mm. The numbers by the peaks refer to the retention time in minutes. The mobile phase was methanol with 0.02 % glacial acetic acid and 0.01 % triethylamine (v/v). The flow rate was 2.0 mL min^{-1} at ambient temperature (23 °C).

which made it more difficult to predict the retention time on coupled columns in this mode. In reversed phase, the individual column of lower selectivity again presented a diluting effect when it was coupled with other column(s) of higher selectivity (Fig. 2-8). However, coupled columns always had higher selectivity than the individual column of the lowest selectivity.

Similar to the new polar organic mode, the retention of analytes in normal phase is not difficult to predict. For all the compounds, the average of the retention on individual columns is fairly close to the retention on the coupled columns. The selectivity of most compounds on coupled columns is an average of the selectivities of individual columns (Fig. 2-9). However, it was found that the elution order for some compounds was reversed on ristocetin A and teicoplanin or vancomycin. As a result,

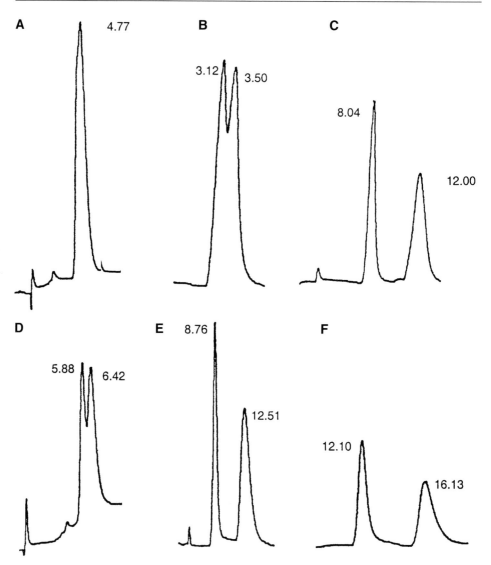

Fig. 2-8. Chromatograms of ketoprofen in reversed phase on: vancomycin (A), teicoplanin (B), ristocetin A (C), vancomycin + teicoplanin (D), ristocetin A + teicoplanin (E), and ristocetin A + vancomycin
+ teicoplanin (F). All columns were 100 × 4.6 mm. The numbers by the peaks refer to the retention time
in minutes. The mobile phase was methanol: 0.1 % triethylammonium acetate (25/75 v/v) pH 6.0. The
flow rate was 1.0 mL min^{-1} at ambient temperature (23 °C).

the selectivity could be lost on the coupled R+T or R+V columns. Therefore, it is the
preferred tactic to couple only vancomycin and teicoplanin in normal phase.

Column coupling proves to be a rapid screening approach in identifying chiral
selectivity in the most efficient and economical way. In addition to the potential for
the simultaneous analysis of a mixture, the coupling practice offers the advantages

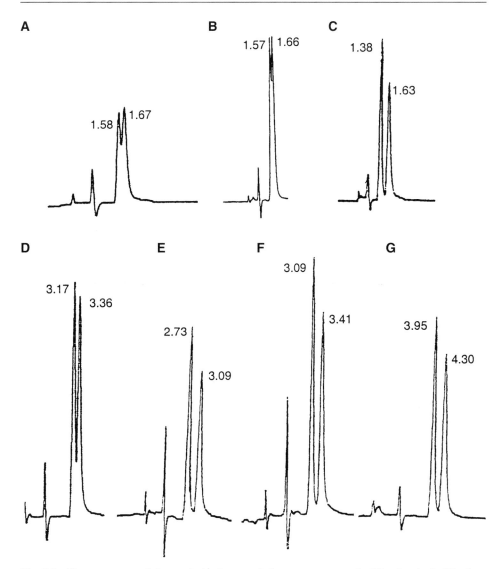

Fig. 2-9. Chromatograms of phensuximide in normal phase on: vancomycin (A), teicoplanin (B), ristocetin A (C), vancomycin + teicoplanin (D), ristocetin A + vancomycin (E), ristocetin A + teicoplanin (F), and ristocetin A + vancomycin + teicoplanin (G). All columns were 100×4.6 mm. The numbers by the peaks refer to the retention time in minutes. The mobile phase was ethanol: hexane (60/40 v/v) and the flow rate was 1.5 mL min^{-1} at ambient temperature (23 °C).

of broad selectivity, ease of operation and flexibility of using one, two or three columns for identifying enantioselectivity. The coupled columns R+V+T showed the broadest selectivity in the new polar organic mode and reversed phase, while the coupled columns V+T showed the broadest selectivity in normal phase.

A rule of thumb has been developed after a large number of analytes were tested. Once the selectivity was observed on the coupled column, a baseline separation can always be achieved on a 25 cm column under optimized conditions. Since the screening procedure already indicates the separation conditions, optimization is straightforward and requires a minimum amount of time.

The new polar organic mode provides broad selectivity in the shortest analysis time. Therefore, it is beneficial to start the screening process with the coupled columns R+V+T in the new polar organic mode. The three coupled columns R+V+T can then be directly switched to reversed phase with methanol as the organic modifier. In case of high back-pressure, the coupled columns R+T could be screened and vancomycin could be tested separately with THF as the organic modifier. In normal phase conditions, it is advantageous to couple columns V+T and use ristocetin A as the backup. When the column coupling method is applied, only four to five runs (each run is within 25 min) are needed before a positive indication is reached concerning the feasibility of a certain macrocyclic glycopeptide CSP and the corresponding operating mode. At this point, either the separation can be optimized using an individual CSP, or other types of CSPs can be screened.

2.4   Optimization

2.4.1   Effect of Flow Rate and Temperature on Enantiomeric Separations

A general phenomenon observed with chiral stationary phases having hydrophobic pockets is that a decrease of flow rate results in an increase in resolution. This change has significant impact mostly in reversed-phase mode (see Fig. 2-10).

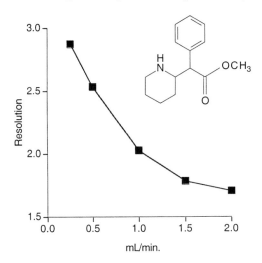

Fig. 2-10. The effect of flow rate on the resolution of methylphenidate enantiomers on vancomycin CSP (250 × 4.6 mm). The mobile phase was methanol: 1.0 % triethyl-ammonium acetate (95/5 v/v) pH 4.1 at ambient temperature (23 °C).

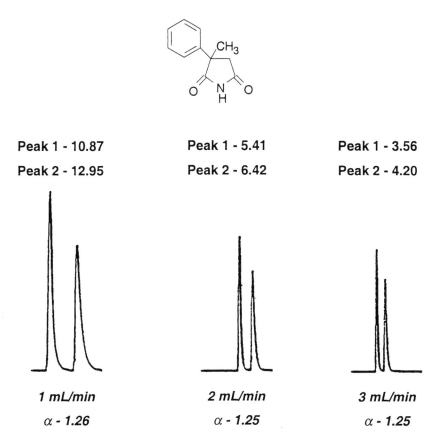

Peak 1 - 10.87   **Peak 1 - 5.41**   **Peak 1 - 3.56**

Peak 2 - 12.95   **Peak 2 - 6.42**   **Peak 2 - 4.20**

1 mL/min   *2 mL/min*   *3 mL/min*

α - *1.26* α - *1.25* α - *1.25*

Fig. 2-11. The effect of flow rate on the selectivity of α-methyl-α-phenyl succinimide on teicoplanin CSP (250×4.6 mm) in normal phase. The mobile phase was ethanol: hexane (20/80 v/v) at ambient temperature (23 °C).

This is because the increased turbulence from higher flow rates decreases the possibility for inclusion complexation, a necessary event for chiral recognition in reversed phase. Some effect has also been observed in the new polar organic mode when k_1' (capacity factor) is small (< 1). Flow rate has no effect on selectivity in the typical normal-phase system, even at flow rates up to 3 mL min^{-1} (see Fig. 2-11).

Changes in temperature have a dramatic effect with all three mobile phase systems on these glycopeptide columns. This is because the binding constant of a solute to the macrocycle involves several interactive mechanisms that change dramatically with temperature. Inclusion complex formation is effectively prevented for most solutes in the temperature range of 60–80 °C. Lowering the temperature generally enhances the weaker bonding forces, resulting in better chiral separation. An example demonstrating the typical effects of temperature is shown in Table 2-5. Retention, selectivity and resolution all decrease when temperature increases. In a recent study [31], it was found that when temperature changes, the retention behavior for

Table 2-5. The effect of temperature on the separation of N-carbamyl-phenylalanine enantiomers on vancomycin CSP.[a]

T (°C)	k'	α	R_s
0	0.51	1.39	1.5
5	0.39	1.34	1.3
15	0.38	1.23	1.0
22	0.31	1.20	0.8
35	0.27	1.11	0.7
45	0.22	1.00	0.0

[a] The column was 250×4.6 mm (i.d.). The mobile phase was acetonitrile: 1 % triethylammonium acetate (10/90, v/v), pH 4.1. The flow rate was 1 mL min⁻¹.

the individual enantiomers of some racemic compounds changes in different ways. Consequently, the elution order of two enantiomers may be reversed on a CSP by simply increasing the operating temperature.

2.4.2 Optimization of Enantiomeric Separations in the New Polar Organic Mode

Since the glycopeptide CSPs contain ionizable groups, ionic interactions play the key role in the chiral recognition mechanism in the new polar organic mode. The ratio of acid/base controls the selectivity due to the fact that the changes in the ratio of acid and base affect the degree of charges on both glycopeptides and the analytes. It is this subtle difference that differentiates the binding energy between two enantiomers and the CSP. It also allows salts of bases to be freely chromatographed. As shown in Fig. 2-12, the effect of acid/base ratio on the separation of sotalol enantiomers on teicoplanin CSP is significant. Generally, the acid/base ratio can be manipulated from 4:1 to 1:4 depending on the sample charge. In a few rare instances the ratios were higher but typically 2:1 is the standard for screening methodology.

The concentration of acid and base controls the retention. In general, the retention time is inversely proportional to acid and base concentrations, which can range from 0.001 % to 1 % (v/v). When intramolecular interaction such as H-bonding overwhelms the potential interaction between CSPs and analytes, higher concentrations of acid and base are needed to alleviate this effect and to promote the necessary interaction for chiral recognition. Usually, glacial acetic acid (HOAc) and triethylamine (TEA) are used as effective acid/base components. In some cases, smaller amounts of trifluoroacetic acid (TFA) can be used instead of acetic acid, whereas ammonia can be used as the alternative base. TFA is advantageous in that it enhances the peak shape and efficiency for some polar compounds, and its higher volatility is more desirable in liquid chromatography – mass spectrometry (LC – MS) operation. However, care must be taken in using TFA, as it may deteriorate the column when it co-exists with water in the column.

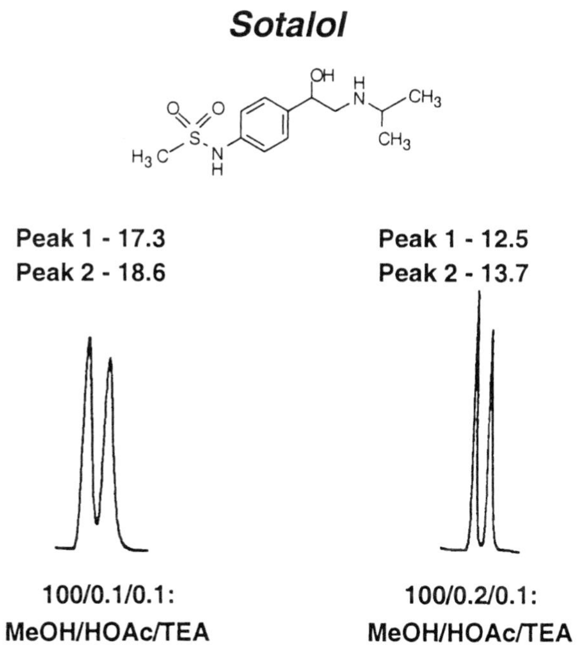

Sotalol

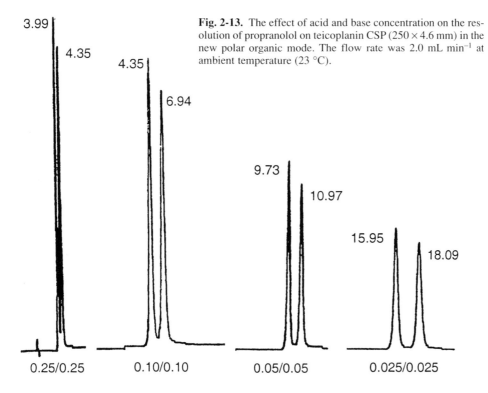

Fig. 2-12. The effect of the acid/base ratio on the selectivity of sotalol on teicoplanin CSP (250 × 4.6 mm) in the new polar organic mode. The flow rate was 1.0 mL min⁻¹ at ambient temperature (23 °C).

Peak 1 - 17.3
Peak 2 - 18.6

Peak 1 - 12.5
Peak 2 - 13.7

100/0.1/0.1:
MeOH/HOAc/TEA

100/0.2/0.1:
MeOH/HOAc/TEA

3.99
4.35

4.35
6.94

9.73
10.97

15.95
18.09

Fig. 2-13. The effect of acid and base concentration on the resolution of propranolol on teicoplanin CSP (250 × 4.6 mm) in the new polar organic mode. The flow rate was 2.0 mL min⁻¹ at ambient temperature (23 °C).

0.25/0.25   0.10/0.10   0.05/0.05   0.025/0.025

An example of the effect of acid and base concentration on the separation of propranolol is shown in Fig. 2-13. In this case, the baseline separation is achieved by adjusting the concentration without changing the acid/base ratio.

2.4.3 Optimization of Enantiomeric Separations in Reversed Phase

In addition to temperature and flow rate, the retention and selectivity in reversed phase are controlled by: (i) the concentration and type of organic modifier; and (ii) the type, concentration and pH of the buffer.

2.4.3.1 Effect of Organic Modifier on Enantiomeric Separations

Various organic modifiers have been used on the glycopeptide CSPs. Methanol, ethanol, isopropanol, acetonitrile and tetrahydrofuran are the most common solvents which give good selectivities for various analytes. Chiral separations on glycopeptide CSPs arc affected by organic modifiers in two ways: (i) the percentage of modifier in the mobile phase; and (ii) the nature of the modifier. Among the chiral compounds that have been resolved in reversed phase on glycopeptide CSPs, especially on teicoplanin and ristocetin A, a large number are α and β amino acids, imino acids and small peptides [32–35]. For these molecules, alcohols (methanol, ethanol, and isopropanol) prove to be effective modifiers. At the same percentage, the three alcohols have shown somewhat different selectivity and resolution with different amino acids (Table 2-6). Unlike traditional reversed phase operations, an increase in the concentration of alcohols results in longer retention and thus, better resolution (Fig. 2-14). Therefore, most amino acids and small peptides are resolved at medium percentage (around 50 %) of alcohols as organic modifiers.

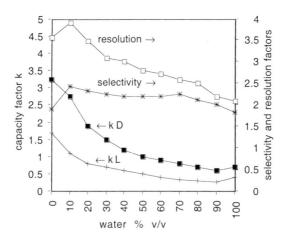

Fig. 2-14. The effect of organic modifier on retention, selectivity and resolution of methionine on teicoplanin CSP (250×4.6 mm). The flow rate was 1.0 mL min^{-1} at ambient temperature (23 °C).

Table 2-6. The effects of organic modifier on the separation of amino acids on teicoplanin CSP.[a]

Compound	Methanol			Ethanol			Isopropanol[b]		
	k'_1	α	R_s	k'_1	α	R_s	k'_1	α	Rs
o-Thiophenylglycine	1.63	2.4	4.7	0.57	4.9	5.4	0.80	4.9	10.1
3-(1-Naphthyl) alanine	2.79	1.3	1.8	1.0	1.6	1.3	1.04	1.8	3.8
Pipecolic acid	2.25	1.6	1.7	1.12	2.0	1.9	1.32	2.2	4.0
Threonine	1.40	1.1	1.0	0.99	1.5	1.6	0.62	1.3	1.6

[a] These data were generated with a 250×4.6 mm teicoplanin column. The mobile phase consisted of alcohol: water (60/40, v/v), and the flow rate was 1 mL min⁻¹ at ambient temperature (23 °C).
[b] Flow rate was 0.5 mL min⁻¹ due to the viscosity of the propanol: water mobile phase that produced a high back-pressure.

For the compounds other than amino acids and peptides [36–39], the retention-versus mobile phase composition usually has a characteristic U-shape. An example is shown in Fig. 2-15 for the separation of 5-methyl-5-phenylhydantoin enantiomers [16]. Increasing the concentration of acetonitrile modifier from 0 % to 50 % caused the retention and enantioselectivity to decrease. At acetonitrile concentrations between 50 % and 80 %, there is little enantioselective retention and the analytes elute near the dead volume of the column. In neat acetonitrile, the retention increases and enantioselectivity returns. The type, as well as the percentage, of organic modifiers have dramatic effects on chiral separations. The effect of organic modifier on the resolution of fluoxetine on vancomycin CSP is demonstrated in Fig. 2-16. When a low percentage (10 %) of organic modifiers is used, THF gave the best resolution. At 50 %, methanol resolved the enantiomers best, while at 90 % the performance of isopropanol exceeded that others. For each organic modifier, different percentages resulted in varying resolutions.

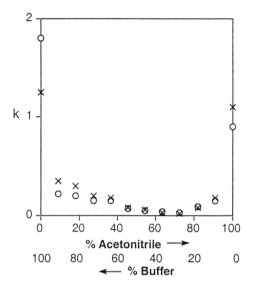

Fig. 2-15. Reversed-phase retention of the first eluted and the second eluted enantiomers of 5-methyl-5-phenylhydantoin as a function of mobile phase composition. The column was a 250×4.6 mm vancomycin CSP. The buffer was triethylammonium acetate at pH 7.0. The flow rate was 1.0 mL min⁻¹ at ambient temperature (23 °C).

In reversed phase, many chiral separations (excluding amino acids) on glycopeptide CSPs are achieved at relatively low percentages of organic modifiers (10–20 %); hence, the screening conditions are set up this way. Alcohols generally require higher concentrations than acetonitrile or THF for comparable retention times. Empirically, THF works best with vancomycin in terms of broad selectivity and high efficiency, while methanol works best for teicoplanin and ristocetin A. Nevertheless, the final choice and percentage of organic modifier is eventually determined by the nature of the analyte and should be optimized accordingly.

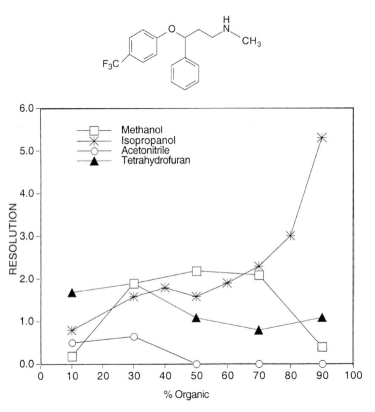

Fig. 2-16. The effect of organic modifiers on the resolution of fluoxetine enantiomers on vancomycin CSP (250 × 4.6 mm). The flow rate was 1.0 mL min^{-1} at ambient temperature (23 °C). (Courtesy of Scott Sharpe, Eli Lilly & Co.)

This observation is important in the study of the chiral recognition mechanism in this system. This may be a practical matter when determining the trace amount of one enantiomer in the presence of its dominant antipode. The smaller peak is always desired to be eluted first for best quantitation.

2.4.3.2 Effect of Aqueous Buffer on Chiral Separations

For most free amino acids and small peptides, a mixture of alcohol with water is a typical mobile phase composition in the reversed-phase mode for glycopeptide CSPs. For some bifunctional amino acids and most other compounds, however, aqueous buffer is usually necessary to enhance resolution. The types of buffers dictate the retention, efficiency and – to a lesser effect – selectivity of analytes. Triethylammonium acetate and ammonium nitrate are the most effective buffer systems, while sodium citrate is also effective for the separation of profens on vancomycin CSP, and ammonium acetate is the most appropriate for LC/MS applications.

Buffer pH is the most important parameter in chiral selectivity on all glycopeptide CSPs. The stability of the analyte–CSP complex depends on the degree of interaction between both entities. Therefore, retention and selectivity of molecules possessing ionizable (acidic or basic) or even neutral functional groups can be affected by altering the pH value. The effect of pH on the resolution of the neutral coumachlor enantiomers is shown in Fig. 2-17. Typically, the use of lower pH has the effect of suppressing the nonchiral retention mechanisms of the stationary phase, which in turn enhances the chiral interactions, and leads to higher resolution. Glycopeptide CSPs are suitable for working at the following pH ranges: 4.0–7.0 for vancomycin, 3.8–6.5 for teicoplanin, and 3.5–6.8 for ristocetin A.

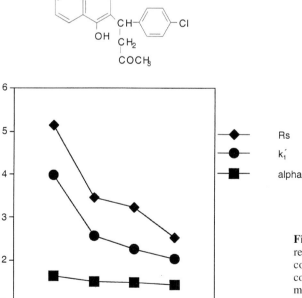

Fig. 2-17. The effect of pH on the retention, selectivity and resolution of coumachlor enantiomers on vancomycin CSP (250 × 4.6 mm). The mobile phase was acetonitrile: 1 % triethylammonium acetate (10/90 v/v). The flow rate was 1.0 mL min⁻¹ at ambient temperature (23 °C).

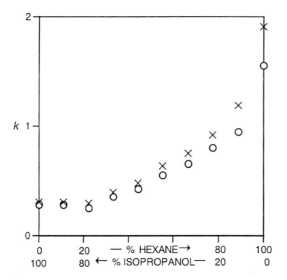

Fig. 2-18. Normal phase retention of the first eluted and second eluted enantiomer of mephenytoin on vancomycin CSP (250 × 4.6 mm). The flow rate was 1.0 mL min^{-1} at ambient temperature (23 °C).

2.4.4 Optimization of Enantiomeric Separations in Normal Phase

Typical normal-phase operations involved combinations of alcohols and hexane or heptane. In many cases, the addition of small amounts (< 0.1 %) of acid and/or base is necessary to improve peak efficiency and selectivity. Usually, the concentration of polar solvents such as alcohol determines the retention and selectivity (Fig. 2-18). Since flow rate has no impact on selectivity (see Fig. 2-11), the most productive flow rate was determined to be 2 mL min^{-1}. Ethanol normally gives the best efficiency and resolution with reasonable back-pressures. It has been reported that halogenated solvents have also been used successfully on these stationary phases as well as acetonitrile, dioxane and methyl *tert*-butyl ether, or combinations of the these. The optimization parameters under three different mobile phase modes on glycopeptide CSPs are summarized in Table 2-7.

Table 2-7. Summary of optimization parameters on glycopeptide CSPs.

New polar organic mode	a. Type of acid and base
	b. Acid/base ratio
	c. Concentration of acid and base
	d. Flow rate
Reversed phase	a. Type of organic modifier
	b. Concentration of organic modifier
	c. Type of aqueous buffer
	d. Concentration of aqueous buffer
	e. pH of aqueous buffer
	f. Flow rate
	g. Temperature
Normal phase	a. Type of polar solvent
	b. Concentration of polar solvent
	c. Acid and base as modifiers
	d. Temperature

2.5 Concluding Remarks

The macrocyclic glycopeptides vancomycin, teicoplanin and ristocetin A have proved to be powerful chiral stationary phases. The unique structures and the variety of functional groups on the macrocycles utilize a large number of interactions possible for chiral recognition. Covalently bonded to silica gel through multiple linkages, these CSPs are multi-modal and it is possible to switch from one mobile phase to another without deleterious effects. The enantiomeric separations are sometimes predictable for structure-related racemates. One important characteristic for the glycopeptide CSPs is their complementary effect, which makes column coupling an efficient and economical screening methodology. The method development with these CSPs is discussed in detail in the new polar organic mode, and reversed-phase and normal-phase modes. The enantiomeric separations with glycopeptide CSPs can be optimized by controlling flow rate, temperature, the acid/base ratio and concentration, the type and amount of organic modifier, the type, concentration and pH of aqueous buffer. As a result of extensive research on these CSPs by scientists worldwide, the summarized method development and optimization protocols may help to stimulate further investigation and understanding of the chiral recognition mechanism, as well as rapid and efficient resolution of greater numbers of chiral compounds.

Acknowledgment

The authors would like to thank Vicki Sutter and Leslie Wrenn for their help with the preparation of this manuscript.

References

[1] Armstrong D. W. (1997), *LC-GC May (Supplement)* : S20–S28.
[2] Okamoto Y., Honda S., Okamoto H., Yuki H., Murata S., Noyori R., Takaya H. (1981), *J. Am. Chem. Soc.* 103 : 6911.
[3] Hermansson J. (1983), *J. Chromatogr.* 269 : 71.
[4] Lindner K. R., Mannschreck A. (1980), *J. Chromatogr.* 193 : 308.
[5] Okamoto Y., Kawashima M., Yakamoto Y., Hatada K. (1984), *Chem. Lett.* 739.
[6] Allenmark S. (1999), *Enantiomer* 4 : 67.
[7] Roumeliotis P., Unger K. K., Kurganov A. A., Davankov V. A. (1982), *Angew. Chem.* 94 : 928.
[8] Mikes F., Boshart G., Gil-Av E. (1976), *J. Chromatogr.* 122 : 205.
[9] Pirkle W. H., Finn J. M., Schreiner J. L., Hamper B. C. (1981), *J. Am. Chem. Soc.* 103 : 3964.
[10] Helgeson R., Timko J., Moreau P., Peacock S., Mayer J., Cram D. J. (1974), *J. Am. Chem. Soc.* 96 : 6762.
[11] Armstrong D. W., Ward T. J., Armstrong R. D., Beesley T. E. (1986), *Science* 232 : 132.
[12] Stalcup A. M., Faulkner J. R., Tang Y., Armstrong D. W., Levy L. W., Regalado E. (1991), *Biomed. Chromatogr.* 15 : 3.
[13] Stalcup A. M., Chang S. C., Armstrong D. W. (1991), *J. Chromatogr.* 540 : 113.
[14] Armstrong D. W., Stalcup A. M., Hilton M. L., Duncan J. D., Faulkner J. R., Chang S. C. (1990), *Anal. Chem.* 62 : 1610.
[15] Stalcup A. M., Gahm K. H. (1996), *Anal. Chem.* 68 : 1369.
[16] Armstrong D. W., Tang Y., Chen S., Zhou Y., Bagwill C., Chen J.-R. (1994), *Anal. Chem.* 66 : 1473.
[17] Ekborg-Ott K. H., Liu Y., Armstrong D. W. (1998), *Chirality* 10 : 434.
[18] Armstrong D. W., Liu Y., Ekborg-Ott K. H. (1995), *Chirality* 7 : 474.
[19] Ekborg-Ott K. H., Kullman J. P., Wang X., Gahm K., He L., Armstrong D. W. (1998), *Chirality* 10 : 627.
[20] Beesley T. E., *Chirobiotic Handbook*, 3rd Edition, Advanced Separation Technologies Inc., 1999.
[21] Chen P., Xing Z., Liu M., Liao Z., Huang D. (1999), *J. Chromatogr. A*, 839 : 239.
[22] Bjarnason B., Chimuka L., Ramstroem O. (1999), *Anal. Chem.* 71 : 2152.
[23] Barnett S. A., Frick L. W. (1979), *Anal. Chem.* 51 : 641.
[24] Benedict C. R. (1987), *J. Chromatogr.* 385 : 369.
[25] Bonanne L. M., Denizot L. M., Tchoreloff P. C., Pierre C., Puisieux F., Cardot P. J. (1992), *Anal. Chem.* 64 : 371.
[26] Chu K. M., Shieh S. M., Wu S. H., Oliver Y. P. (1992), *J. Chromatogr. Sci.* 30 : 171.
[27] Kristensen K., Angelo K. R., Blemmer T. (1994), *J. Chromatogr. A* 666 : 283.
[28] Naidong W., Lee J. W., Hulse J. D. (1994), *J. Liq. Chromatogr.* 17 : 3747.
[29] Johnson D. V., Wainer I. W. (1996), *Chirality* 8 : 551.
[30] Wang A. X., Lee J. T., Beesley T. E. (2000), LC–GC, June: 626–639.
[31] Scott R. P. W., Beesley T. E. (1999), *Analyst* 124 : 713.
[32] Bethod A., Liu Y., Bagwill C., Armstrong D. W. (1996), *J. Chromatogr. A* 731 : 123.
[33] Peter A., Torok G., Armstrong D. W. (1998), *J. Chromatogr. A* 793 : 283.
[34] Tesarova E., Bosakova Z., Pacakova V. (1999), *J. Chromatogr. A* 838 : 121.
[35] Lehotay J., Hrobonova K., Krupcik J., Cizmarik L. (1998), *Pharmazie* 53 : 12.
[36] Joyce K. B., Jones A. E., Scott R. J., Biddlecombe R. A., Pleasance S. (1998), *Rapid Commun. Mass Spectrom.* 12 : 1899.
[37] Fried K. M., Koch P., Wainer I. W. (1998), *Chirality* 10 : 484.
[38] Aboul-Enein H. Y., Serignese V. (1998), *Chirality* 19 : 358.
[39] Tesarova E., Zaruba K., Flieger M. (1999), *J. Chromatogr. A* 844 : 137.

3 Combinatorial Approaches to Recognition of Chirality:
Preparation and Use of Materials for the Separation of Enantiomers

František Švec, Dirk Wulff and Jean M. J. Fréchet

3.1 Introduction

The continuing trend to replace racemic drugs, agrochemicals, flavors, fragrances, food additives, pheromones, and some other products with their single enantiomers is driven by their increased efficiency, economic incentive to avoid the waste of the inactive enantiomer, and regulatory action resulting from the awareness that individual enantiomers have different interactions with biological systems. There are several methods to obtain enantiomerically pure compounds. The most important are: (i) syntheses based on chiral starting materials from natural sources such as amino acids; (ii) enantioselective reactions; and (iii) separations of mixtures of enantiomers using methods such as crystallization via diastereoisomers, enzymatic or chemical kinetic resolution, and chromatographic separation [1–3].

Although very efficient, the broad application of the direct preparation is restricted due to the limited number of pure starting enantiomers. The design of a multistep process that includes asymmetric synthesis is cumbersome and the development costs may be quite high. This approach is likely best suited for the multi-ton scale production of "commodity" enantiomers such as the drugs ibuprofen, naproxen, atenolol, and albuterol. However, even the best asymmetric syntheses do not lead to products in an enantiomerically pure state (100 % enantiomeric excess). Typically, the product is enriched to a certain degree with one enantiomer. Therefore, an additional purification step may be needed to achieve the required enantiopurity.

The chromatographic methods that include gas and liquid chromatography, as well as electrophoresis fit into the third group of methods designed for the separation of individual enantiomers from their mixtures. These techniques, characterized by the use of chiral stationary phases (CSP) and chiral additives, respectively, emerged more than three decades ago as valuable methods for analytical assays in academic laboratories and clinical testing. Since chromatography can be used in virtually any scale, it has also been used for preparative- and production-scale separations due to its high efficiency and ease of operation. For example, preparative liquid chromatography implemented in the simulated moving bed (SMB) format enables the isolation of sufficient quantities of pure chiral compounds to carry out early pharmacological and toxicological studies providing both enantiomers for comparative biological testing [4]. Chromatographic separations continue to be used for the process and quality controls, even after a large-scale asymmetric technology had been developed and implemented.

3.2   Engineering of a Chiral Separation Medium

A chiral separation medium is a complex system. Ideally, interactions that lead to enantioseparation are maximized while nonspecific interactions should be completely suppressed. Typically, a medium for chromatographic separations involves the solid support, the selector, and the linker connecting the two, as shown in scheme 3-1.

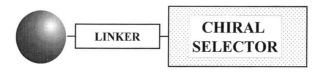

Scheme 3-1

Most commercial CSPs contain chiral selectors (vide infra) supported by porous silica beads. Silica-based chromatographic supports have numerous advantages such as broad range of different porosities, high mechanical stability, and resistance to swelling. Columns packed with these materials generally exhibit high efficiencies. However, residual silanol groups on the surface of the silica may contribute to non-specific interactions with the separated enantiomers, thereby decreasing the overall selectivity of the separation medium. This was demonstrated by the improved enantioselectivities measured for CSPs that had their residual silanol groups capped after the attachment of a chiral selector [5]. In contrast to silica particles, synthetic organic polymers are more seldom used as a platform for the preparation of chiral stationary phases. Their excellent stability over the entire range of pH, the variety of available chemistries, and the more accurate control of both the functionality and the porous properties make them a good alternative to the well-established silica matrices. Although porous polymer beads have been used successfully for a wide variety of chromatographic separations including the size-exclusion chromatography of synthetic polymers, the normal-phase or reversed-phase separations of small molecules, and the ion-exchange and hydrophobic interaction chromatography of biopolymers, there are only a few examples of polymer-based chiral separation media with attached selectors [6–10]. In contrast, a larger number of chiral separations has been demonstrated using polymer-based molecular imprinted separation media [5, 11–13].

The use of a polymeric support also affords a unique opportunity to control independently the variables that may affect the chiral recognition process, which is hard to achieve with silica. For example, the type and number of reactive sites can be easily adjusted with a polymer support. We recently reported an extensive study of the effects of support chemistry, surface polarity, length and polarity of the tether, and selector loading [14]. Our group also demonstrated that separation media based on an organic polymer support provide enhanced enantioselectivities and reduced retention times when compared to analogous silica-based chiral stationary phases, mostly as a result of substantially decreased nonspecific interactions (Fig. 3-1) [8].

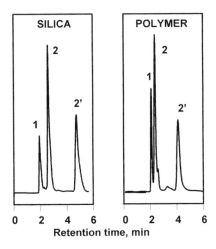

Fig. 3-1. Separation of racemic 3,5-dinitrobenzamido leucine *N,N*-diallylamide on silica and polymer-based chiral stationary phases. Conditions: column size 150 × 4.6 mm i.d.; mobile phase 20 % hexane in dichloromethane; flowrate 1 mL min^{-1}; injection 7 μg. Peaks shown are: 1,3,5-tri-*tert.*-butylbenzene (1), *R*-enantiomer (2); *S*-enantiomer (2'). (Reprinted with permission from ref. [8]. Copyright 1997 American Chemical Society.)

3.3   Chiral Selectors

Enantioseparation is typically achieved as a result of the differences in interaction energies $\Delta(\Delta G)$ between each enantiomer and a *selector*. This difference does not need to be very large, a modest $\Delta(\Delta G) = 0.24$ kcal/mol is sufficient to achieve a separation factor α of 1.5. Another mechanism of discrimination of enantiomers involves the preferential inclusion of one into a "cavity" or within the helical structure of a polymer. The selectivity of a selector is most often expressed in terms of retention of both enantiomers using the separation factor α that is defined as:

$$\alpha = k'_2/k'_1 \tag{1}$$

where k'_1 and k'_2 are the retention factors of the first and the second peak, respectively, defined as

$$k' = (t_r - t_0)/t_0 \tag{2}$$

where t_r and t_o are the retention times for the analyte and the unretained compound (void volume marker), respectively.

Chiral selectors are the most important part of the separation system. This is why most attention during the development of new chiral separation media has always been devoted to selectors. As a result of the growing interest in chiral chromatography, a large number of phases and additives have emerged to meet the challenge of enantiomer separations [5, 15, 16]. For example, more than 90 CSPs were commercially available for separations in the liquid chromatographic mode in the early 1990s [17].

The majority of currently known selectors can be divided into the following categories:

1. *Proteins.* A chiral stationary phase with immobilized α_1-acid glycoprotein on silica beads was introduced by Hermansson in 1983 [18, 19]. Several other proteins such as chicken egg albumin (ovalbumin), human serum albumin, and cellohydrolase were also used later for the preparation of commercial CSPs. Their selectivity is believed to occur as a result of excess of dispersive forces acting on the more retained enantiomer [17]. These separation media often exhibit only modest loading capacity.

2. *Modified polysaccharides.* Although derivatives of microcrystalline cellulose have been used for chiral separations since the 1970s [20], materials useful in high-performance liquid chromatography (HPLC) were only developed by Okamoto in the mid-1980s. CSPs involving various esters and carbamates of cellulose and amylose coated on wide-pore silica are currently the most frequently used chiral media for chromatographic separations in both analytical and preparative scales [21, 22]. Although they often do not exhibit very high selectivities, they separate an extremely broad range of different racemates.

3. *Synthetic polymers.* In the 1970s, Blaschke prepared several crosslinked gels from N-acryloylated L-amino acids and a small percentage of ethylene dimethacrylate or divinylbenzene and used them for the low-pressure chromatographic resolution of racemic amino acid derivatives and mandelic acid [23]. Another polymer-based CSP was later prepared by Okamoto from isotactic poly(triphenylmethyl methacrylate). This material is the prototypical polymeric selector with a well-defined one-handed helical structure [24]. This polymer was prepared by anionic polymerization using a chiral organolithium initiator, and then coated onto porous silica beads. While these columns were successful in the separation of a broad variety of racemates, their relative lack of chemical stability and high cost make them less suitable for large-scale applications.

4. *Macrocyclic glycopeptides.* The first of these CSPs – based on the "cavity" of the antibiotic vancomycin bound to silica – was introduced by Armstrong [25]. Two more polycyclic antibiotics teicoplanin and ristocetin A, were also demonstrated later. These selectors are quite rugged and operate adequately in both normal-phase and reversed-phase chromatographic modes. However, only a limited number of such selectors is available, and their cost is rather high.

5. *Cyclic low molecular weight compounds.* Chiral separations using chiral crown ethers immobilized on silica or porous polymer resins were first reported in the

mid-1970s [6]. These highly selective and stable selectors have found only a limited application for the separation of atropoisomers. In contrast, modified cyclic glucose oligomers – cyclodextrins – have proven to be very universal chiral selectors for chiral separations in electrophoresis, gas chromatography, and liquid chromatography [26]. In addition to the formation of reversible stereoselective inclusion complexes with the hydrophobic moieties of the solute molecules that fit well into their cavity, they are often functionalized to further enhance hydrogen bonding and dipolar interactions [27]. Attached covalently to porous silica beads, they afford very robust CSPs with modest selectivities for a number of racemates.

6. *Metal ion complexes.* These "classic" CSPs were developed independently by Davankov and Bernauer in the late 1960s. In a typical implementation, copper (II) is complexed with L-proline moieties bound to the surface of a porous polymer support such as a Merrifield resin [28–30]. They only separate well a limited number of racemates such as amino acids, amino alcohols, and hydroxyacids.

7. *Small chiral molecules.* These CSPs were introduced by Pirkle about two decades ago [31, 32]. The original "brush"-phases included selectors that contained a chiral amino acid moiety carrying aromatic π-electron acceptor or π-electron donor functionality attached to porous silica beads. In addition to the amino acids, a large variety of other chiral scaffolds such as 1,2-disubstituted cyclohexanes [33] and cinchona alkaloids [34] have also been used for the preparation of various brush CSPs.

3.3.1 Design of New Chiral Selectors

CSPs with optically active polymers, such as modified cellulose, polyacrylates, and proteins, have been used successfully for a variety of enantioseparations [5, 13, 15, 17]. Despite extended studies, the mechanism of separation for these CSPs is not yet completely understood, which makes it difficult to develop new media of this type. In contrast, bonded natural and synthetic chiral selectors such as substituted cyclodextrins, crown ethers, and brush-type selectors have several advantages including well-defined molecular structures and sufficiently developed enantiomer "recognition" models. For example, the separation of enantiomers with brush-type stationary phases is based on the formation of diastereoisomeric adsorbate "complexes" between the analyte and the selector. According to the Dalgliesh's 3-point model [35], enantiomer recognition is achieved as a result of three simultaneous attractive interactions (donor–acceptor interactions such as hydrogen bonding, π-stacking, dipole–dipole interactions, etc.) between the selector and one of the enantiomers being separated. At least one of these interactions must be stereochemically dependent [36–39]. Compared to all other selectors, brush-type systems afford the most flexibility for the planned development of a variety of different chiral stationary phases suitable for the separation of a broad range of analyte types [40, 41].

The majority of the original chiral selectors for brush-type CSPs were derived from natural chiral compounds. Selectors prepared from amino acids, such as phenyl glycine and leucine, [42–45] and quinine [46, 47], are just a few examples of the most common chiral moieties. However, further development of highly selective CSPs requires the design of new types of synthetic receptors that will also make use of compounds outside the pool of natural chiral building blocks.

3.4   In Pursuit of High Selectivity

According to Equation 3, the resolution R_s of two peaks in column separation is controlled by three major variables: retention defined in terms of the retention factor k'; column efficiency expressed as the number of theoretical plates N; and selectivity characterized by the selectivity factor α [48]:

$$R_s = \frac{\sqrt{N}}{4}(\alpha - 1)\frac{k_1'}{1 + k_1'} \tag{3}$$

In this equation, k_1' is the retention factor of the first peak. The most significant contribution to the overall resolution has the selectivity term $(\alpha - 1)$ since the resolution is a linear function of the selectivity factor. Obviously, an excellent separation can also be achieved on columns with a high efficiency. However, the dependency of resolution on efficiency is not linear, and levels off at high efficiencies thus making the quest for a further increase less useful. Since the technology of packed analytical columns is well established and columns with very high efficiencies can be produced, baseline enantioseparations are achieved even with selectors that have low selectivity factors α close to 1. This is why many commercial columns are very successful despite their modest selectivity factors for most racemates that typically do not exceed $\alpha = 3$. In fact, very high selectivity factors are often not desirable for analytical separations since the second peak would elute much later and the time required for the separation would be extended unnecessarily [49]. A highly desirable feature for chiral columns is their broad selectivity, i.e. their ability to separate a large number of various enantiomers.

Most of the criteria and features outlined above for liquid chromatography media also apply to the development of selectors for electrodriven separations such as electrophoresis and electrochromatography.

Chromatographic separations in preparative columns and on preparative and process scale are based on the same concepts. However, packing large-scale columns to achieve efficiencies matching those of analytical columns remains a serious challenge. Typically, preparative columns have much lower efficiencies even if they are packed with analytical grades of stationary phases. Therefore, preparative columns have to be much longer in order to obtain the same number of theoretical plates that enable separations similar to those achieved in smaller columns. Unfortunately, the use of longer columns substantially contributes to the costs of the equipment, and their ultimate length is limited by the overall pressure drop that can be tolerated by the system. SMB technology helps to solve both these difficulties [50].

A better solution for preparative columns is the development of separation media with substantially increased selectivities. This approach allows the use of shorter columns with smaller number of theoretical plates. Ultimately, it may even lead to a batch process in which one enantiomer is adsorbed selectively by the sorbent while the other remains in the solution and can be removed by filtration (single plate separation). Higher selectivities also allow overloading of the column. Therefore, much larger quantities of racemic mixtures can be separated in a single run, thus increasing the throughput of the separation unit. Operation under these "overload" conditions would not be possible on low selectivity columns without total loss of resolution.

Another important issue that must be considered in the development of CSPs for preparative separations is the solubility of enantiomers in the mobile phase. For example, the mixtures of hexane and polar solvents such as tetrahydrofuran, ethyl acetate, and 2-propanol typically used for normal-phase HPLC may not dissolve enough compound to overload the column. Since the selectivity of chiral recognition is strongly mobile phase-dependent, the development and optimization of the selector must be carried out in such a solvent that is well suited for the analytes. In contrast to analytical separations, separations on process scale do not require selectivity for a broad variety of racemates, since the unit often separates only a unique mixture of enantiomers. Therefore, a very high key-and-lock type selectivity, well known in the recognition of biosystems, would be most advantageous for the separation of a specific pair of enantiomers in large-scale production.

Despite continuing progress in the design of new selectors, the process is slow as it mostly involves a traditional one-selector/one-column-at-a-time methodology.

3.5 Acceleration of the Discovery Process

3.5.1 Reciprocal Approach

The first approach to the accelerated development of chiral selectors reported by Pirkle's group in the late 1970s relied on the "principle of reciprocity" [51]. This is based on the concept that if a molecule of a chiral selector has different affinities for the enantiomers of another substance, then a single enantiomer of the latter will have different affinities for the enantiomers of the identical selector. In practice, a separation medium is prepared first by attaching a single enantiomer of the target compound to a solid support that is subsequently packed into a HPLC column. Racemates of potential selectors are screened through this column to identify those that are best separated. The most promising candidate is then prepared in enantiopure form and attached to a support to afford a CSP for the separation of the target racemate. This simple technique was used by several groups for the screening of various families of compounds [52–55] The reciprocal method is particularly suitable for situations in which the target enantiomer is known and its separation from a racemate is required. Since chromatographic techniques are readily automated, a broad variety of novel chiral ligands may be considered.

3.5.2 Combinatorial Chemistry

Although the reciprocal approach potentially enables the screening of large numbers of compounds, only the advent of combinatorial chemistry brought about the tools required for the synthesis of large libraries of potential selectors in a very short period of time. In addition, using the methods of combinatorial chemistry, novel strategies different from those of the reciprocal approach could also be developed.

Combinatorial chemistry and high-throughput parallel synthesis are powerful tools for the rapid preparation of large numbers of different compounds with numerous applications in the development of new drugs and drug candidates [56–58], metal-complexing ligands and catalysts [59–63], polymers [64], materials for electronics [65, 66], sensors [67], supramolecular assemblies [68–70], and peptidic ligands for affinity chromatography [71]. The essence of combinatorial synthesis is the ability to generate and screen or assay a large number of chemical compounds – a "library" – very quickly. Such an approach provides the diversity needed for the discovery of lead compounds and, in addition, allows their prompt optimization. The fundamentals of combinatorial chemistry, including rapid screening methodologies, are reviewed in numerous papers and books [72, 73]. The following sections of this chapter will describe a variety of different combinatorial methods that have led to selectors for the recognition of chirality aiming mainly at the development of robust media for the separation of enantiomers.

3.6 Library of Cyclic Oligopeptides as Additives to Background Electrolyte for Chiral Capillary Electrophoresis

Enantioresolution in capillary electrophoresis (CE) is typically achieved with the help of chiral additives dissolved in the background electrolyte. A number of low as well as high molecular weight compounds such as proteins, antibiotics, crown ethers, and cyclodextrins have already been tested and optimized. Since the mechanism of retention and resolution remains ambiguous, the selection of an additive best suited for the specific separation relies on the one-at-a-time testing of each individual compound, a tedious process at best. Obviously, the use of a mixed library of chiral additives combined with an efficient deconvolution strategy has the potential to accelerate this selection.

The power of a combinatorial approach to chiral additives for CE was first demonstrated by Jung and Schurig who used a library of cyclic hexapeptides [74]. Since the number of hexapeptides representing all possible combinations of 20 natural L-amino acids is 64×10^6, the first study involved only mixed libraries of hexapeptides of the type c(OOXXXO) consisting of three fixed positions O and three randomized positions X represented by any of 18 natural amino acids (cysteine and tryptophan were not included into the scheme). Three cyclopeptide libraries c(L-Asp-L-Phe-XXX-D-Ala), c(L-Arg-L-Lys-XXX-D-Ala), and c(L-Arg-L-Met-XXX-D-Ala), each consisting of 5832 members, were prepared and tested in chiral CE. When dissolved in an electrolyte to form 10 mmol/L solutions, all three libraries enabled the separation of racemates. For example, the first library facilitated the baseline separation of racemic Tröger's base in a 67 cm-long capillary with a selectivity factor α of 1.01 and column efficiency of 360 000 plates. Similarly, the second library helped to resolve the *N*-2,4-dinitrophenyl (DNP) derivative of glutamic acid in a capillary with the same length affording a selectivity factor of 1.13 and a column efficiency of 79 000 plates. These results indicate the presence of useful selectors in the mixed library. However, this brief study did not attempt the deconvolution of the mixture and did not identify the best selector.

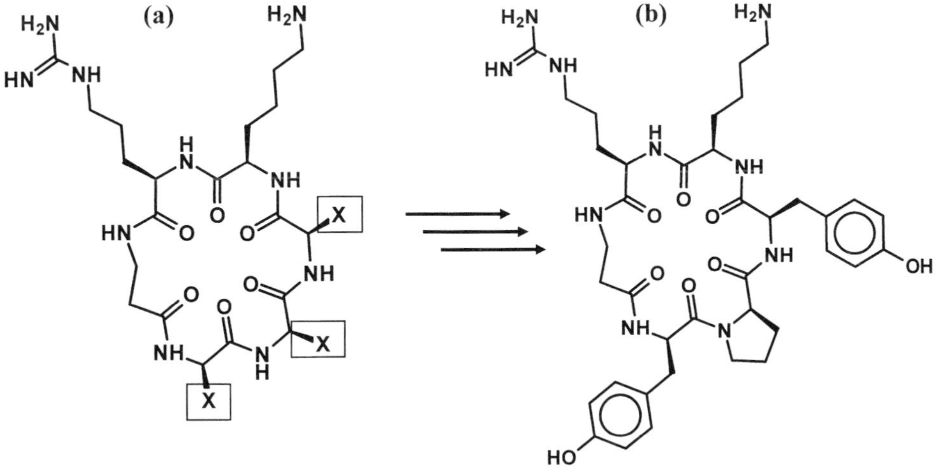

Scheme 3-2.

The deconvolution of a cyclic hexapeptide library to specify the best selector for the target racemate has recently been reported by Chiari et al. [75]. Several libraries of linear hexapeptides with protected lateral chains were prepared using solid-phase synthesis on Merrifield resin, and the cyclization reaction was carried out after cleavage in solution. The study also started with a mixed library of 5832 compounds consisting of cyclic c(OOXXXO) hexapetides with 3 fixed "O" positions consisting of L-arginine, L-lysine, and β-alanine and 3 randomized positions (X) occupied by any of the 18 L-amino acids (Scheme 3-2a). Once again, cysteine and tryptophan were not included. The substitution of D-alanine originally used by Jung and Schurig

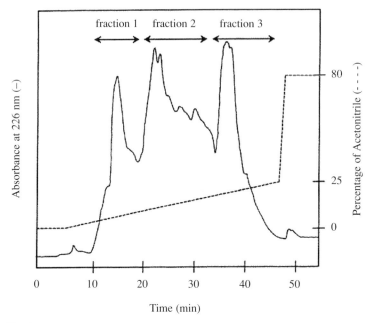

Fig. 3-2. Semipreparative RP-HPLC profile of cyclo(Arg-Lys-X-Pro-X-Ala). The crude sublibrary (160 mol) was dissolved in 0.1 % (v/v) TFA and applied to a Whatman Partisil 10 μm ODS-2 (1 × 50 cm) column. The peaks were eluted using a 40-min linear gradient of 0–25 % acetonitrile in water at a flowrate of 7 mL min^{-1}. Fractions were collected every 2 min and pooled in three fractions as indicated by arrows; 130 μmol of peptides was recovered (yield 81 %). (Reprinted with permission from ref. [75]. Copyright 1998, American Chemical Society.)

with β-alanine led to improved resolution of DNP-glutamic acid enantiomers achieved with the complete mixture.

Since the proline residue in peptides facilitates the cyclization, 3 sublibraries each containing 324 compounds were prepared with proline in each randomized position. Resolutions of 1.05 and 2.06 were observed for the CE separation of racemic DNP-glutamic acid using peptides with proline located on the first and second random position, while the peptide mixture with proline preceding the β-alamine residue did not exhibit any enantioselectivity. Since the c(Arg-Lys-O-Pro-O-β–Ala) library afforded the best separation, the next deconvolution was aimed at defining the best amino acid at position 3. A rigorous deconvolution process would have required the preparation of 18 libraries with each amino acid residue at this position.

However, the use of a HPLC separation step enabled a remarkable acceleration of the deconvolution process. Instead of preparing all of the sublibraries, the c(Arg-Lys-O-Pro-O-β-Ala) library was fractionated on a semipreparative C$_{18}$ HPLC column and three fractions as shown in Fig. 3-2 were collected and subjected to amino acid analysis. According to the analysis, the least hydrophobic fraction, which eluted first, did not contain peptides that included valine, methionine, isoleucine, leucine, tyrosine, and phenylalanine residues and also did not exhibit any separation ability for the tested racemic amino acid derivatives (Table 3-1).

Table 3-1. Values of enantiomeric resolution of DNP-amino acids in a running electrolyte containing the three fractions 1, 2, and 3 of the cyclo(Arg-Lys-X-Pro-X-β Ala) sublibrary separated by preparative HPLC.

Analyte	Resolution			
	c(Arg-Lys-X-Pro-X-β Ala)	Fraction 1	Fraction 2	Fraction 3
DNP-D,L-Glu	2.05	1.09	4.05	1.69
DNP-D,L-Ala	0.80	0	5.54	2.45
DNP-D,L-Leu	0	0	3.53	0.89

This led to the conclusion that these amino acids were essential for the resolution capability and only 6 new libraries of 18 compounds had to be synthesized with these amino acid residues to define the position 3. Surprisingly, the separation abilities of all six libraries were very similar. Therefore, tyrosine was chosen for continuing deconvolution, since it is convenient as its aromatic ring can easily be detected by UV spectrometry. The last step, defining position 5, required the synthesis and testing of 6 individual hexapeptides.

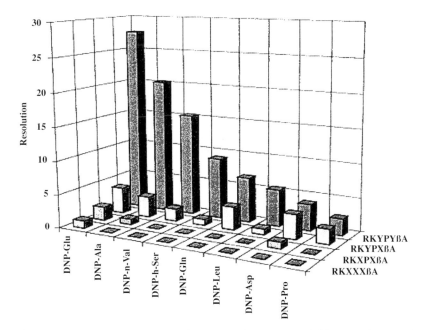

Fig. 3-3. Comparison of the values of enantiomeric resolution of different DNP-D,L-amino acids at different deconvolution stages of a cyclic hexapeptide sublibrary. Resolution values in a cyclo(Arg-Lys-X-X-X-β-Ala) sublibrary, in the first line, are compared to those obtained in sublibraries with a progressively increasing number of defined positions. All the sublibraries were 30 mM in the running buffer while the completely defined cyclo(Arg-Lys-Tyr-P-Tyr-β–Ala) peptide is used at 10 mM concentration. Conditions: cyclopeptide sublibrary in 20 mM sodium phosphate buffer, pH 7.0; capillary, 50 μm i.d., 65 cm total length, 57 cm to the window; V = –20 kV, I = 40; electrokinetic injection, –10 kV, 3 s; detection at 340 nm. (Reprinted with permission from ref. [75]. Copyright 1998, American Chemical Society.)

The improvements in resolution achieved in each deconvolution step are shown in Figure 3-3. While the initial library could only afford a modest separation of DNB-glutamic acid, the library with proline in position 4 also separated DNP derivatives of alanine and aspartic acid, and further improvement in both resolution and the number of separable racemates was observed for peptides with hydrophobic amino acid residues in position 3. However, the most dramatic improvement and best selectivity were found for c(Arg-Lys-Tyr-Pro-Tyr-β-Ala) (Scheme 3-2a) with the tyrosine residue at position 5 with a resolution factor as high as 28 observed for the separation of DNP-glutamic acid enantiomers.

In addition to the development of the powerful chiral additive, this study also demonstrated that the often tedious deconvolution process can be accelerated using HPLC separation. As a result, only 15 libraries had to be synthesized instead of 64 libraries that would be required for the full-scale deconvolution. A somewhat similar approach also involving HPLC fractionations has recently been demonstrated by Griffey for the deconvolution of libraries screened for biological activity [76]. Although demonstrated only for CE, the cyclic hexapeptides might also be useful selectors for the preparation of chiral stationary phases for HPLC. However, this would require the development of non-trivial additional chemistry to appropriately link the peptide to a porous solid support.

3.6.1 Library of Chiral Cyclophanes

Inspired by the separation ability of cyclic selectors such as cyclodextrins and crown ethers, Malouk's group studied the synthesis of chiral cyclophanes and their intercalation by cation exchange into a lamellar solid acid, α-zirconium phosphate aiming at the preparation of separation media based on solid inorganic-organic conjugates for simple single-plate batch enantioseparations [77–80].

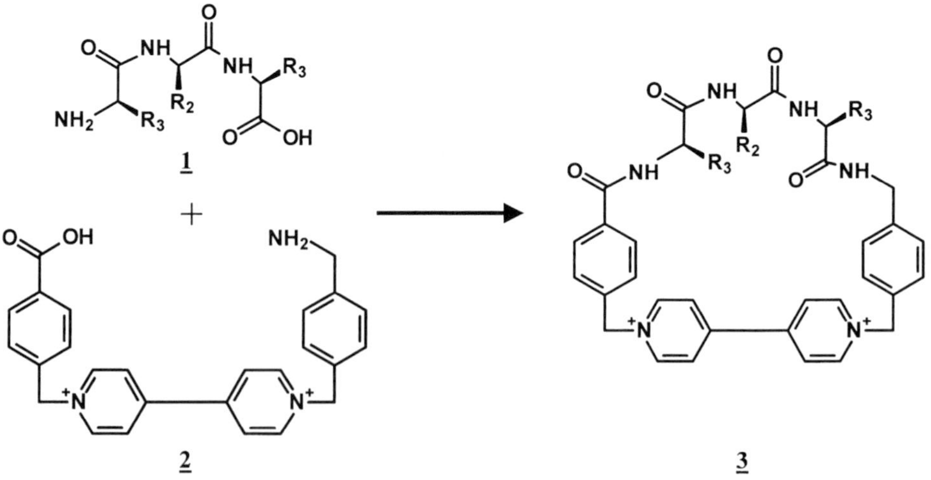

Scheme 3-3.

An example of the modular preparation of the cyclophane **3** from the substituted bipyridine **2** and a general tripeptide **1** is shown in Scheme 3-3. The host molecule **3** contains a pre-organized binding pocket. The overall basicity of such molecules also facilitates their intercalation within the lamellas of acidic zirconium phosphate, thus making this chemistry well suited for the desired application.

While the bipyridinium part of the cycle is fixed, the peptidic module is amenable to combinatorial variation. The ability of a library of cyclophanes **3** containing 20 dipeptides and one tripeptide to recognize chiral compounds was studied in deuterium oxide solutions using the facile ^{1}H NMR titration technique that required only a small amount of the selector [77–79]. The chemical shifts of both protons of the –CH$_2$– group linking the bipyridinium unit with the phenyl groups of the cyclophane were monitored as a function of concentration of the added guest molecules (Fig. 3-4) and used to calculate of binding constants K.

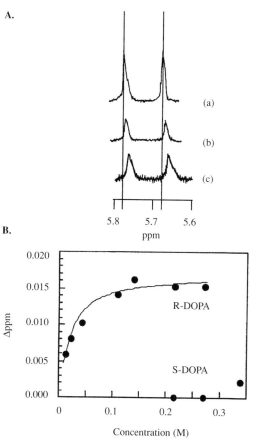

Fig. 3-4. (A) Changes in chemical shift of protons of cyclophane –CH$_2$– groups between bipyridinium and phenyl in ^{1}H NMR spectra of **3** as a function of (R)-DOPA concentration (a) 0, (b) 0.111, and (c) 0.272 mol L^{-1}. (B) Change in chemical shift plotted against the analytical concentration of (R)- and (S)-DOPA. The solid line is calculated for 1:1 host – guest complexation. (Reprinted with permission from ref. [79]. Copyright 1998, American Chemical Society.)

For example, cyclophane **3** containing (*L*)-val-leu-ala tripeptide showed significant association with (*D*)-3,4-dihydroxyphenylalanine **4** (DOPA) and the drug nadolol **5** with *K* values of 39 ± 6 and 23 ± 3 mol^{-1}, respectively, demonstrating rather high selectivity of this cyclic selector. In contrast, only very low binding constants were observed for (*L*)-DOPA (*K* = 3 mol^{-1}), *D* and *L* tryphtophan (*K* = 5 and 6 mol^{-1}), and *N*-2-naphthylalanine (*K* = 10 mol^{-1}) [79]. A further increase in the hydrophobicity and, perhaps, enhanced π-acceptor ability of the lateral functionalities of the oligopeptide moiety may help substantially to improve the selectivity that would be required for their successful application in intercalated solids. It should be noted that this category of selectors is exceptional since, in contrast to the vast majority of typical selectors, it operates in environment friendly aqueous media. Their solubility in water results from the cationic nature of the cyclophane.

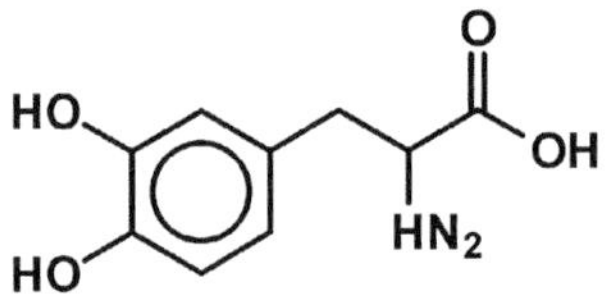

4 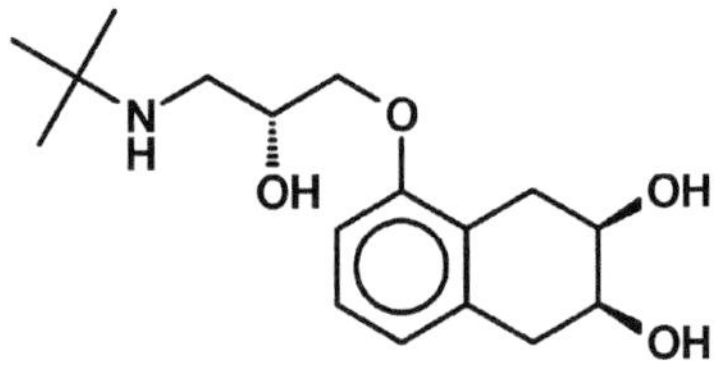   **5**

3.6.2 Modular Synthesis of a Mixed One-Bead – One-Selector Library

The screening of libraries of compounds for the desired property constitutes an essential part of the combinatorial process. The easier and the faster the screening, the higher the throughput and the more compounds can be screened in a unit of time. This paradigm has led Still's group to develop a combinatorial approach to chiral selectors that involves a visual screening step by optical microscopy that enables the manual selection of the best candidates [81].

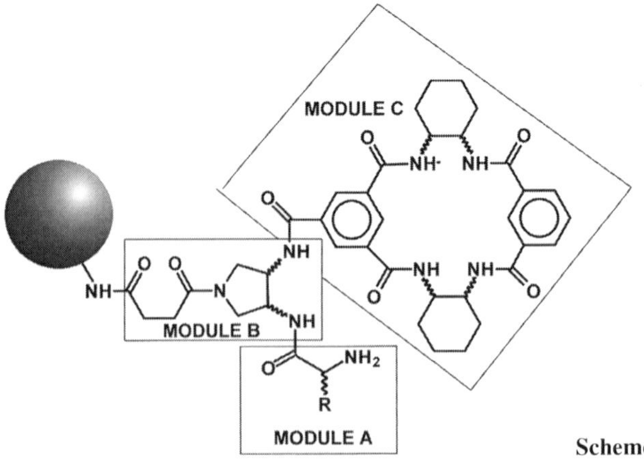

Scheme 3-4.

In order to prove the concept, they prepared a library of 60 selectors using three different building modules A, B, and C (Scheme 3-4). Module A consists of 15 different D- and L-amino acids, while module C was a cyclic amide formed by the condensation of two *RR* or *SS* 1,2-diaminohexane molecules with one isophthalic acid and one trimesic acid unit. Two different stereoisomers (*RRRR* and *SSSS*) were prepared. These two modules A and C were linked through module B (3,4-diaminopyrrolidine) that serves as a turn element directing modules A and C toward one another. This module was also used for the attachment of the selector to the solid support. Aminomethylated Merrifield resin (100 μm polystyrene beads crosslinked with 2 % divinylbenzene) modified with ε-aminocaproic acid was chosen as the support, and the library was prepared using a split synthesis process that led to a mixture of different beads with each bead containing only one selector [72]. Briefly, a batch of reactive beads was split into four parts and reacted with small amounts of tag acids to encode the stereoisomers used in the first reaction step and to enable decoding the successful selectors. Each pool was then treated in a separate flask with modules (*RR*)-B-(*RRRR*)-C, (*RR*)-B-(*SSSS*)-C, (*SS*)-B-(*RRRR*)-C, and (*SS*)-B-(*SSSS*)-C, respectively. Once these reactions were completed, beads from all flasks were combined and then split into 15 reaction vessels. Each set contained beads with all four combinations of modules B–C. In the next step, the beads in each reactor were pooled, tagged and treated with 15 different activated amino acids (module A). Once this reaction was complete, the beads were combined again and used for the screening.

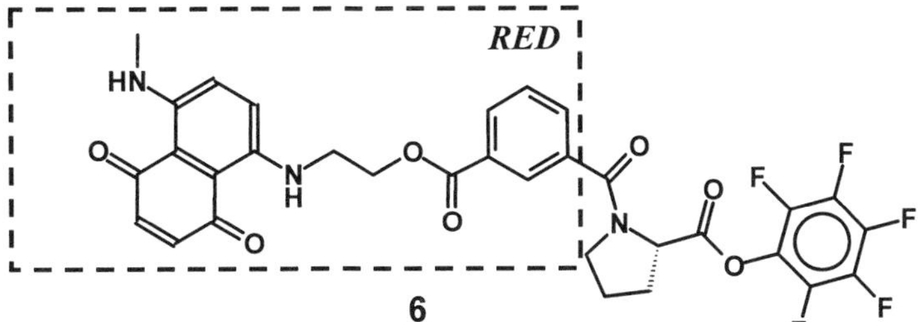

6

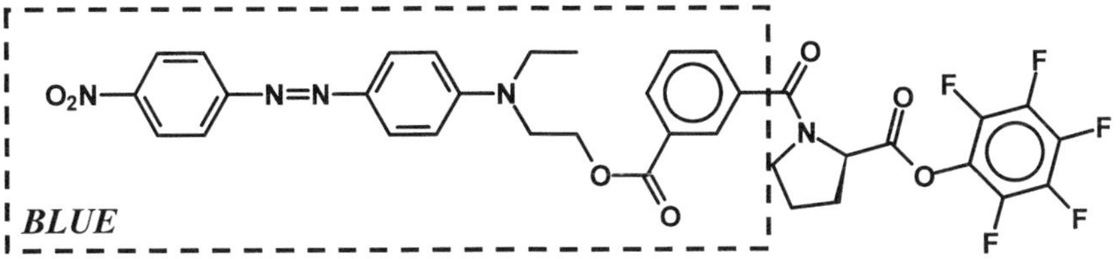

7

To find the most efficient selectors in the library, blue and red dye-labeled enantiomeric probe molecules **6** and **7** were prepared by linking pentafluorophenyl esters of L- and D-proline with Disperse Blue 3 and Disperse Red 1, respectively, through an isophthaloyl (shown in structures **6** and **7**) or a succinyl moiety. For detection, a

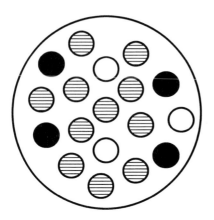

Fig. 3-5. Schematic of the visual screening of colored beads in the field of optical microscope. Open circles red beads, close circles blue beads, dashed circles brown beads.

portion of beads was added to an excess of a mixture containing both proline – dye conjugates and left to interact for 4 h. The beads were then washed and observed by optical microscopy. A typical picture showing the desired effect is presented schematically in Fig. 3-5.

Differently colored beads could be seen in the field of observation. Those beads that exhibited higher affinity to respective colored D- and L-proline – dye conjugates were red and blue, while beads with no selectivity were brown. Beads with the purest blue or red color were manually picked and decoded to determine the structures of their selectors. Once decoded, gram quantities of beads bearing these selectors were prepared and treated again with the mixture of dye-proline conjugates. Subsequent release of adsorbed molecules and determination of the enantiomeric excess (ee.) enabled direct comparison of the enantioselectivities of these selectors (Table 3-2). Enantioselectivities similar to those shown in Table 3-2 were also found for the racemic proline pentafluorophenyl ester, and confirm that there is no positive or negative contribution to selectivity entailed by the large dye moiety.

Table 3-2. Enantiodiscrimination of selected library members using the two-color assay with proline derivatives.

Library member	Enantiomeric excess (% ee)
L-His-(SS)-B-(RRRR)-C	49 for D
D-His-(RR)-B-(SSSS)-C	51 for L
L-Asp-(SS)-B-(RRRR)-C	44 for D
D-Asp-(RR)-B-(SSSS)-C	48 for L
L-Asn-(SS)-B-(SSSS)-C	51 for L
D-Asn-(RR)-B-(RRRR)-C	39 for D

Although the preparation of the quite complex selector modules prior to the synthesis of the library represented a rather significant synthetic effort, this study showed clearly the potential of combinatorial chemistry in the early development stage of a chiral separation medium and demonstrated a novel approach to rapid screening that might be amenable to full automation in the future.

3.7 Combinatorial Libraries of Selectors for HPLC

3.7.1 On-Bead Solid-Phase Synthesis of Chiral Dipeptides

Several attempts to prepare efficient chiral stationary phases using Merrifield's solid-phase peptide synthesis have been reported in the past. For example, in 1977 Gruska [82] prepared a tripeptide bound to a solid support using a sequence of protection – coupling – deprotection reactions. This approach appeared to suffer from incomplete conversion in the coupling steps, and the stationary phase exhibited only a modest selectivity. A similar stationary phase was later prepared by attaching the pure tripeptide L-val-ala-pro, prepared in solution, to porous silica beads [83]. This CSP exhibited higher selectivity than that of Gruska thus indicating the possible detrimental effect of undesired or uncontrolled functionalities on the recognition ability of the stationary phase. Recently, Welch analyzed these results and realized that the primary reason for the relative failure of the early approaches to on-bead solid-phase synthesis of oligopeptide selectors could be traced to the relatively low reactivity of the functional silane reagent, 1-trimethoxysilyl-2-(4-chloromethyl-phenyl)ethane, used for the preparation of the original chiral stationary phases [84]. As a result of steric constrains, selector surface coverage of only 0.3 mmol g^{-1} could be achieved after activation using this bulky silane reagent. In contrast, Welch easily obtained CSPs containing at least twice as much selector using aminopropyltriethoxysilane activation. This group also optimized the reaction conditions to afford silica beads with a high amine surface coverage and to realize their essentially quantitative functionalization in the subsequent reaction step. This more successful approach enabled study of the effects of various variables on the separation properties of chiral stationary phases thus prepared.

CSP 1   **CSP 3**

First, they compared CSPs 1 and 3 prepared by the two-step solid-phase methodology with their commercially available counterparts (CSPs 2 and 4) obtained by direct reaction of the preformed selector with a silica support. Although no exact data characterizing the surface coverage density for these phases were reported, all of the CSPs separated all four racemates tested equally. These results shown in Table 3-3 subsequently led to the preparation of a series of dipeptide and tripeptide CSPs 5–10 using a similar synthetic approach. Although the majority of these phases exhibited selectivities lower or similar to those of selectors built around a single amino acid (Table 3-3), this study demonstrated that the solid-phase synthesis was a

Table 3-3. Enantioseparation of 2-methylnaphthoyl-*N,N*-diethylamide and naproxene methyl ester using CSPs 1-11.

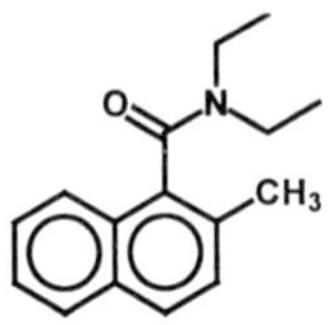

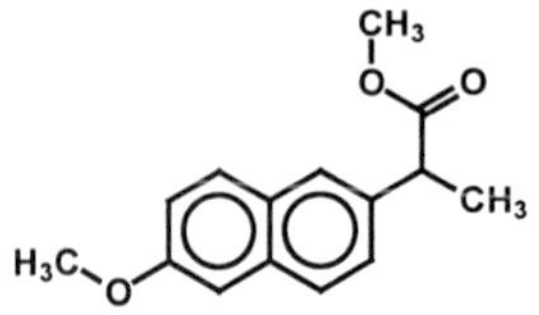

CSP	k'_1	α	k'_1	α
1	5.67	1.10	3.19	1.33
2[a]	6.48	1.08	3.64	1.33
3	10.43	1.10	3.16	1.00
4[a]	9.08	1.07	2.54	1.00
5	6.64	1.15	1.81	1.00
6	7.60	1.10	2.18	1.00
7	5.28	1.12	1.45	1.00
8	4.95	1.03	5.30	1.29
9	7.77	1.11	1.87	1.00
10	4.81	1.13	1.35	1.00
11	3.15	2.22	8.06	1.84

[a] Commercial CSPs 2 and 4 from Regis Technologies, Inc. (Morton Grove, IL) have selectors close to those of CSPs 1 and 3, respectively.
Conditions: mobile phase 10 % 2-propanol in hexane, column 250 × 4.6 mm i.d., flow rate 2 mL min⁻¹, UV detection at 254 nm.

viable alternative to the more traditional approaches and opened the access to libraries of chiral separation media.

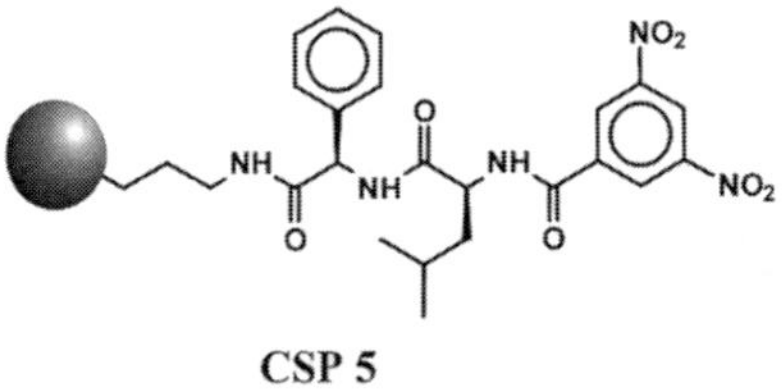

CSP 5

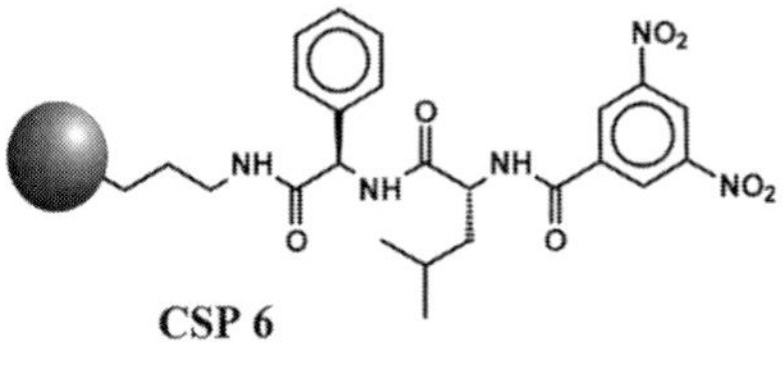

CSP 6

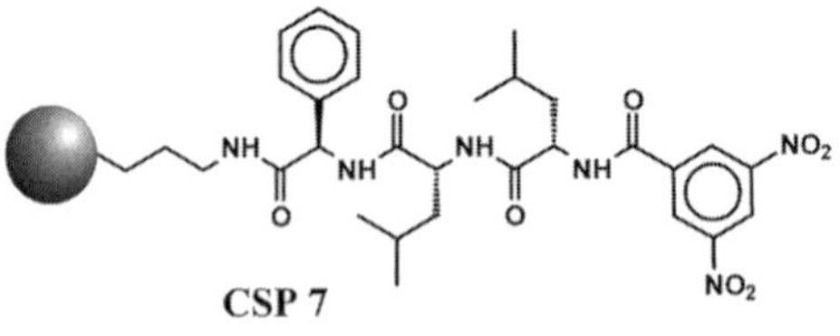

CSP 7

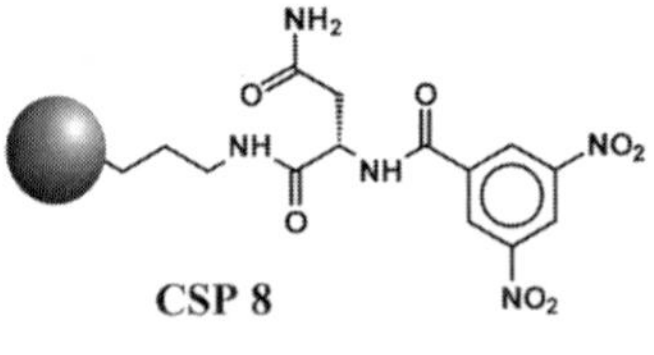

CSP 8

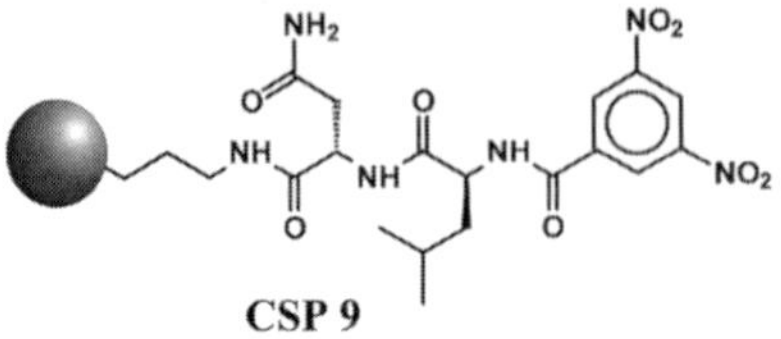

CSP 9

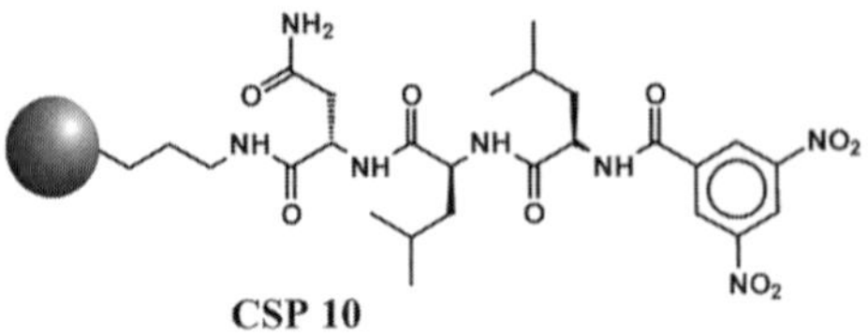

CSP 10

The experiments of the initial study were performed on a 5 g scale. Although fully feasible, this quantity of functionalized beads required the use of very substantial amounts of reagents and solvents for both the preparation and the chromatographic testing. Therefore, the same group later developed a microscale methodology for screening with only 50 mg of the CSP [85]. Their approach assumed that if an efficient CSP is placed in a solution of a racemate, it should adsorb preferentially only one of the enantiomers, thus depleting it from the solution. To confirm this assumption, they placed a few milligrams of the beads in a vial and added a dilute solution of a less than equimolar amount of the target racemate. After equilibration, a sample of the supernatant liquid was injected into a commercial chiral HPLC column and the peak areas were determined for both separated enantiomers to calculate the selectivity. If the areas were equal, the CSP did not exhibit any selectivity, while any change in the amount of an enantiomer represented some level of selectivity.

After this feasibility test, a library consisting of 50 types of beads each containing a different dipeptide selector attached through its C-terminal group was prepared and screened (Fig. 3-6) [84]. The first amino acid residue (aa 1) was chosen from a

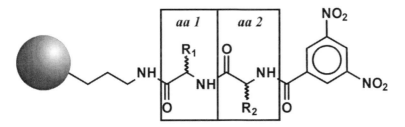

Fig. 3-6. General structure of dipeptide CSPs.

group consisting of D enantiomers of phenylglycine, valine, leucine, glutamine, and phenylalanine. Both D and L enantiomers of the same set of amino acids were then used as the second residue (aa 2), thus affording 50 different dipetides (5 × 10). The N-termini of these dipeptides were then capped with the π-acidic dinitrobenzoyl (DNB) group. The selectivity of each member of this parallel library was screened against the model racemate N-(2-naphthyl)alanine diethylamide (**8**).

8

A number of interesting conclusions could be drawn from the screening. For example, the amide hydrogen atom at the second amino acid residue close to the

DNB group appeared to be essential for high selectivity. Similarly, a large group affording steric hindrance at the same amino acid also improved the resolution. In contrast, proline at that position did not lead to CSPs with good selectivities. Although the substituted amino acid-based selectors were thoroughly studied earlier using various techniques including X-ray diffraction and NMR, this study brought about unexpected results. For example, the glutamine residue (gln) at the initial position was beneficial for selectivity. Similarly, homochiral (D–D) dipeptides afforded better selectivity than many heterochiral sequences. The best selectivity of this selector library was observed for (L)-gln-(L)-val-DNB. Although successfully demonstrated, the "in-batch" screening is less sensitive than the direct separation in HPLC mode and its use appears limited to the discovery of selectors with selectivity factors of at least 1.5. In addition, this evaluation allows only relative comparisons and exact numerical values for the selectivity factors cannot be calculated easily.

To further extend this study, the authors expanded their selection of amino acids including hydrogen bonding residues (L-isomers of glutamine, asparagine, serine, histidine, arginine, aspartic acid, and glutamic acid) in position 1 closest to the surface of the support and both D and L amino acids with bulky substituents (leucine, isoleucine, t-leucine, valine, phenylalanine, and tryptophan) in the position 2 [86]. The terminal amine functionalities of these dipeptides were again capped by dinitrobenzyl groups. One sublibrary of 39 attached selectors could be prepared directly, while the second sublibrary involving 32 selectors required the Fmoc lateral chain protection during its preparation. Hence, the complete library used in this study incorporated 71 dipeptide selectors out of 98 possible structures. All of these CSPs were tested for the resolution of **8** using the batch approach. Evaluation of results shown in Fig. 3-7 indicated that glutamic acid, aspartic acid, and histidine in position 1 and leucine, isoleucine, and phenylalanine in position 2 afforded selectors with enantioselectivity far better than that of the gln-val-DNB selector lead identified in the original library [84].

The usefulness of this solid-phase synthesis/screening was finally validated by synthesizing 5 g of beads with the (L)-glu-(L)-leu-DNB selector. These were packed into a 250 × 4.6 mm i.d. HPLC column and evaluated using normal-phase chromatographic conditions. The separation of racemic **8** shows Fig. 3-8. This separation was remarkable for several reasons: first, for its excellent selectivity factor (α = 20.74) enabling an outstanding separation of both enantiomers with an isocratic mixture of 2-propanol-hexane; second, the k' value for the second peak was 78.99 and the peak did not elute until after almost 2 h, indicating that a rather strong interaction is involved in the recognition process; and finally, the column afforded a high selectivity factor of over 18 even in pure ethyl acetate that might be a better solvent for many racemates than the hexane mixture and can easily be recycled. The large "distance" between the peaks of both enantiomers resulting from the high selectivity was found extremely useful for separations under overload conditions. For example, Fig. 3-9 shows a remarkable enantioseparation of 100 mg of the model racemate on the analytical size column that produced both enantiomers in optical purity of 98.4 and 97 % ee, respectively [86].

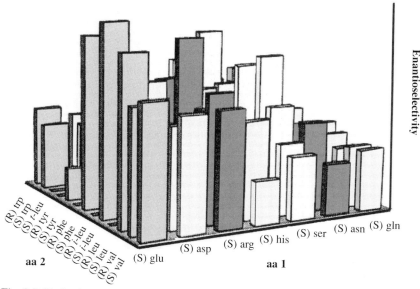

Fig. 3-7. Evaluation of a focused library of 71 DNB-dipeptide CSPs for enantioseparation of the test racemate **8.** (Reprinted with permission from ref. [86]. Copyright 1999, American Chemical Society.)

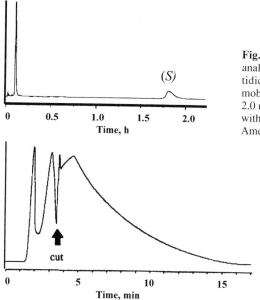

Fig. 3-8. HPLC evaluation of a 250×4.6 mm i.d. analytical column packed with the selected dipeptidic (S)-Glu-(S)-Leu-DNB CSP. Conditions: mobile phase 20 % 2-propanol in hexane, flowrate 2.0 mL min^{-1}, UV detection at 280 nm. (Reprinted with permission from ref. [86]. Copyright 1999, American Chemical Society.)

Fig. 3-9. Preparative HPLC of 100 mg of the test racemate **8** in a single 2 mL injection using a 250×4.6 mm i.d. column containing (S)-Glu-(S)-Leu-DNB CSP. Conditions: mobile phase ethyl acetate, flowrate 2.0 mL min^{-1}, UV detection at 380 nm. Injection 2 mL of 50 mg mL^{-1} racemate solution. Fractions collected before and after the indicated cut point were 98.4 % ee and 97 % ee pure, respectively. (Reprinted with permission from ref. [86]. Copyright 1999, American Chemical Society.)

The use of silica beads as a support for the preparation of peptide libraries appears to be somewhat problematical. While Welch could demonstrate the very satisfactory results described above [84–86], Li was not able to achieve complete coupling of the first amino acid to all of the surface functional groups of modified silica, thus leaving behind a number of functionalities that might be detrimental for the desired chiral separations [87]. Therefore, Li's group used the organic polymer supports that are most common in the field of solid-phase synthesis: aminomethylated Merrifield resin (2 % crosslinked polystyrene beads) and amine-terminated polyethylene glycol modified polystyrene resin (NovaSyn TG amino resin). They prepared two 4 × 4 parallel libraries using these supports, each consisting of 16 selectors obtained from four amino acids (L-leucine **9**, L-alanine **10**, glycine **11**, and L-proline **12**) and four carboxylic acids providing π blocks (3,5-dinitrobenzyl **13**, benzyl **14**, naphthyl **15**, and anthryl **16**) with varied aromatic moiety. The achiral glycine unit **11** served as a negative internal control. The solid-phase synthesis of this 16-member library on resin was performed using a Hi-top filter plate manual synthesizer and Fmoc strategy.

The screening was performed in a way similar to that of Welch, except that it involved the use of a spectropolarimeter instead of chiral chromatography to determine the selectivity. Equal amounts of the target racemate **17** were added into each of the 16 wells containing beads and the ellipticity of the supernatant liquid in each well was measured after equilibrating for 24 h at the wavelength of the maximum adsorption (260 nm). Knowing the specific ellipticity of one enantiomerically pure analyte and the total concentration of enantiomers in the solution determined from UV adsorption, the enantiomeric ratio of components in the supernatant liquid could easily be calculated. The results are summarized in Fig. 3-10. Obviously, no selectivity is expected for the glycine-based selectors. Thus the readings for these selectors set the accuracy limits of the circular dichroism method. Comparison of ellipticities measured with the same selector prepared on both different resins indicated that the more polar and hydrophilic TG resin afforded lower selectivities. However, the effect of the matrix did not change the fact that the highest ellipticity was found for beads with the same selectors. This suggested that the support chemistry might affect the screening, a fact, that was also demonstrated quite convincingly by our research group [8, 10].

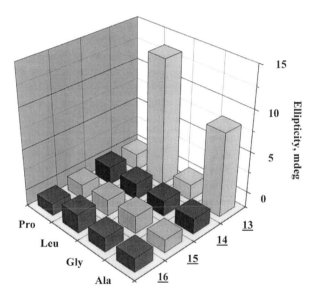

Fig. 3-10. Ellipticities measured at 260 nm for separations achieved with the members of the parallel library of 16 dipeptide CSPs (Reprinted with permission from ref. [87]. Copyright 1999, American Chemical Society.)

The two best selectors resulting from Li's screening, DNB-L-ala and DNB-L-leu, were then prepared on a larger scale, attached to silica beads modified with 3-amino-propyl-triethoxysilane, and the CSPs were packed into columns. Respective separation factors of 4.7 and 12 were found for the separation of racemic naphthyl leucine ester **17** using these CSPs.

3.7.2 Reciprocal Screening of Parallel Library

In our own research we have demonstrated the power of the reciprocal approach with the screening of a library of substituted dihydropyrimidines. These compounds represent a new family of potential selectors not derived from natural chiral blocks [55, 88]. The library of dihydropyrimidines was prepared using a multicomponent condensation strategy [89]. Multicomponent reactions involve two or more reactants in a single-step process that provides durable core structures and highly variable side chains from simple starting materials. Therefore, these strategies are very useful for combinatorial syntheses of libraries of small molecules. The Biginelli dihydropy- rimidine synthesis, first reported more than 100 years ago [90] is one of such multi- component reactions. It involves a one-pot cyclocondensation of β-keto esters **18,** aldehydes **19,** and ureas **20** that affords dihydropyrimidine heterocycles **21** (Scheme 3-5) [91]. This simple approach was applied in our laboratory to create a library of over 160 diverse racemic dihydropyrimidines [55]. Although this library was not very large, its diversity was sufficient to obtain an efficient selector and to establish a correlation between structural features and enantioselectivity. The approach is also notable for its direct applicability to libraries of racemic molecules.

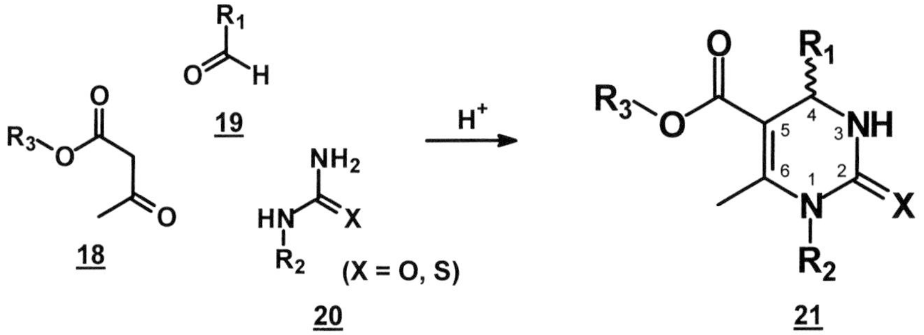

Scheme 3-5.

In order to perform such a correlation, our library was "screened" using a "recip- rocal" CSP with an arbitrary bound chiral target (*L*)-(3,5-dinitrobenzoyl) leucine (Fig. 3-11).

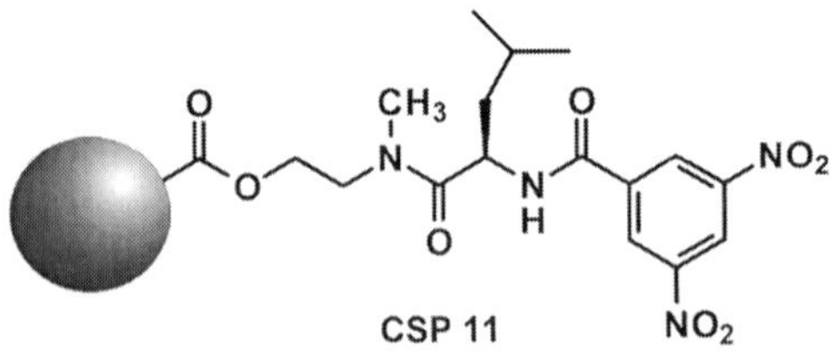

CSP 11

The target was immobilized on monodisperse macroporous poly ((*N*-methyl)aminoethyl methacrylate-*co*-methyl methacrylate-*co*-ethylene di-

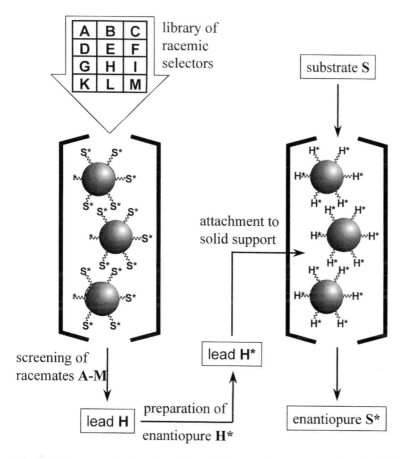

Fig. 3-11. Concept of reciprocal combinatorial approach to the preparation of chiral stationary phase. (Reprinted with permission from ref. [55]. Copyright 1999, American Chemical Society.)

methacrylate) beads [10] affording CSP **11.** Some results of the screening of the library are shown in Fig. 3-12 and 3-13. They revealed that many racemic dihy-dropyrimidines are not resolved at all (separation factor $\alpha = 1.0$), while rather high α values of up to 5.2 were achieved for top candidates such as 4-(9-phenanthryl)-dihydropyrimidine **22.** Figure 3-14A shows the chromatographic separation of this racemate.

We also performed a single-crystal X-ray structure analysis of this lead compound. The solid state structure of this compound depicted in Fig. 3-15 shows a half-boat-like ("sofa") conformation with the 9-phenanthryl group in a *quasi*-axial or *quasi*-flagpole position, and the α,β-unsaturated exocyclic ester in a *s-cis* conformation. This cleft-like conformation is advantageous for the creation of centers with a high recognition ability, since one enantiomer "fits" in better than the other thus leading to selectivity.

Inspection of the large body of data collected from the separation experiments also revealed several structural requirements necessary for good chiral recognition.

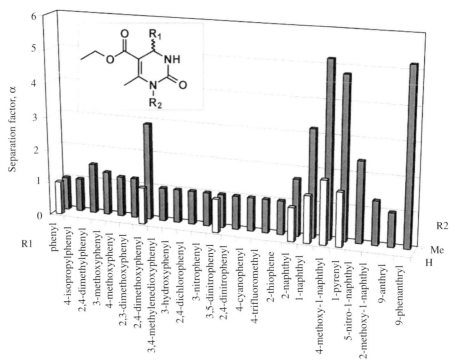

Fig. 3-12. Selectivity factors for the separations of sublibraries of racemic ethyl (6-methyl-) and (ethyl 1,6-dimethyl) 2-oxo-4-substituted-1,2,3,4-tetrahydropyrimidine-5-carboxylates. (Reprinted with permission from ref. [55]. Copyright 1999, American Chemical Society.)

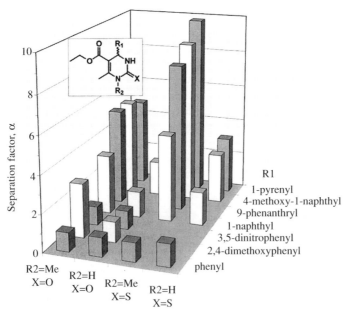

Fig. 3-13. Selectivity factors for the separations of sublibraries of racemic (ethyl 6-methyl-2-oxo-), (ethyl 1,6-dimethyl-2-oxo-), (ethyl 6-methyl-2-thio-), and (ethyl 1,6-dimethyl-2-thio) 4-substituted-1,2,3,4-tetrahydropyrimidine-5-carboxylates. (Reprinted with permission from ref. [55]. Copyright 1999, American Chemical Society.)

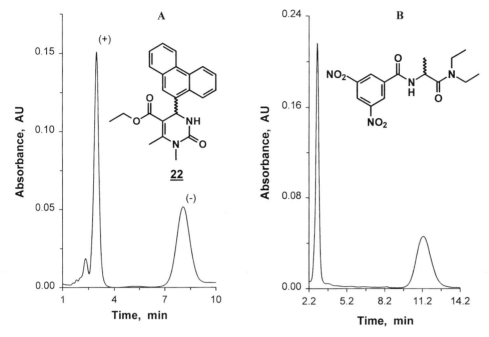

Fig. 3-14. Separation of (A) (±)-4-(9-phenanthryl)-dihydropyrimidine **22** on chiral stationary phase CSP **11** and (B) racemic 3,5-dinitrobenzamidoalanine-*N,N*-diethylamide on chiral stationary phase CSP **12**. Conditions: column 150 × 4.6 mm i.d.; mobile phase dichloromethane; flowrate 1 mL min^{-1}.

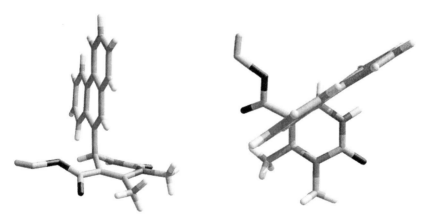

Fig. 3-15. Spatial structure of (±)-4-(9-phenanthryl)-dihydropyrimidine **22** determined by X-ray diffraction. The hydrogen atoms are not shown for clarity.

For example, only those dihydropyrimidines that contained a hydrogen-bonding donor at position 3 next to the chiral center were separated. Remarkably, dihydropyrimidines with non-substituted nitrogen atoms at positions 1 and 3 resulted in separations with longer retention times and decreased separation factors α. Increasing the π-basicity of the aromatic group at C4 resulted in higher separation factors due to a stronger interaction with the π-acidic 3,5-dinitrobenzoyl group of CSP **11** used for the screening. Obviously, the substitution pattern as well as the π-basicity of the aromatic group at the C4 atom play essential roles. Dihydropyrimidines with *ortho*-substituted aromatic groups show much higher enantioselectivities compared to *meta*- and *para*-substituted groups. For example, the observed enantioselectivity for the 1-naphthyl-dihydropyrimidines was almost twice that of the corresponding 2-naphthyl derivative (see Fig. 3-12). Addition of a second ortho substituent in the aromatic ring of the dihydropyrimidines led to a dramatic deterioration of enantioseparation as observed with the 9-anthryl or 2-methoxy-1-naphthyl substituted compounds. Finally, another important effect on enantioselectivity was observed when a thiourea was used instead of urea for the preparation of **21**. In general, the selectivity factor for the 2-thio analogs increased two-fold compared to the corresponding 2-oxo-DHPMs (see Fig. 3-13). For example, 4-(9-phenanthryl)-2-thiodihydropyrimidine exhibited a selectivity of $\alpha = 11.7$ instead of 5.2 for the best oxygenated analogue. However, the lability of the thio-containing compounds compared to the oxo-analogue led us to select the latter for the preparation of more rugged CSPs.

In the next step, the best candidate from the series 2-oxo-4-(9-phenanthryl)-dihydropyrimidine **22** was prepared and isolated in enantiomerically pure form, then attached to a macroporous polymer support. To attach the isolated selector to the amino functionalized macroporous polymethacrylate support, a suitable reactive "handle" had to be introduced into the dihydropyrimidine. We chose to functionalize the methyl group at the C6 carbon atom by a simple bromination to afford (-)-**22**. Coupling of this compound to the amino functionalized support then gave the desired chiral stationary phase CSP **12** (Scheme 3-6) containing 0.20 mmol g^{-1} of the selector.

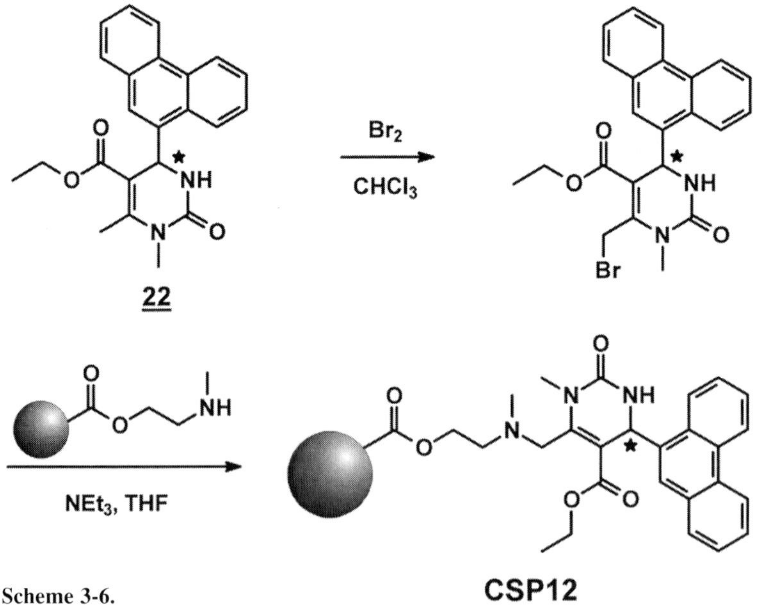

Scheme 3-6.   **CSP12**

CSP **12** afforded good separations for a variety of racemic α-amino acid derivatives. Figure 3-14B shows an example of a typical separation of 3,5-dinitrobenzamidoalanine-*N,N*-diethylamide enantiomers with a separation factor α = 7.7. Even though CSP **12** was designed for the separation of derivatized amino acids, other classes of compounds such as dihydropyrimidines and profens could also be resolved. In addition, this phase could be utilized under reversed-phase separation conditions. Despite the suppression of the hydrogen bonding interactions between the CSP and analyte, rather good enantioselectivities (α = 3.5) were obtained even in a mobile phase consisting of 50 % water. This confirmed that CSP **12** was a versatile phase capable of enantioseparations in either reversed-phase or normal-phase mode.

The reciprocal strategy is best suited for typical situations encountered in the industry that require the preparation of a highly selective CSP for the separation of only a single racemic product such as a drug. Since single enantiomers of that compound must be prepared for the testing, the preparation of a "reciprocal" packing with a single enantiomer attached to the support does not present a serious problem. Once this CSP is available, a broad array of libraries of potential racemic selectors can easily be screened. Although we selected a simple Biginelli three-component condensation reaction to prepare the library of selectors based on "non-natural" compounds, many other libraries of chiral organic compounds could also be screened as potential selectors for the chiral recognition of specific targets. This approach provides one more benefit: when such an extensive study is carried out with structurally related families of compounds typical of chemical libraries, a better understanding of chiral recognition may quickly be generated and used for design of even more successful selectors.

3.7.3 Reciprocal Screening of Mixed Libraries

The reciprocal screening of a mixed library described by Li's group very recently [92] is an interesting variation of the approach outlined in the previous text. Following a standard procedure, a L-naphthylleucine CSP **13** containing the target analyte was prepared first by attaching L-**17** onto silica using a standard hydrosilylation procedure.

CSP 13

A mixed 4 × 4 peptide library consisting of 16 members was again prepared from the earlier-shown two families of building blocks **9–12** (all L enantiomers) and

13–16 shown earlier, each terminated with a glycine unit, using a solid-phase synthesis followed by cleavage. This complete library was injected on a short column packed with CSP **13** and eluted in a gradient of 2-propanol in hexane (Fig. 3-16a). The chromatogram features a number of peaks. Obviously the retention times depended on the interaction of individual library members with the immobilized target as well as on possible interactions with both the support and the mobile phase. However, the retention times alone did not provide any information that might be related to the chiral recognition since only one enantiomer of each potential selector is present in the library. Therefore, an identical library was also prepared using all D enantiomers of the amino acid building blocks and the separation process was repeated on the same column (Fig. 3-16b).

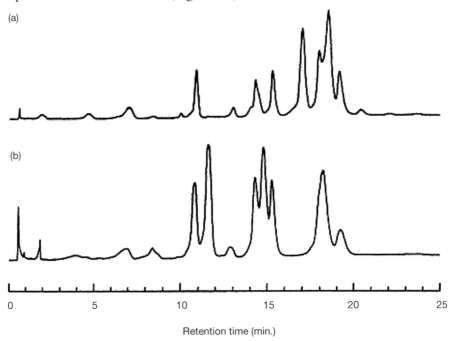

Retention time (min.)

Fig. 3-16. Chromatograms of mixed libraries of 16 L (a) and 16 D (b) selectors using reciprocal stationary phase CSP **13**. Conditions: column 50 × 4.6 mm i.d.; mobile phase gradient of 5–20 % 2-propanol in hexane; flowrate, 1.2 mL min^{-1}; UV detection at 254 nm. (Reprinted with permission from ref. [92]. Copyright 1999, American Chemical Society.)

The substantial difference between these two chromatograms was a clear proof that CSP **13** interacted differently with the mixtures of L and D enantiomers. This also indicated the presence of at least one pair of enantiomers that interacted selectively with the CSP. Unfortunately, a tedious synthesis of 16 sublibraries (eight L and eight D) containing decreasing numbers of blocks had to be prepared to deconvolute the best selector. A comparison of the chromatograms obtained from these sublibraries in each deconvolution step was used again, and those selectors for which no difference was observed were eliminated. This procedure enabled the identification of two powerful selectors, DNB-L-leu-gly and DNB-L-ala-gly in agreement with the results of the parallel screening [87].

These two selectors terminated with a glycine were then prepared on a larger scale, their carboxyl groups reacted with 3-aminopropyltriethoxysilane, and the conjugate immobilized onto silica. Each CSP was packed into columns and used for the separation of racemic (1-naphthyl)leucine ester **17**. Separation factors of 6.9 and 8.0 were determined for the columns with DNB-ala-gly and DNB-leu-gly selector respectively. These were somewhat lower than those found for similar CSPs using the parallel synthesis and attached through a different tether [87].

The major weakness of this method appears to be the limited size of the library that can be screened. Although the authors believe that their method is well suited to screen medium-sized libraries with up to a few hundred members, the most important limit is the requirement of having equal sets of both L and D building units. This is easily achieved with amino acids that provide relatively large diversity at a reasonable cost. However, this may be a serious problem with many other families of chiral compounds. The other drawback of the current implementation is the large number of sublibraries that must be synthesized and screened to specify the best selector. In fact, the number of sublibraries in the published procedure equaled the number of members of the original library thus making the expected acceleration effect of this combinatorial approach questionable. However, the use of the mass spectroscopic detection during the first two parallel screening separations of Fig. 3-16 would afford molecular weights of the separated compounds that are specific for each individual selector. If the retention of any of the injected compounds is the same for both D- and L-libraries, no selectivity occurs. In contrast, different retention times in both runs indicate selectivity and even allow an estimation of selectivity factors. Such an approach might totally avoid the tedious multistep deconvolution process and accelerate the screening procedures.

3.7.4 Library-On-Bead

Our group also demonstrated another combinatorial approach in which a CSP carrying a library of enantiomerically pure potential selectors was used directly to screen for enantioselectivity in the HPLC separation of target analytes [93, 94]. The best selector of the bound mixture for the desired separation was then identified in a few deconvolution steps. As a result of the "parallelism advantage", the number of columns that had to be screened in this deconvolution process to identify the single most selective selector CSP was much smaller than the number of actual selectors in the library.

Our strategy consisted of the following steps: A mixture of potential chiral selectors is immobilized on a solid support and packed to afford a "complete-library column", which is tested in the resolution of targeted racemic compounds. If some separation is achieved, the column should be "deconvoluted" to identify the selector possessing the highest selectivity. The deconvolution consisted in the stepwise preparation of a series of "sublibrary columns" of lower diversity, each of which constitute a CSP with a reduced number of library members.

The feasibility of this approach was demonstrated with a model library of 36 compounds prepared from a combination of three Boc protected L-amino acids (valine **23**, phenylalanine **24**, and proline **25**) and 12 aromatic amines (3,4,5-trimethoxyaniline (**26**), 3,5-dimethylaniline (**27**), 3-benyloxyaniline (**28**), 5-aminoindane (**29**), 4-*tert*-butylaniline (**30**), 4-biphenylamine (**31**), 1-3-benyloxyaniline (**28**), 5-aminoindane (**29**), 4-*tert*-butylaniline (**30**), 4-biphenylamine (**31**), 1-aminonaphthalene (**32**), 4-tritylaniline (**33**), 2-aminoanthracene . (**34**), 2-aminofluorene (**35**), 2-aminoanthraquinone (**36**), 3-amino-1-phenyl-2-pyrazolin-5-one (**37**)). The complete library was prepared by a two-step procedure that includes the activation and coupling of the N-Boc-protected α-amino acids with the various amines followed by deprotection of the resulting protected amides (Scheme 3-7). The mixture of deprotected amino acid derivatives in solution was then immobilized onto a polymeric solid support, typically activated 5-μm macroporous poly(hydroxyethyl methacrylate-*co*-ethylene dimethacrylate) beads, to afford the chiral stationary phases with a multiplicity of selectors. Although the use of columns

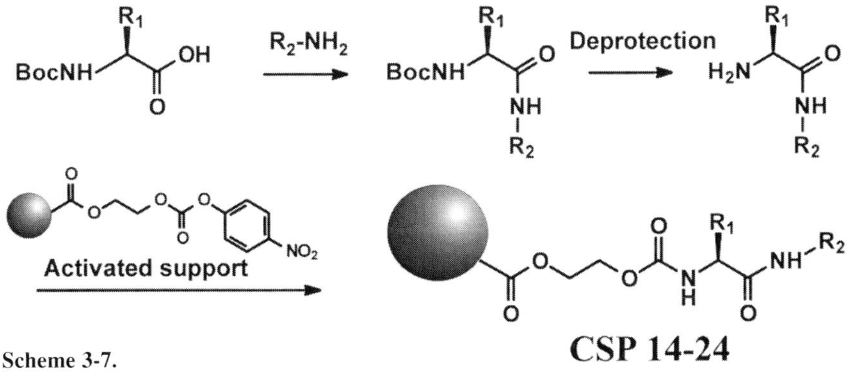

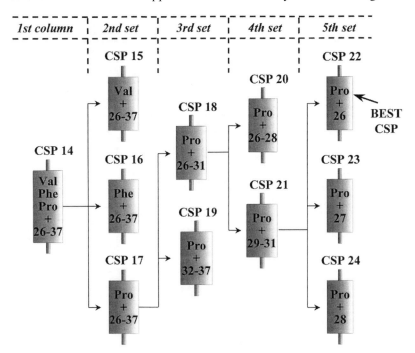

Scheme 3-7. **CSP 14-24**

with mixed selectors has not been recommended for actual enantioseparations [95], it is ideally suited for our combinatorial approach to optimized selectors.

As expected from the design of the experiment, the HPLC column packed with CSP **14** containing all 36 members of the library with π-basic substituents separated π-acid substituted amino acid amides. Although encouraging since it suggested the presence of at least one useful selector, this result did not reveal which of the numerous selectors on CSP **14** was the most powerful one. Therefore, a deconvolution process involving the preparation of series of beads with smaller numbers of attached selectors was used. The approach is schematically outlined in Fig. 3-17.

Fig. 3-17. Schematic of the deconvolution process used in the library-on-bead approach.

In the next step, each single amino acid was coupled separately with the 12 amines resulting in three new CSPs (CSP **15**–CSP **17**) each containing mixed ligands. Once packed into HPLC columns these CSPs were evaluated. The highest selectivity factor of 13.7 in the first deconvoluted series of columns was found for the proline-based CSP **17** while the α-values of the two other columns were close to 5. Further deconvolution of the proline-based column was carried out by splitting the 12 members of the amine building blocks **26–37** in two subgroups to afford a third set of columns (CSP **18** and CSP **19**). Thus, the first proline-based sublibrary column (CSP **18**) was prepared using 6 amines with smaller aromatic substituents (amines **26–31**), and the second column (CSP **19**) with amines having larger substituents (**32–37**). These columns exhibited selectivity factors of 13.6 and 7.3, respectively. In the next step, the 6 amine-based groups present in the more selective column CSP **18** were divided again into 2 groups (**26–28** and **29–31**), containing 3 amines with mainly *meta*-substituted aromatic amines and another 3 with *para*-substituted amines. The respective columns CSP **20** and CSP **21** exhibited rather high α-values of 17.4 and 14.9 for racemic (3,5-dinitrobenzoyl)leucine diallylamide and indicated that both groups involved at least one selector with a very high selectivity. This was not surprising for CSP **20,** since we have previously demonstrated that the selector substituted with 3,5-dimethylaniline (**27**) that was intentionally used in our experimental design, was quite powerful [8]. Since the performance of these two columns was similar, we decided to further deconvolute CSP **21** as it involved an entirely new set of selectors. Three columns CSP **22**–CSP **24** packed with beads containing only individual selectors were prepared with the amines **29, 30,** and **31,** respectively. Although all these columns exhibited rather high selectivities, an α-value of 23.1 was achieved with CSP **22** featuring 5-aminoindane **29** as a part of the proline selector. Figure 3-18 shows the changes in selectivity factors determined for (3,5-dinitrobenzoyl)leucine diallylamide on CSP **14–24.**

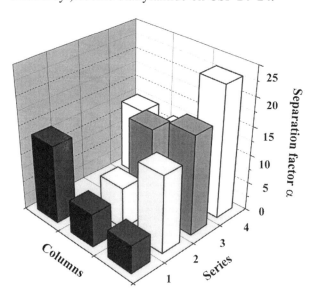

Fig. 3-18. Selectivity factors α determined for (3,5-dinitrobenzoyl)leucine diallylamide on CSP 15–24. Conditions: analyte (3,5-dinitrobenzoyl)leucine diallylamide; column 150 × 4.6 mm i.d.; mobile phase 20 % hexane in dichloromethane; flowrate 1 mL min⁻¹, UV detection at 254 nm.

Since this method of screening initially operated by selecting groups of molecules rather than individual compounds, and since the difference between both CSP **20** and CSP **21** was small, it is indeed possible that our "best" CSP **22** was not actually the most efficient selector of the original mixture. To confirm this, as well as to satisfy our curiosity to uncover which other selector was very powerful, we prepared three additional columns CSP **25–27** containing single proline-based selectors with amines **26–28** as a control experiment. As expected from the previous work [8], CSP **26** prepared with amine **27** also exhibited a very high selectivity ($\alpha = 24.7$ for (3,5-dinitrobenzoyl)leucine diallylamide) similar to that of CSP **22**. Surprisingly, CSP **24** and CSP **27**, prepared with amines **26** and **28** respectively, afforded only modest α-values of less than 4.

The rapid increase in the separation factors observed for the individual series of columns reflected not only the improvement in the intrinsic selectivities of the individual selectors but also the effect of increased loading with the most potent selector. Although the overall loading determined from nitrogen content remained virtually constant at about 0.7 mmol g^{-1} for all CSPs, the fractional loading of each selector increased as the number of selectors in the mixture decreased. Thus, the whole method of building block selection and sublibrary synthesis can be also viewed as an amplification process.

In the classical one-column-one-selector approach, the number of columns that have to be tested equals the number of selectors. Using the chemistry described above, this would require the preparation, packing, and testing of 36 CSPs. In contrast, our combinatorial scheme allowed us to obtain a highly selective CSP from the same group of 36 selectors using only 11 columns (less than one-third). A simple theoretical calculation reveals that the use of all 20 natural amino acids with 12 amines would lead to a library of 240 selectors. While the preparation and testing of 240 columns would be time consuming, a mixture of these selectors could be deconvoluted using our approach with only 15 columns or just 1/16 of the total number of columns that would otherwise be required. The parallelism advantage of the "library-on-bead" approach with mixed selector column would be even more impressive with much larger libraries of selectors for which the deconvolution by splitting the library in each step to two or three sublibraries would rapidly lead to the most selective CSP. Obviously, this approach can dramatically decrease the time required for the development of novel CSPs.

Although the power of this combinatorial approach was clearly demonstrated, our method also has some limitations. For example, in a hypothetical situation in which only a single selector is active and all members of a much larger library are attached to the beads in equal amounts, the percentage of the active selector in the mixture is low. Despite its possibly of high specific selectivity (selectivity per unit of loading), the actual selectivity of a mixed selector CSP may be rather small because of the low loading of the specific selector. Accordingly, the peaks for both enantiomers may elute close to each other and the actual separation may become impossible to observe within the limits of experimental errors. Thus the sensitivity of the chromatographic screening may somewhat limit this approach. However, the number of selectors that can be screened in a single column remains impressive.

In contrast, there are fewer limitations from the chemical point of view. The preparation of large, well-defined, libraries that involve amino acid building blocks has been demonstrated many times. Carefully optimized reaction conditions for the preparation of other mixed libraries can also ensure that each desired compound is present in sufficient amount. However, the reaction rates of some individual selectors with the activated solid support may be lower than that of others. As a result, the more reactive selectors would occupy a majority of the sites within the beads. Since the most reactive selectors may not be the most selective, testing of a slightly larger number of specifically designed CSPs may be required to reduce the effect of false-negative results.

While the reciprocal approach is best suited for the development of a CSP for a single, well-known racemic target, the library-on-bead technique is more useful for the initial scanning of various targets to find a lead selector. It is easy to imagine a development laboratory with a number of columns with immobilized libraries of selectors used to screen target racemates in very rapid fashion. Such pre-screening would suggest the type of selector chemistry that may be best suited for a specific target. The next step would either involve deconvolution of the library on bead or reciprocal testing of parallel libraries of selectors with analogous core chemistries.

3.8   Conclusion

Combinatorial chemistry, a powerful tool in many areas such as drug discovery, materials research, and catalysis, can also be used effectively in the area of molecular recognition to discover new selectors for the recognition of chirality. To date, only a few combinatorial strategies leading to chiral selectors have been demonstrated in the literature. Other approaches such as combinatorial molecular imprinting [96–98] may soon emerge to expand the scope of combinatorial recognition processes. In the future, it is very likely that combinatorial methods will become a widely used tool, even for the development of effective selectors for specific targets.

The power of combinatorial chemistry resides in both the large numbers of compounds that can be prepared within a very short period of time and the rapid assay and deconvolution techniques that may be used for testing to discover the optimal or near-optimal selector within the library. This availability of libraries encompassing a broad diversity of ligand types enables rapid identification of suitable selector families, their comparative screening, and the rapid preparation of custom-made separation media for the resolution of specific racemates [99]. As an additional benefit, studies carried out with broad arrays of structurally related families of selectors can further improve the general understanding of chiral recognition.

Chiral separation media are quite complex systems. Therefore, neither combinatorial methods nor even the identification of the best selector can ensure that an outstanding chiral separation medium will be prepared. This is because some other variables of the system such as the support, spacer, and the chemistry used for their connection to the selector must also be taken into the account. It is likely, that combinatorial methods will also soon be used for the optimization of these subunits of the system.

Acknowledgments

Support of this research by a grant of the National Institute of General Medical Sciences, National Institutes of Health (GM-44885) is gratefully acknowledged. This work was also partly supported by the Division of Materials Sciences of the U.S. Department of Energy under Contract No. DE-AC03-76SF00098.

References|

[1] Seebach, D.; Sting, A. R.; Hoffmann, M. *Angew. Chem., Intl. Ed.* **1996**, *35*, 2708–2747.

[2] Gawley, R. E.; Aubé, J. *Principles of Asymmetric Synthesis;* Pergamon: New York, 1996.

[3] Abdoul-Enein, H. Y.; Wainer, I. W. *The Impact of Stereochemistry on Drug Development and Use;* Wiley-VCH: New York, 1997.

[4] Adachi, S. *J. Chromatogr. A* **1994**, *658*, 271–282.

[5] Ahuja, S. *Chiral Separations: Applications and Technology;* American Chemical Society: Washington D.C., 1997.

[6] Sogah, G. D. Y.; Cram, D. *J. Am. Chem. Soc.* **1976**, *98*, 3038.

[7] Hosoya, K.; Yoshizako, K.; Tanaka, N.; Kimata, K.; Araki, T.; Haginaka, J. *Chem. Lett.* **1994**, 1437–1438.

[8] Liu, Y.; Svec, F.; Fréchet, J. M. *J. Anal. Chem.* **1997**, *69*, 61–65.

[9] Sinibaldi, M.; Castellani, L.; Federici, F.; Messina, A.; Girelli, A. M.; Lentini, A.; Tesarova, E. *J. Liq. Chromatogr.* **1995**, *18*, 3187–3203.

[10] Lewandowski, K.; Murer, P.; Svec, F.; Fréchet, J. M. *J. Anal. Chem.* **1998**, *70*, 1629–1638.

[11] Wulff, G. *Angew. Chem., Intl. Ed. Eng.* **1995**, *34*, 1812–1832.

[12] Cormack, P. G.; Mosbach, K. *React. Funct. Polym.* **1999**, *41*, 115–124.

[13] Allenmark, S. *Chromatographic Enantioseparations: Methods and Applications;* Ellis Horwood: New York, 1991.

[14] Lewandowski, K.; Murer, P.; Svec, F.; Fréchet, J. M. *J. Anal. Chem.* **1998**, *70*, 1629–1638.

[15] Jinno, K. *Chromatographic Separations Based on Molecular Recognition;* Wiley-VCH; New York, 1997.

[16] Beesley, T. E.; Scott, R. P. *Chiral Chromatography;* Wiley-VCH: New York, **1998**.

[17] Siret, L.; Bargmann-Leyder, N.; Tambute, A.; Caude, M. *Analysis* **1992**, *20*, 427–435.

[18] Hermansson, J. *J. Chromatogr.* **1983**, *26*, 71.

[19] Hermansson, J. *Trends Anal. Chem.* **1989**, *8*, 251.

[20] Hesse, G.; Hagel, R. *Chromatographia* **1973**, *6*, 277–280.

[21] Okamoto, Y.; Aburatani, M.; Hatada, K. *Bull. Chem. Soc. Jap.* **1990**, *63*, 955–957.

[22] Okamoto, Y.; Yashima, E. *Angew. Chem., Intl. Ed.* **1998**, *37*, 1021–1043.

[23] Blaschke, G.; Schwanghart, A. D. *Chem. Ber.* **1976**, *109*, 1967–1975.

[24] Okamoto, Y.; Nakano, T. *J. Am. Chem. Soc.* **1980**, *102*, 6358.

[25] Armstrong, D. W.; Tang, Y.; Chen, S.; Zhou, Y.; Bagwill, C.; Chen, J. R. *Anal. Chem.* **1994**, *66*, 1473–1484.

[26] Harada, A.; Furue, M.; Nozakura, S. *J. Pol. Sci.* **1978**, *16*, 189.

[27] Armstrong, D. W.; DeMond, W. *J. Chromatogr. Sci.* **1984**, *22*, 411.

[28] Rogozhin, S. V.; Davankov, V. A. *Dokl. Akad. Nauk SSSR* **1970**, *192*, 1288.

[29] Humbel, F.; Vonderschmitt, D.; Bernauer, K. *Helv. Chim. Acta* **1970**, *53*, 1983.

[30] Yamskov, I. A.; Berezin, B. B.; Davankov, V. A.; Zolotarev, Y. A.; Dostavalov, I. N.; Myasoedov, N. F. *J. Chromatogr.* **1981**, *217*, 539–543.
[31] Pirkle, W. H.; Finn, J. M.; Schreiner, J. L.; Hamper, B. C. *J. Am. Chem. Soc.* **1981**, *103*, 3964–3966.
[32] Pirkle, W. H.; Hyun, M. H. *J. Org. Chem.* **1984**, *49*, 3043–3046.
[33] Gargaro, G.; Gasparrini, F.; Misiti, D.; Palmiery, G.; Pierini, M.; Villani, C. *Chromatographia* **1987**, *24*, 505–509.
[34] Lämmerhofer, M.; Lindner, W. *J. Chromatogr.* **1996**, *741*, 33–48.
[35] Dalgliesh, C. E. *J. Chem. Soc.* **1952**, 3940–3942.
[36] Pirkle, W. H.; Burke, J. A.; Wilson, S. R. *J. Am. Chem. Soc.* **1989**, *111*, 9222–9223.
[37] Pirkle, W. H.; Pochapsky, T. C. *J. Am. Chem. Soc.* **1987**, *109*, 5975–5982.
[38] Pirkle, W. H.; Pochapsky, T. C. *Chem. Rev.* **1989**, *89*, 347–362.
[39] Davankov, V. A. *Chromatographia* **1989**, *27*, 475–482.
[40] Pirkle, W. H.; Bowen, W. E. *HRC-J. High Resolut. Chromatogr.* **1994**, *17*, 629–633.
[41] Pirkle, W. H.; Murray, P. G.; Burke, J. A. *J. Chromatogr.* **1993**, *641*, 21–29.
[42] Pirkle, W. H.; Welch, C. J.; Lamm, B. *J. Org. Chem.* **1992**, *57*, 3854–3860.
[43] Gasparrini, F.; Misiti, D.; Villani, C.; Borchardt, A.; Burger, M. T.; Still, W. C. *J. Org. Chem.* **1995**, *60*, 4314–4315.
[44] Cuntze, J.; Diederich, F. *Helv. Chim. Acta* **1997**, *80*, 897–911.
[45] Zimmerman, S. C.; Kwan, W. S. *Angew. Chem. Intl. Ed. Eng.* **1995**, *34*, 2404–2406.
[46] Lämmerhofer, M.; Lindner, W. *J. Chromatogr. A* **1998**, *829*, 115–125.
[47] Lämmerhofer, M.; Lindner, W. *J. Chromatogr.* **1999**, *839*, 167–182.
[48] Snyder, L. R.; Kirkland, *J. J. Introduction to Modern Liquid Chromatography*; Wiley: New York 1979.
[49] Pirkle, W. H.; Pochapsky, S. *J. Chromatogr.* 1986, *369*, 175–177.
[50] Strube, J.; Jupke, A.; Epping, A.; Smidt-Traub, H.; Schulte, M.; Devant, R. *Chirality* **1999**, *11*, 440–450.
[51] Pirkle, W. H.; House, D. W.; Finn, J. M. *J. Chromatogr.* **1980**, *192*, 143–158.
[52] Pirkle, W. H.; Burke, J. A. *J. Chromatogr.* **1991**, *557*, 173–185.
[53] Pirkle, W. H.; Koscho, M. E. *J. Chromatogr.* **1999**, *840*, 151–158.
[54] Welch, C. J. *J. Chromatogr. A* **1994**, *666*, 3–26.
[55] Lewandowski, K.; Murer, P.; Svec, F.; Fréchet, J. M. J. *J. Comb. Chem.* **1999**, *1*, 105–112.
[56] Terrett, N. K.; Gardner, M.; Gordon, D. W.; Kobylecki, R. J.; Steele, J. *Chemistry-A European Journal* **1997**, *3*, 1917–1920.
[57] Thompson, L. A.; Ellman, J. A. *Chem. Revs.* **1996**, *96*, 555–600.
[58] Gordon, E. M.; Gallop, M. A.; Patel, D. V. *Acc. Chem. Res.* **1996**, *29*, 144–154.
[59] Menger, F. M.; Eliseev, A. V.; Migulin, V. A. *J. Org. Chem.* **1995**, *60*, 6666–6667.
[60] Francis, M. B.; Finney, N. S.; Jacobsen, E. N. *J. Am. Chem. Soc.* **1996**, *118*, 8983–8984.
[61] Gilbertson, S. R.; Wang, X. *Tetrahedr. Lett.* **1996**, *37*, 6475–6478.
[62] Cole, B. M.; Shimizu, K. D.; Krueger, C. A.; Harrity, J. A.; Snapper, M. L.; Hoveyda, A. H. *Angew. Chem., Intl. Ed. Eng.* **1996**, *35*, 1668–1671.
[63] Cole, B. M.; Shimizu, K. D.; Krueger, C. A.; Harrity, J. A.; Snapper, M. L.; Hoveyda, A. H. *Angew. Chem. Intl. Ed. Eng.* **1996**, *35*, 1668–1671.
[64] Brocchini, S.; James, K.; Tangpasuthadol, V.; Kohn, J. *J. Am. Chem. Soc.* **1997**, *119*, 4553–4554.
[65] Schultz, P. G.; Xiang, X. D. *Current Opinion in Solid State & Materials Science* **1998**, *3*, 153–158.
[66] Xiang, X. D.; Sun, X. D.; Briceno, G.; Lou, Y.; Wang, K. A.; Chang, H.; Wallace-Freedman, W. G.; Chen, S. W.; Schultz, P. G. *Science* **1995**, *268*, 1738–1740.
[67] Dickinson, T. A.; Walt, D. R.; White, J.; Kauer, J. S. *Anal. Chem.* **1997**, *69*, 3413–3418.
[68] Burger, M. T.; Still, W. C. *J. Org. Chem.* **1995**, *60*, 7382–7383.
[69] Goodman, M. S.; Jubian, V.; Linton, B.; Hamilton, A. D. *J. Am. Chem. Soc.* **1995**, *117*, 11610–11611.
[70] Carrasco, M. R.; Still, W. C. *Chemistry & Biology* **1995**, *2*, 205–212.
[71] Huang, P. Y.; Carbonell, R. G. *Biotechnol. Bioeng.* **1995**, *47*, 288–297.
[72] Wilson, S. R.; Czarnik, A. W. *Combinatorial Chemistry: Synthesis and Application;* Wiley: New York, 1997.
[73] Terrett, N. K. *Combinatorial Chemistry*; Oxford University Press: Oxford, 1998.
[74] Jung, G.; Hofstetter, H.; Feiertag, S.; Stoll, D.; Hofstetter, O.; Wiesmuller, K. H.; Schurig, V. *Angew. Chem. Intl. Ed.* **1996**, *35*, 2148–2150.

[75] Chiari, M.; Desperati, V.; Manera, E.; Longhi, R. *Anal. Chem.* **1998**, *70*, 4967–4973.

[76] Griffey, R. H.; An, H. Y.; Cummins, L. L.; Gaus, H. J.; Haly, B.; Herrmann, R.; Cook, P. D. *Tetrahedron* **1998**, *54*, 4067–4076.

[77] Garcia, M. E.; Gavin, P. F.; Deng, N. L.; Andrievski, A. A.; Mallouk, T. E. *Tetrahedr. Lett.* **1996**, *37*, 8313–8316.

[78] Gavin, J. A., Deng, N. L.; Alcala, M.; Mallouk, T. E. *Chem. Mater.* **1998**, *10*, 1937–1944.

[79] Gavin, J. A.; Garcia, M. E.; Benesi, A. J.; Mallouk, T. E. *J. Org. Chem.* **1998**, *63*, 7663–7669.

[80] Mallouk, T. E.; Gavin, J. A. *Acc. Chem. Res.* **1998**, *31*, 209–217.

[81] Weingarten, M. D.; Sekanina, K.; Still, W. C. *J. Am. Chem. Soc.* **1998**, *120*, 9112–9113.

[82] Kitka, E. J.; Gruska, E. *J. Chromatogr.* **1977**, *135*, 367–376.

[83] Howard, W. A.; Hsu, T. B.; Rogers, L. B.; Nelson, D. A. *Anal. Chem.* **1985**, *57*, 606–610.

[84] Welch, C. J.; Bhat, G.; Protopopova, M. N. *Enantiomer* **1998**, *3*, 463–469.

[85] Welch, C. J.; Protopopova, M. N.; Bhat, G. *Enantiomer* **1998**, *3*, 471–476.

[86] Welch, C. J.; Bhat, G.; Protopopova, M. N. *J. Comb. Chem.* **1999**, *1*, 364–367.

[87] Wang, Y.; Li, T. *Anal. Chem.* **1999**, *71*, 4178–4182.

[88] Lewandowski, K.; Murer, P.; Svec, F.; Fréchet, J. M. *J. Chem. Commun.* **1998**, 2237–2238.

[89] Armstrong, R. W., Combs, A. P.; Tempest, P. A.; Brown, S. D.; Keating, T. A. *Acc. Chem. Res.* **1996**, *29*, 123–131.

[90] Biginelli, P. *Gazz. Chim. Ital.* **1893**, *23*, 360–416.

[91] Kappe, C. O. *Tetrahedron* **1993**, *49*, 6937–6963.

[92] Wu, Y. Q.; Wang, Y.; Yang, A. L.; Li, T. *Anal. Chem.* **1999**, *71*, 1688–1691.

[93] Murer, P.; Lewandowski, K.; Svec, F.; Fréchet, J. M. *J. Anal. Chem.* **1999**, *71*, 1278–1284.

[94] Murer, P.; Lewandowski, K.; Svec, F.; Fréchet, J. M. *J.Chem. Commun.* **1998**, 2559–2560.

[95] Pirkle, W. H.; Welch, C. J. *J. Chromatogr. A* **1996**, *731*, 322–326.

[96] Lanza, F.; Sellergren, B. *Anal. Chem.* **1999**, *71*, 2092–2096.

[97] Takeuchi, T.; Fukuma, D.; Matsui, J. *Anal. Chem.* **1999**, *71*, 285–290.

[98] Sabourin, L.; Ansell, R. J.; Mosbach, K.; Nicholls, I. A. *Anal. Commun.* **1998**, *35*, 285–287.

[99] Stinson, S. C. *Chem. Eng. News* **1999**, October 11, p. 110.

4 CHIRBASE: Database Current Status and Derived Research Applications using Molecular Similarity, Decision Tree and 3D "Enantiophore" Search

Christian Roussel, Johanna Pierrot-Sanders, Ingolf Heitmann and Patrick Piras

4.1 Introduction

The past two decades have seen remarkable advances in chiral chromatography, as only 20 years ago, the direct resolution of enantiomers by chromatography was still considered to be an impressive technical achievement.

The enormous increase in the number of groups working in this domain, in concert with the advances in the fundamental techniques of chromatography and laboratory automation (screening technologies) have led to the rapid and unprecedented accumulation of data [1].

Despite the difficulties caused by the rapidly expanding literature, the use of chiral stationary phases (CSPs) as the method of choice for analysis or preparation of enantiomers is today well established and has become almost routine. It results from the development of chiral chromatographic methods that more than 1000 chiral stationary phases exemplified by several thousands of enantiomer separations have been described for high-performance liquid chromatography (HPLC).

The intended fields of usage of these methods include a broad range of R&D and control applications (pharmacokinetics, asymmetric synthesis, enzymatic resolution, simulated moving bed technology, etc.) in the pharmaceutical and agrochemical, as well as the food and biotechnology industries. This gives rise to one important outcome for the application of chiral technology: an increasing number of scientists with different scientific cultures have to select the correct analytical tools.

If the end-users of these tools have different objectives and requirements depending on the field of their activity, they do have in common the same issue: which CSP and working conditions should be selected for the enantiomeric separation of a given pair of enantiomers?

Such an achievement of a chiral separation often requires large numbers of expensive and time-consuming laboratory experiments, even for experienced groups and organizations. Without considering the resulting loss of time and money for those who repeat negative experiments already mentioned in the literature, substantial savings in resources can be expected by strategies in which a substance's behavior on a given CSP can be evaluated from a database search of its molecular structure.

The need for an overall and combined chemical structure and data search system became clear to us some time ago, and resulted in the decision to build CHIRBASE, a molecular-oriented factual database. The concept utilized in this database approach is related to the importance of molecular interactions in chiral recognition mechanisms. Solely a chemical information system permits the recognition of the molecular key fingerprints given by the new compound among thousands of fingerprints of known compounds available in a database.

The first system designed to perform these basics was developed in 1989 [2]. In early 1990, a first version of 5000 entries was made available to the scientific community on an IBM PC DOS-based program: ChemBase. Now, some 10 years later, the database has increased to several tens of thousands of entries, and the entire system can be searched with the powerful ISIS software [3] as rapidly as it was searched on the old computer systems.

Since the original version was released, there have been significant changes [4] and improvements following the advent of new generations of computers and chemical substructure systems [5], but the main goal of the project has not changed: to gather together comprehensive structural, experimental and bibliographic information on successful or unsuccessful chiral separations which have been obtained on CSPs by liquid chromatography, and with an emphasis on molecular structures searches.

Besides the rapidly growing accumulation of data, the need to systematize and analyze the data becomes correspondingly ever more demanding. Hence today, the challenge resides not only in the management of this huge quantity of information, but also in its investigation in order to reveal the knowledge implicit within the data. It must be emphasized here that CHIRBASE data can be directly exported from ISIS under various formats; consequently, a number of different computational methods and tools to handle and analyze these data can be readily employed.

In this chapter, we will discuss the present status of CHIRBASE and describe the various ways in which two (2D) or three-dimensional (3D) chemical structure queries can be built and submitted to the searching system. In particular, the ability of this information system to locate and display neighboring compounds in which specified molecular fragments or partial structures are attached is one of the most important features because this is precisely the type of query that chemists are inclined to express and interpret the answers. Another aspect of the project has been concerned with the interdisciplinary use of CHIRBASE. We have attempted to produce a series of interactive tools that are designed to help the specialists or novices from different fields who have no particular expertise in chiral chromatography or in searching a chemical database.

Finally – and previously defined as a long-term goal of our development effort – data collection has reached the point where meaningful data mining studies are beginning to emerge. The extraction of implicit, previously unknown, and potentially useful patterns from data is facilitated by the link between 2D or 3D database searching, automated molecular property calculation, data analysis and visualization tools (Fig. 4-1). These developments will certainly play an important role in our objective to build an information system that provides not only a data collection but also some knowledge about the processes of the chiral separation, and also help to identify the research that is needed to fill the gaps in our knowledge.

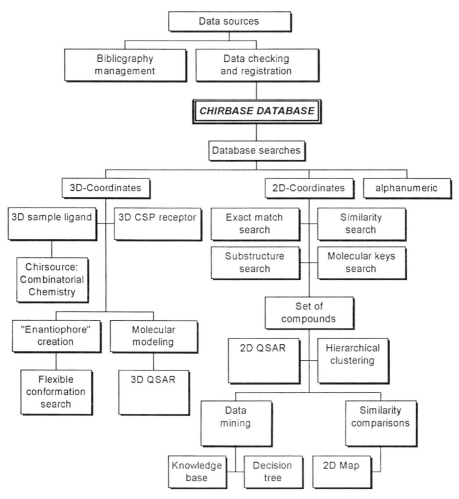

Fig. 4-1. General organization of the CHIRBASE project.

4.2 Database Status, Content and Structure

As we have already pointed out, the first objectives of CHIRBASE project are to:

- compile a comprehensive bibliography;
- offer a diagnostic resource for resolution of newly prepared compounds; and
- supplement existing entries by new structural and bibliographic information as they become available (the database is updated quarterly).

The sources of the data comprise not only the primary chemical literature from international journals and patents (including the synthetic chemistry literature), but also the secondary and the gray literature: manufacturer catalogues, posters or lectures from congresses and unpublished experiments sent by the authors.

Some details of the database are available on the Web [6] and consist today of about 61 000 different chiral separations (Table 4-1). Between 12 000 and 15 000 new separations are stored continuously each year in CHIRBASE.

Table 4-1. Overall statistics.

Number of entries (unique sample-CSP combinations)	40 000
Number of experiments (different chiral separations)	61 000
Number of samples with 2D coordinates	18 000
Number of samples successfully converted into 3D coordinates	16 277
Number of CSPs (2D co-ordinates)	1077
Number of solvents or modifiers (2D co-ordinates)	205
Number of entries with data corrected/completed by the authors	6110
Number of new chiral separations per update (each 4 months)	3–5000

As already stated above, the database has been developed using ISIS software. The program operation is very simple, and about 30 min to learn the particular commands of this structure-searching program. ISIS provides both storage and retrieval of chemical structures. It is also possible to store text and numeric data into database entries. Because molecular structures are searchable in many ways, ISIS software is an excellent tool for exploiting data, and not simply archiving it.

ISIS databases are hierarchical, so CHIRBASE was designed to incorporate about 60 data fields on several levels of detail (the main fields are listed in Table 4-2). The first level contains the molecular structure of the sample combined to the molecular structure of the CSP, producing a unique location or entry for a specific sample-CSP couple. Consequently, in the current version of CHIRBASE, which contains 40 000 entries, one entry corresponds to the separation of one sample on one CSP and contains in different sublevels a compilation of all the references and the various analytical conditions available for this separation.

If some fields may be empty in the sublevels, all the fields in the main level are required for each entry. A new chiral separation record can be added in CHIRBASE solely if the authors correctly identify both sample and CSP. Since the beginning of the project, our policy has been to contact the authors of all publications containing incomplete, ambiguous or inconsistent data and to ask for additional information. Providing the separations with unique case numbers helps us considerably in this essential task, and also facilitates avoiding redundancies in the database. When chiral separations are reported for the second time in a new publication with exactly the same chromatographic conditions, this is stated in a footnote added in the field « comments ». In this field, miscellaneous information that cannot appear elsewhere are listed (detection limit, description of a reported chromatogram, racemization study, mobile phase limitations, etc.).

Table 4-2. Main CHIRBASE fields.

Category	Field names	Category	Field names
REFERENCE	Authors	CONDITIONS	Method
	Journal		Detection
	Volume		Scale
	Page		Injected amount
	Year		Flow-rate
SAMPLE	Sample_name		Temperature
	Sample_chirality		pH
	Sample_molstructure		Mobile_phase
	Sample_molformula	RESULTS	First_eluted
	Sample_molecular_weight		Second_eluted
CSP	CSP_ame		k1
	CSP_molstructure		k2
	CSP_molecular_weight		Rt1
	CSP_molformula		Rt2
	CSP_trade_name		Alpha
	CSP_supplier		Resolution
	CSP_type_of_column		
	CSP_particle_size		Comments

4.3   Data Registration

New entries are built via an in-house-developed ISIS application called REGISTER.

This application contains customized forms (Fig. 4-2) and dialog boxes which assist the user during registration.

The forms comprise several sections (REF, CSP, COND, SAMP, and DATA) which follow the ordered way chiral separations are planned in an organized laboratory. The user first puts in the section REF the full information related to a given reference. Then, in CSP, SAMP and COND he or she will enter all the columns, samples and operating conditions. The separations are actually created in the last windows DATA, where each sample will be connected to all the columns which have been used, and each column will be connected to all the tested conditions; in front of these conditions are entered the experimental results of each separation. This hierarchical structure thus enables a rapid and easy registration of several chiral separations for a given compound.

All the sections must be completed by the user and then submitted to a single administrator for addition to the database. Upon completion of the form, the user has the option of making a check submission, which processes the data and performs error checks as normal, but displays the verdict on screen for the user rather than sending the data to the administrator. A variety of errors are checked, including missing data and inconsistent data, invalid molecular structures or numeric data outside the normal range. When the user is satisfied with the form data, they can be submitted to the administrator via the "export" button. Upon submission, the data are stored in a file in an archive directory of submissions. A copy of the files is then sent to the administrator. The export function of the REGISTER program causes the data to be read and converted into a database of the same structure as the final database storing all the existing chiral separations data. For each record, some fields (such as the solvent or modifier molecular structures) are automatically calculated from the transferred information. The administrator can then choose to add each imported record to the master CHIRBASE database and, if necessary, add further data or comments to each record, thus ensuring the homogeneity of the available information on each separation.

It is important to stress that in contrast to computer-generated derived databases, a great deal of human involvement is necessary for the creation and updating of entries.

The most complicated aspect of the CHIRBASE project effort is the actual incorporation and validation of data. It is largely due to the complexity of the problem and to the difficulty of extracting and interpreting the relevant information, since the vast majority of all useful data is disseminated in the papers rather than in a "user-readable" or a "computer-readable" form.

We are providing these tools to enter data into the database and trying to convince chemists that it is in their own interest to participate with their data for use by the community. This is a very important aspect of the project, as an increasing number of results obtained as a result of analytical chiral methods are published without adequate description of these tools.

The REGISTER application can also be easily modified to meet the requirements of other organizations, which may use it for corporate data registration and storage as a central depository of proprietary chiral separations.

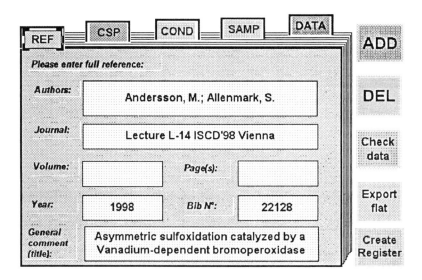

Fig. 4-2. Screen of REGISTER (data registration tool for enantioseparations).

4.4   Searching the System

ISIS provides means for manipulation, comparison, search and retrieval of records. Searches through the database can be carried out in a number of ways using:

- Molecule queries (one or two associated molecules) for sample, CSP or solvent:
  – Exact match
  – Substructure
  – Similarity
  – Name
  – Formula

- Data queries for chromatographic data, literature references, analytical conditions and any other text or numeric data that is associated with a separation:
  – String, substring or numeric data

- Multiple component queries for any combination of structures (one or two molecules) and text or numeric data:
  – Combined molecule and data components

As we have already mentioned, graphical molecule-oriented searching strategies are the best tools to solve a chiral separation problem. Chemical substructure searching (SSS) is the ability to search a structure database using any structure fragment. The results of a SSS are the chemical structures in the database which contain the query structure fragment. One of the major advantages of this type of search in the chiral chromatography domain is that it allows the retrieval of a given sample of all the derivatives, and thus a choice of the best derivatization reagent to achieve a good resolution on a specific CSP.

Molecular similarity searching provides the possibility of finding unrelated but functionally analogous molecules. This is a very nice feature because many distinct structures in contact with a CSP often share the same active sites. The compounds which have a structure similar to the structure of the sample query can be displayed automatically in order of their similarity. The degree of similarity is measured by a numerical value on a scale of 0 to 100 that may be included in the output form. An example of a similarity search is shown in Fig. 4-3. In this example, a search is being performed for the AZT with a similarity value >65 %.

The stereochemistry shown in CHIRBASE includes relative configurations, absolute configurations, and geometric isomers. A chiral flag on the chiral selector of a CSP identifies a specific enantiomer of a molecule. Up- and down- stereo bonds represent relative configurations, and together with a chiral flag, they represent the absolute configuration of the CSP. Polymeric supports are stored as generic monomer structures in the database. In such molecules, the pseudoatom "[" defines the linkage that connects all the monomers of the polymer. Other searchable pseudoatoms may be connected to a portion of the polymer structure such as a methylene group. For instance, the pseudoatoms "coated" or "Beads" will characterize the preparation mode of the CSP.

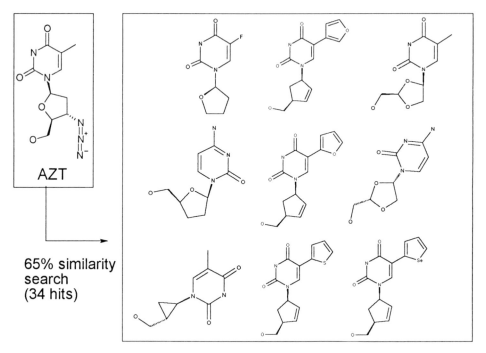

Fig. 4-3. Molecular similarity searching of AZT in CHIRBASE. (Compounds reported in Refs. [7–12].)

In addition to addressing the practicalities of database maintenance, we also aim to enhance the range of the facilities available, to make the information within CHIRBASE more readily accessible to users. CHIRBASE contains two form-based applications for query building designed to help novice or expert users to formulate queries: the query menu and the automatic search tool.

4.4.1 The Query Menu

CHIRBASE provides integrated responses from single questions, as well as from combinatorial questions constructed on the basis of any specific query correspond-ing to one or several field(s) occurring in the database. With the molecular structure of a sample in hand, the search can be conducted interactively from the query menu form.

By clicking the appropriate buttons on the form, the user can combine molecular structure queries of sample, CSP and solvent, using operators AND, OR, NOT with data queries in one search. A query for the search of chiral separations of alpha-aro-matic acids on any polysaccharide phases coated on silica gel providing an alpha value superior to 1.2 is shown in Fig. 4-4.

The query menu is simple to use for answering most of the questions commonly addressed to the database:

We have one specific sample. We want to find all possible CSPs for its analytical or preparative separation and closely related molecules.

We want to explore the application of specific CSPs. We want to find all separations that use them or that use structurally similar CSPs and specific data that are associated with them.

We have specific types of compounds of interest. We want to include or exclude that use these specific solvents or types of solvents.

We want to find all separations that have a given combination of structural features and data in common.

We want to retrieve the literature citations to specific structures or family of structures.

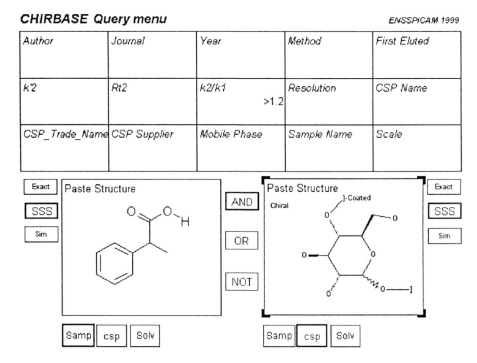

Author	Journal	Year	Method	First Eluted
k'2	Rt2	k2/k1 >1.2	Resolution	CSP Name
CSP_Trade_Name	CSP Supplier	Mobile Phase	Sample Name	Scale

CHIRBASE Query menu   *ENSSPICAM 1999*

Fig. 4-4. The query menu form: search of chiral separations of alpha-aromatic acids on any polysaccharide CSPs with $\alpha > 1.2$.

4.4.2 The Automatic Search Tool

The entire formulation of a chemical question can involve a number of issues, some of which are often obscure to the user. Questions such as what components can be substituted at a given atom site, what types of bonds are favorable for a given chemical fragment query (single, double, ring, chain, aromatic, etc.), and should the query contain explicit or implicit substitution, are just a few of the issues facing a user.

For the inexperienced or infrequent users it is even more of a problem. Inexperienced users should have the ability to exploit easily the wide variety of structural search techniques.

To address these issues regarding adequate structural searching in CHIRBASE, some facilities have been recently added to the user interface. The result is the automatic generation and search of strategic 2D query structures defined with the help of the following commands (Fig. 4-5):

- Auto Search: This button initiates from a structure query two or three automated series of search: exact and substructure searches in local desktop versions; exact, substructure and similarity searches in network version (under ISIS/Host). All the result lists are saved in CHIRBASE using « exact-auto », « SSS-auto » and « SIMXX %-auto » names. XX is the highest similarity search value (from 80 % to 40 %) allowing to retrieve hits in CHIRBASE. The records in all the lists are unique. The SSS-auto list does not include records that are in the exact-auto list. The SIMXX %-auto list does not include records that are in exact and SSS-auto lists.
- Auto Build: This application has been built upon the knowledge and experiences of our working group in order that beginner has access to all pertinent information. Auto Build automatically adds in the sample query structure the appropriate atom and bond query features. Then clicking on SSS button initiates a substructure search. Some examples of the query features, which may be added to a query structure, are:
  – I, Cl, F, Br atoms are replaced by a list of halogens [I, Cl, Br, F].
  – Cycles or chains are respectively specified with the "Rn" or "Ch" query bond features.
  – The query atom «A» (Any atoms) is added in phenyl, naphthyl, etc. aromatic cycles to allow finding any hetero-aromatic cycles.
  – S/D (Single/Double bonds) query bond features added if tautomers can be built from sample structure.

When a search of any sort is completed, the number of hits is reported to the user. The search can be further narrowed by restricting the range of structural features to be retrieved. Specific substituents can be added and substring/numerical searches can also be included on fields such as "alpha" and "CSP trade name". Here, the rapidity with which results to the queries are provided is decisive, because the answers often constitute the support of a subsequent query. This feedback loop is extremely resourceful and any use of chemical database should be based around this inventive approach. A highly interactive system which supplies immediately the answer to a question enables the user to detect the defaults in a question and then successively to adjust the query until he or she finds the best questions which are often not well-known in the beginning of the search. Thus, the crucial point in searching CHIRBASE is not only how to formulate a question, but also to find the best questions. That is why different output forms have been developed to visualize the results of database searches: either full data for each matching record, concise data comprising selected fields (the summary form is displayed in Fig. 4-6) or tables are available.

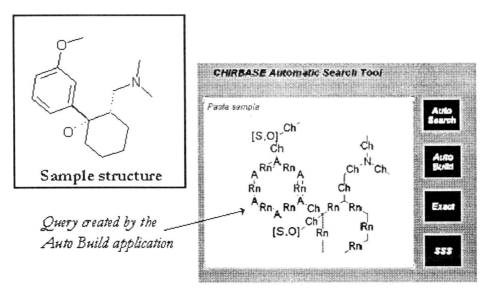

Fig. 4-5. Use of the Auto Build command to create a query structure. A: any atoms; Ch: chain bond; Rn: ring bond; [S,O]: oxygen or sulfur atom.

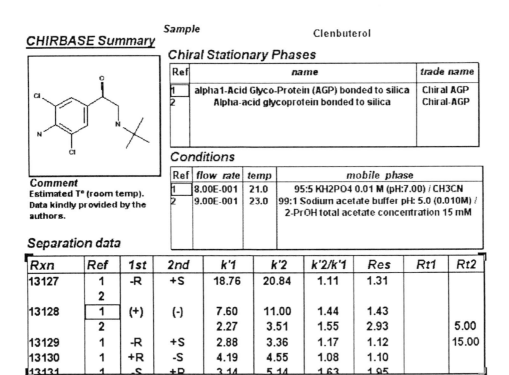

CHIRBASE Summary

Sample   Clenbuterol

Chiral Stationary Phases

Ref	name	trade name
1	alpha1-Acid Glyco-Protein (AGP) bonded to silica	Chiral AGP
2	Alpha-acid glycoprotein bonded to silica	Chiral-AGP

Comment
Estimated T° (room temp).
Data kindly provided by the
authors.

Conditions

Ref	flow rate	temp	mobile phase
1	8.00E-001	21.0	95:5 KH2PO4 0.01 M (pH:7.00) / CH3CN
2	9.00E-001	23.0	99:1 Sodium acetate buffer pH: 5.0 (0.010M) / 2-PrOH total acetate concentration 15 mM

Separation data

Rxn	Ref	1st	2nd	k'1	k'2	k'2/k'1	Res	Rt1	Rt2
13127	1	-R	+S	18.76	20.84	1.11	1.31		
	2								
13128	1	(+)	(-)	7.60	11.00	1.44	1.43		
	2			2.27	3.51	1.55	2.93		5.00
13129	1	-R	+S	2.88	3.36	1.17	1.12		15.00
13130	1	+R	-S	4.19	4.55	1.08	1.10		
13131	1	-S	+R	3.14	5.14	1.63	1.95		

Fig. 4-6. CHIRBASE Summary output form.

4.5 3D Structure Database Searches

Today, 3D databases, which provide the means for storing and searching for 3D information of compounds, are proven to be useful tools in drug discovery programs. This is well exemplified with the recent discovery of novel nonpeptide HIV-1 protease inhibitors using pharmacophore searches of the National Cancer Institute 3D structural database [13–15].

In 3D searching strategies, the relevant molecules will be found by searching the fitting 3-D molecular properties around the chirality center instead of similar functional groups, as it can be searched using 2D query structures. The facilities of ISIS/3D fulfil this ability to import and search 3D structures in CHIRBASE. The 3D structures were built with the program CORINA developed by Gasteiger's group [16] and included in TSAR [17], a fully integrated quantitative structure-activity relationship (QSAR) package. For each 2D structure, only a single conformation was stored in CHIRBASE because ISIS/3D allows conformationally flexible substructure (CFS) searching [18]. In a CFS search, the conformational fitting is the process of rotating single bonds in 3D structures to fit the constraints of the query. Actually, searching a database of flexible conformations is quite efficient and does not require the storage of accurate models as produced from X-ray crystallographic data, as long as we always build the database with models derived from the same force field calculations and therefore control the errors.

A very interesting approach in the prediction of a chiral resolution is the study of the solute binding to the stationary phase receptor. This habitually requires the utilization of sophisticated docking and molecular dynamics techniques, and much computer time. However, numerous reasonable predictions of CSP-solute binding have been made in the literature simply by specifying the spatial arrangement of a small number of atoms or functional groups [19]. Here, we will call such an arrangement "enantiophore". A key concept of CHIRBASE 3D searches is to find molecules from a 3D query of a compound reduced to a description of the chemical environment that corresponds to the enantiophore hypothesis. This new approach now permits us to consider the molecules in terms of interaction centers of the solute and the geometric relationship between those centers. For example, an "enantiophore" may be defined by three atoms separated by a set of distances with some tolerances (usually 0.5 A). Then a search in ISIS/3D will find only the structures that satisfy the enantiophore query. In this work, we will explore two kinds of strategies.

4.5.1 Queries Based on CSP Receptor

The enantiophore query used in the search is derived from the CSP and directly built from a 3D structure model of the target CSP molecule, as it can be used today for the determination of new lead compounds [20, 21]. This procedure does not need an important modeling expertise. One can easily recognize the different center types in the receptor in question. These can be hydrogen-bond donors and acceptors, charged centers, aromatic ring centers, lipophile centers, etc. These will be assumed to represent the types of interactions that the sample is likely to have at the CSP receptor sites. Once potential CSP active sites have been identified, the user can propose different enantiophores, which are complementary to the sites and since, may be considered as CSP receptor-based queries. The main purpose of this enantiophore building is then to locate in CHIRBASE the potential structures that can fit into a given CSP receptor.

This technique has been used to track conceivable enantiophores of the Whelk-O 1 CSP ((3R,4S)-4-(3,5-dinitrobenzamido)-3-[3-(dimethylsilyloxy)propyl]-1,2,3,4-tetrahydrophenanthrene) which was prepared in 1992 by Pirkle and Welch [22]. As seen in Fig. 4-7, a Whelk enantiophore should contain two ore more of the following molecular properties:

- aromatic center (π-π interaction)
- hydrogen-bond donor center
- hydrogen-bond acceptor center
- dipole-dipole stacking

The hypothetical enantiophore queries are constructed from the CSP receptor interaction sites as listed above. They are defined in terms of geometric objects (points, lines, planes, centroids, normal vectors) and constraints (distances, angles, dihedral angles, exclusion sphere) which are directly inferred from projected CSP receptor-site points. For instance, the enantiophore in Fig. 4-7 contains three point attachments obtained by:

- projecting a point at a 3 Å distance along a line that is perpendicular to an aromatic ring plane. This point is surrounded by an exclusion sphere (1.5 Å in diameter) to exclude atoms within this sample space.
- projecting a point along the direction of a hydrogen at a 3 Å distance from the hydrogen bond donor site.

The distances then retrieved between the enantiophore points are exactly based on the actual receptor-site information, and so well justify the expression of "CSP receptor-based query" quoted above. It is well recognized today that samples that are resolved on a common CSP receptor do not systematically bind with analogous functional groups. Many CSPs often present multiple modes of chiral recognition mechanisms. For this Whelk CSP study, we have elaborated the following simple enantiophore queries (Fig. 4-8) using the exposed strategy:

Query 1: contains two aromatic groups. The «A» (any) atoms mean that we allow the retrieval of heterocyclic systems.
Query 2: an aromatic group and a hydrogen-bond acceptor center. R1 is a complex generic group (not detailed here) which delimits the search to all groups that provides lone-pair electrons to the hydrogen bond.
Query 3: an aromatic group and a hydrogen-bond donor center.
Query 4: one aromatic group and an amide group.

All these queries have at least one lipophilic aromatic group.

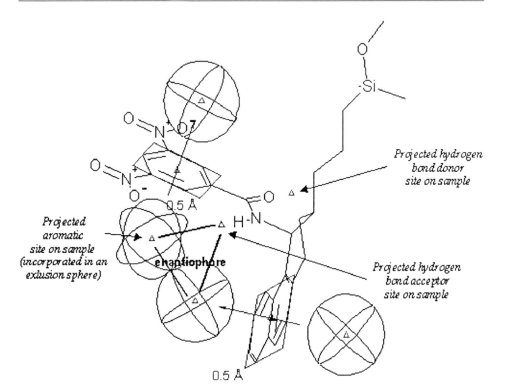

Fig. 4-7. One possible three interaction points enantiophore built from the Whelk-O 1 CSP.

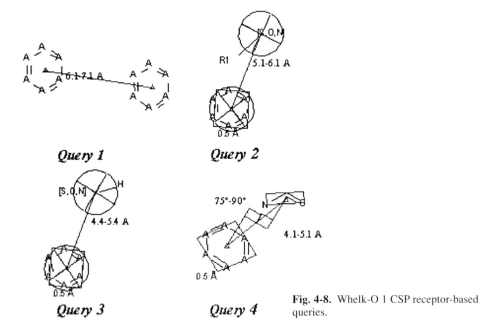

Query 1

Query 2

Query 3

Query 4

Fig. 4-8. Whelk-O 1 CSP receptor-based queries.

One may be surprised by the simplicity of those enantiophore queries. Nevertheless, in respect of the stereochemical constraints, such combinations of interactions involving hydrogen-bond donor or acceptor groups and π-stacking are sufficiently specific and directional to provide adequate chiral recognition models. A certain tolerance has been attached to the distances (± 0.5Å) or angles because a precise coincidence of atoms is not necessary as specific interactions may take place over a range of angles and distances.

The four queries were examined against a list of samples tested on Whelk CSP that constitutes our search domain. Search results are summarized in Table 4-3. Of the 616 3D structures in this database list, 370 fit at least one of the query (one sample may fit more than one query) and 335 are given as resolved according to chromatographic data or information reported in the field comment. Query 2 retrieved the largest number of compounds with a high percentage of resolved samples in the hit list. While the number of hits retrieved with Query 1 is lower, this query provided a similar proportion of resolved samples (93 %).

Table 4-3. Search results from CHIRBASE 3D with enantiophore queries built from the Whelk CSP.

	Number of hits	Number of resolved samples	% of resolved samples in the hit list
Query 1	141	131	93
Query 2	286	266	93
Query 3	116	76	65
Query 4	129	109	84
Total	370	335	90

One approach to dealing with the problem of the consistency of the distance ranges attached to the queries is to perform additional CFS searches with different distance constraints assigned arbitrarily. We have not yet completed these trials, but first results led to a decrease in the number of retrieved hits and the yields of compounds given as resolved. Another way to control the quality of the enantiophore queries is to seek if they really contain the essential features responsible for chiral recognition. This can be done by examining the query embedded within the retrieved structures in their bound conformations as displayed in Fig. 4-9. In this figure, the query embedded in sample 4 indicates that only one aromatic group presumably interacts with the CSP; the other aromatic group seems not involved in sample binding. This assumption was quite consistent with other result hits. For instance, the same query was also able to retrieve *N*-methyl-3-phenyl-3-trimethylsilyl-propanamide, which is also well separated on Whelk CSP and only contains one aromatic group.

From these preliminary results, we can assume that such enantiophore queries could be used to search in a database of compounds with unknown enantioselectivities. With respect to the percentage of resolved samples that are retrieved in CHIRBASE, the resulting list should contain a similar yield of compounds providing favorable specific interactions with the CSP receptor binding sites.

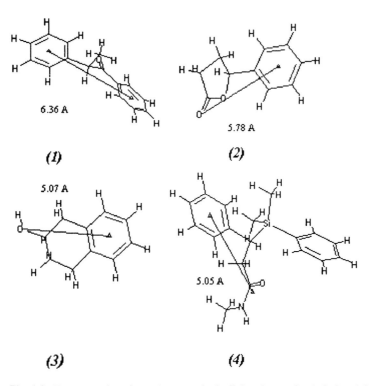

Fig. 4-9. Four examples of sample respectively fitting the queries 1, 2, 3 and 4 of Fig. 4-8.

4.5.2 Queries Based on Sample Ligand

This method represents the most common and traditional application of computational tools to rational drug design. From a list of molecules of known activity, one can establish a 3D-pharmacophore hypothesis that is then transformed into a 3D-search query. This query is then used to search a 3D database for structures that fit the hypothesis within a certain tolerance. If the yield of active molecules is significant, then the query can be used to predict activities on novel compounds. In our situation, the enantiophore is built from the superposition of a list of sample molecules, which are all well separated on a given CSP. Hence, the common features of this series of molecules can become a good enantiophore hypothesis for the enantioresolution on this CSP.

Starting from a collection of samples remarkably well resolved (alpha > 6) on Chiralcel OD (Cellulose tris(3,5-dimethylphenylcarbamate) coated on aminopropyl silica), a putative three-point enantiophore for binding to CSP was derived (Fig. 4-10). This enantiophore query was used to search (CFS 3D search) within a list comprising 4203 compounds tested on Chiralcel OD. From this search domain of CHIRBASE 3D, 191 structures were found to match the enantiophore.

In order to enhance our ligand-based query hypothesis, the structural fragments of the initial query were generalized but linked with the same distance constraints. A search of this final query (see Fig. 4-10) in the same list yielded 690 hits and a statistically significant correlation of the presence of this enantiophore and the enantioselectivity of the compounds was found (94 % of those are well resolved on Chiralcel OD). Note that out of the 4203 compounds of the Chiralcel OD domain search, a 2D search found 1900 structures that contain the substructural features of the generalized query.

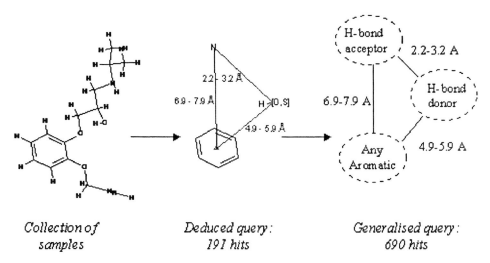

Collection of   *Deduced query:*   *Generalised query:*
samples   *191 hits*   *690 hits*

Fig. 4-10. Chiralcel OD ligand-based queries.

In Fig. 4-11, two different samples are displayed in their original conformations and conformations fitted to the query as they are highlighted by the CFS search process. The CFS process rotates single bonds between two atoms to find the maximum and minimum difference possible with the distance and angle constraints. Then, using a torsional fitter, it attempts to minimize in those conformations the deviations between measured values of 3D constraints and the values that are specified in the 3D-search query.

We have seen here that these simple methods which only rely upon the optimal use of molecular graphics tools can address highly specific receptor-ligand interactions.

These first-created enantiophores are rudimentary, but may serve as useful guidelines for a further design of more sophisticated and efficient search queries in consideration of possible alternative modes of binding and conformational changes in the CSP receptor structure. Undoubtedly, this query optimization will soon take advantage of the backgrounds of our new 3D-database project called CHIRSOURCE.

CHIRSOURCE aims to explore the use of chiral chromatography for combinatorial chemistry approaches. Combinatorial chemistry, as well as parallel synthesis, requires the availability of both enantiomers to address the configurational diversity issues. The availability of both enantiomers is not so common as far as the sources come from the « chiral pool » or implies two separated asymmetric synthesis or resolution, which often require further enantiomeric purification. Preparative (or semipreparative) chiral liquid chromatography is the method of choice for the availability of both enantiomers in high enantiomeric purity in a single shot. Simulated moving bed technology is available for larger-scale separations.

Designed from CHIRBASE-3D, CHIRSOURCE provides 30 000 structures in terms of configurational diversity, most of them easily available by semipreparative scale on corporate installation or in dedicated companies with minor further optimization.

We are today persuaded that CHIRSOURCE can help to reduce the costs and means that are required to launch a new chiral drug to market. CHIRSOURCE will include molecular attributes (dipole, lipophilicity, surface area and volume, HOMO-LUMO, Verloop parameters, molar refractivity) and molecular indices (describing connectivity, shape, topology and electrotopology, atom, ring and group counts). Such 3D molecular descriptors are often used in cluster analysis to identify dissimilar compounds for combinatorial chemistry and high-throughput screening applications.

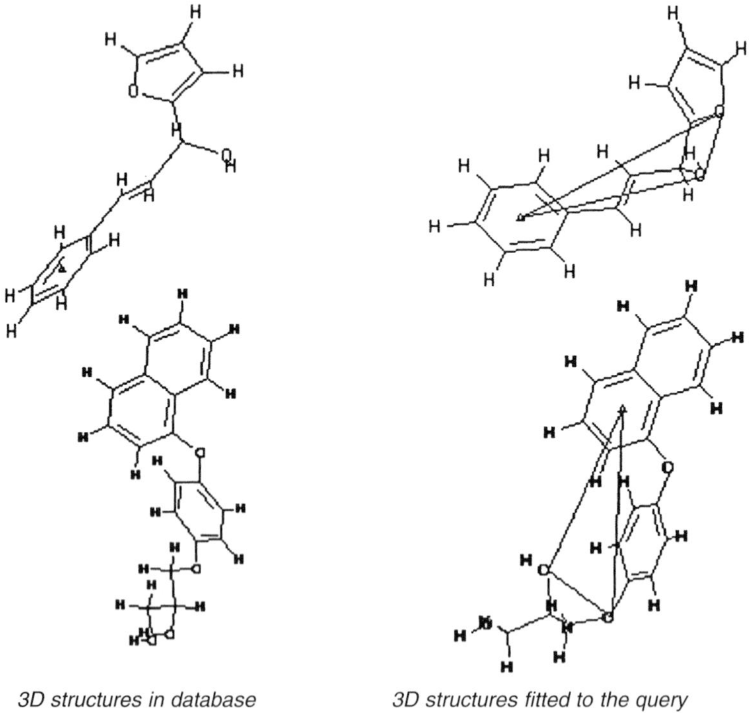

3D structures in database   3D structures fitted to the query

Fig. 4-11. Examples of structure fitting the generalized Chiralcel OD ligand-based query.

4.6 Dealing with Molecular Similarity

Besides 3D structure database searches, molecular similarity is also widely used for drug design by the pharmaceutical industry, as demonstrated by two recent reviews [23, 24]. More particularly, 2D fingerprints used to calculate the 2D topological similarity of molecules were found valid to quantify molecular diversity and thus manage the global diversity of structure databases [25]. In this section, we describe the application of similarity measures, in order to determine some relationships between CSPs by production of a molecular similarity matrix displayed as a dot plot. More precisely, the molecular similarity calculations applied to CHIRBASE provides a way of comparing the samples within a dataset, as well as comparing different datasets using the two following methodologies:

1. Select a set of compounds resolved on a given CSP, calculate the similarity indices between all possible molecule pairs, and then use these indices to build a similarity matrix containing relevant information about the structural diversity within the set of samples separated on this CSP.
2. Select two sets of compounds resolved on two different CSPs, calculate the similarity indices between all possible molecule pairs of these two sets, and then use these indices to build a similarity matrix containing relevant information about the structural affinities of these two CSPs.

The similarity matrices are constructed by one in-house program developed inside CHIRBASE using the application development kit of ISIS. They contain the similarity coefficients as expressed by the Tanimoto method. In ISIS, the Tanimoto coefficients are calculated from a set of binary descriptors or molecular keys coding the structural fragments of the molecules.

These structural key descriptors incorporate a remarkable amount of pertinent molecular arrangements covering each type of interaction involved in ligand-receptor bindings [26]. Since every structure in a database is represented by one or more of the 960 key codes available in ISIS, suppose that two molecules include respectively A and B key codes, then the Tanimoto coefficient is given by:

$$\frac{A \cap B}{[A \cup B] - [A \cap B]}$$

In ISIS, the similarity value is ranging between 0 and 100. A similarity value of 0 means that the two molecules are totally dissimilar, whereas a value of 100 will be obtained when the two molecules are 100 % identical. The matrices are called similarity matrix by convention, as larger numbers indicate more similarity between items. Dot plots of the matrix are produced by another in-house application developed with Visual Basic using the InovaGIS object library [27]. The pixels in the map are color-coded by similarity coefficients, providing a visual representation of similitudes among one or two sets of molecules. Such a representation is a simple but very powerful means for quickly visualizing and finding trends in very large data sets (up to 250 000 points). The present work is preliminary, and it is intended to illustrate one interesting issue that can be addressed with CHIRBASE.

4.6.1 Comparison of Sample Similarities within a Molecule Dataset

In these first studies, similarity measures were investigated to survey the molecular diversity of a set of molecules resolved on a given CSP in order to compare the extent of their application range.

Three types of CSPs were compared:

- Polysaccharide-based CSPs [28]:
  - Chiralcel OD: Cellulose tris(3,5-dimethylphenylcarbamate coated on amino-propyl silica.
  - Chiralpak AD: Amylose tris(3,5-dimethylphenylcarbamate) coated on amino-propyl silica.
- Pirkle-like CSPs [29]:
  - Whelk-O 1: (3R,4S)-4-(3,5-Dinitrobenzamido)-3-[3-(dimethylsilyloxy)pro pyl]-1,2,3,4-tetrahydrophenanthrene).
  - Pirkle DNPG: (R)-N-3,5-Dinitrobenzoyl-phenylglycine covalently bonded to aminopropyl silica.
- Inclusion-based CSPs [28]:
  - Crownpak CR(+): (S)-18-crown-6 ether coated on silica.

In the first category, we have chosen two cellulose- and amylose-based CSPs which provide today a considerable application range in CHIRBASE. In the second category, we have chosen to evaluate the behavior of the Whelk-O 1 CSP because of its well-recognized ability to resolve a broader range of samples than standard Pirkle-like CSPs. In the last category, Crownpak CR is a good example of a CSP offering a limited field of application with a large proportion of amino acids in our molecule library. Today, such qualitative trends can be revealed to the analyst from the published literature, or by manual examination of the structures in CHIRBASE through a time-consuming and biased procedure, but have not yet been clearly determined through a rational and systematic manner.

A set of about 500 molecules was used for each CSP. Figure 4-12 illustrates some results of dot plots. The similarity measures are displayed here according to a gray value gradient (white for 0 to black for 100). As the way the data points are presented to the application is dependent upon the organization of the molecules in CHIR-BASE, dots are put in the maps at random positions by the algorithm in order to provide a good statistical repartition, and thus facilitate the ability to distinguish visually the global diversity of samples.

For each of these images, a mean value of the luminance can then be measured directly using a standard photo-editor software.

As shown in Table 4-4, the image luminance values are found to be in good agreement with the calculated mean similarity values of the molecule sets. Therefore, the level of similarity of the molecule set can be immediately judged based on the aver-

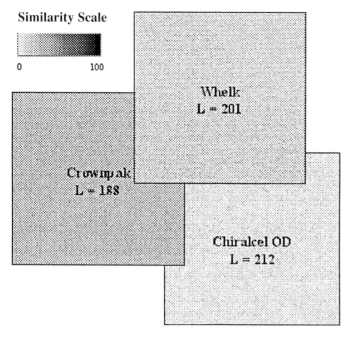

Fig. 4-12. Similarity maps of Crownpak CR, Whelk-O 1 and Chiralcel OD.
L: luminance value of the image.

age luminance of the full image: the smaller similarity between molecules, the higher luminance in the picture will be found. As the luminance values depend on the initial choice of molecules, other experiments were repeated with different populations of molecule. The results obtained with these additional molecule sets showed very little variation of map patterns.

To demonstrate the excellent correlation ($r^2 = 0.99$) between the luminance of the images and molecular diversity, we plotted the luminance values of the map versus the mean similarity values of data sets (Fig. 4-13). From this plot, a scoring scheme for the classification of CSPs from specific to broad application range can be well established: Crownpak CR > Pirkle DNBPG > Whelk > Chiralpak AD > Chiralcel OD.

Table 4-4. Map luminance and mean similarity values of CSP datasets.

	Mean similarity	Luminance
Whelk	21.34	201
Chiralcel OD	16.62	212
Chiralpak AD	18.20	209
Pirkle DNBPG	24.89	192
Crownpak CR	26.30	188

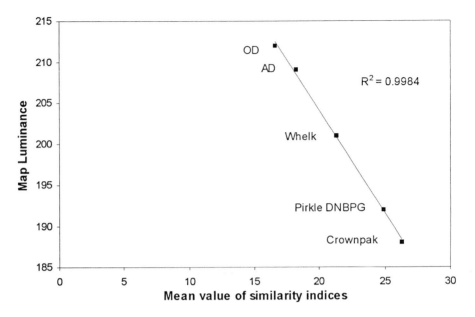

Fig. 4-13. Plot of map luminance versus mean similarity of molecule sets.

The results obtained are in accordance with our previous observations:

- Chiralcel OD and Chiralpak AD are associated with the largest mean values of molecular diversity.
- Whelk-O 1 and Pirkle DNBPG have appreciable differences in terms of application range. This is not surprising since Pirkle DNBPG often requires a prior derivatization of solutes to achieve a separation. This also confirms the atypical character of Whelk-O 1 compared to other Pirkle-like CSPs.
- Crownpak CR(+) exhibits the lowest diversity of structural features.

For comparative purposes, Chiralcel OD and Crownpak CR could be used as an extreme case to delineate the basis of a molecular diversity scale.

4.6.2 Comparison of Molecule Dataset Similarities between two CSPs

In the following studies, the same computational steps have been used in a straight-forward manner as before to compare pairs of CSP. The molecule datasets employed in these studies are the same as that used above. In addition, two protein-based CSP were also compared:

- Chiral-AGP: alpha-acid glycoprotein bonded to silica.
- Ultron ES-OVM: ovomucoid-conjugated bonded to aminopropyl-silica.

A quick inspection of similarity maps in Fig. 4-14, allows one to see at once that Chiralcel OD and Whelk-O 1 molecule sets contain notable structural differences, whereas AGP and OVM data sets contain much more structurally related molecules.

Caution must be emphasized here that this simple method which aims to measure the molecular diversity between two CSP classes does not provide an absolute scale. However, a relative analysis of luminance values (Table 4-5) can show how potentially different are the application range of two CSPs and can also help to select a subset of CSPs that represent the largest scope of applications.

Table 4-5. Map luminance of pairs of CSP datasets.

	Luminance
OD-Whelk	213
AD-Whelk	209
AD-OD	214
Whelk–Pirkle DNBPG	202
AD-Pirkle	206
OD-Pirkle	210
AGP-OVM	197

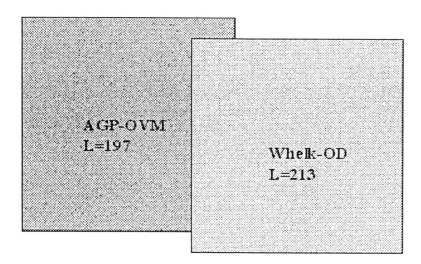

Fig. 4-14. Similarity maps comparing molecular diversity between two couples of CSP.

A data plot, as displayed in Fig. 4-15, may then constitute a useful support for the simple selection of candidate CSPs that should be available in a laboratory. For purposes of comparison, luminance data were scaled by normalizing the data in the range [0,100] by means of the following equation:

$$D_i = [100\,(L_i - L_{min})/(L_{max-}\,L_{min})]$$

where D_i is the scaled value, L_i is the original luminance, and L_{min} and L_{max} are the minimum and maximum values of luminance.

As we have already indicated, the diversity value of molecule sets combining two CSPs is difficult to interpret on an absolute scale. Only the relative position of each set can be useful to compare, and also the arrangement of the points in regard to the molecular diversity inherent to each individual molecule set of CSP.

On this basis, the AD-OD combined set appears to be the most diverse, and its score establishes a practical larger bound. It is interesting to note that this combined set has a larger diversity than each original AD and OD set. This increase of diversity is also observed for the combined OD-Whelk set. This may explain why these CSPs are often good candidates in CSP screening strategies. Comparison of Whelk-Pirkle set reveals no reduction of diversity of the original Whelk set. This result suggests that more specific Pirkle-like CSPs contribute well to augment the diversity space defined by the Whelk-O 1 CSP.

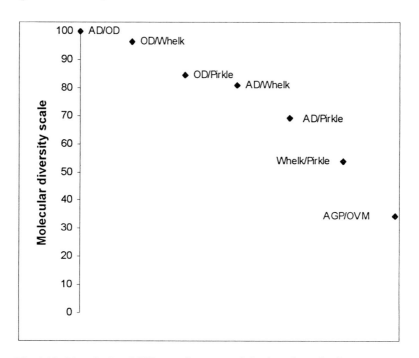

Fig. 4-15. Plot of pairs of CSP according to a scaled value of map luminance.

More surprising is the loss of diversity shown by the AGP-OVM combined set; this result seems to confirm the significant "overlap" of AGP and OVM solutes found in earlier works [30].

As a first conclusion, this work shows that similarity coefficients, which code molecules in terms of chemical substructures, are useful to assess the efficiency of CSPs. The purpose of this work was not to propose a new method for solving the complex task of CSP classification. Indeed, the aim was basically to compare CSP applications through simple and easily interpretable similarity maps for simplifying the analysis of large data matrices. From a practical point of view, it is impossible to test all the existing CSPs. The comparison of maps allows a direct classification of whether a given CSP presents a broad variety of applications or not. For screening purposes, a CSP choice made throughout such studies should be much better than a random selection of CSPs.

Furthermore, this approach can also supply a straightforward procedure to predict the potentialities of newly designed CSPs. Also, similarity maps can serve to depict resemblance between CSPs when there is no information available regarding the structural requirements for interaction with CSP. Compared to other methods such as hierarchical clustering approaches using structure-based fingerprints, our approach requires much less CPU time (less than 1 h to build a map of 250 000 dots). Thus, this rapid diversity analysis process may be proven useful in other areas, such as aiding in investigating diversity in databases of high-throughput screening results.

4.7 Decision Tree using Application of Machine Learning

Machine learning provides the easiest approach to data mining, and also provides solutions in many fields of chemistry: quality control in analytical chemistry [31], interpretation of mass spectra [32], as well prediction of pharmaceutical properties [33, 34] or drug design [35].

Utilization of intelligent systems in chiral chromatography starts with an original project called "CHIRULE" developed by Stauffer and Dessy [36], who combined similarity searching and an expert system application for CSP prediction. This issue has recently been reconsidered by Bryant and co-workers with the first development of an expert system for the choice of Pirkle-type CSPs [37].

Machine learning can analyze a large dataset and determine what information is most pertinent. Such generalized information can then be converted into knowledge through the generation of rule sets that will enable faster and more relevant decisions.

A decision tree is constituted of two types of nodes: *parent* and *leaves*. Each parent node corresponds to a question or an attribute; each leaf node designates a single class. The branches connected to a parent node correspond to a split of the population node according to the answers to the question or the value of the attribute. Each subset of the population is split again, recursively, using different questions or attributes until a subset belong to a single class. In this case, the branch of the tree stops with a leaf node labeled with a single class.

A tree is read from root to leaves. We begin at the root of the tree which contains all the population. Then, following the relevant branches according to the question asked at each branch node, we finally reach a leaf node. The label on that leaf node provides the class which is the resulting conclusion induced from the tree.

The first tree induction algorithm is called ID3 (Iterative Dichotomizer version 3) and was developed by Quinlan [38]. Subsequent improved versions of ID3 are C4.5 and C5. In our study, we used MC4 decision tree algorithm which is available in the MLC++ package [39]. MC4 and C4.5 use the same algorithm with different default parameter settings.

The purpose of this study is only intended to illustrate and evaluate the decision tree approach for CSP prediction using as attributes the 166 molecular keys publicly available in ISIS. This assay was carried out a CHIRBASE file of 3000 molecular structures corresponding to a list of samples resolved with an α value superior to 1.8. For each solute, we have picked in CHIRBASE the traded CSP providing the highest enantioselectivity. This procedure leads to a total selection of 18 CSPs commercially available under the following names: Chiralpak AD [28], Chiral-AGP [40], Chiralpak AS [28], Resolvosil BSA-7 [41], Chiral-CBH [40], CTA-I (microcrystalline cellulose triacetate) [42], Chirobiotic T [43], Crownpak CR(+) [28], Cyclobond I [43], DNB-Leucine covalent [29], DNB-Phenylglycine covalent [29], Chiralcel OB [28], Chiralcel OD [28], Chiralcel OJ [28], Chiralpak OT(+) [28], Ultron-ES-OVM [44], Whelk-O 1 [29], (R,R)-β-Gem 1 [29].

After importing the data file into MLC++ and selecting "gain-ratio" as splitting method, the program builds the full tree shown in Fig. 4-16. The tree has 631 nodes, 316 leaves and 107 attributes. Attributes are molecular key features and leaves are CSPs.

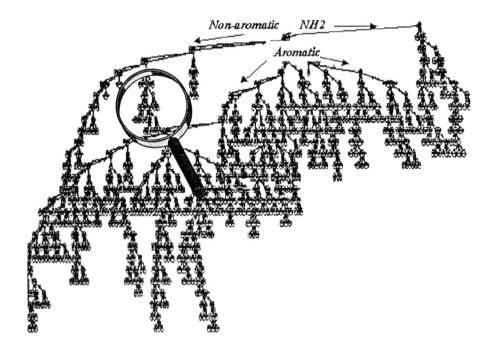

Fig. 4-16. Decision tree built by MLC++ from the analysis of 3000 solutes resolved on 18 commercially available CSPs. The magnifying glass shows the region zoomed in Fig. 4-17.

As the entire tree is complex and cannot be clearly displayed in one screen, we report in Fig. 4-17 an expanded (zoomed) fraction of the "nonaromatic" population set of the tree.

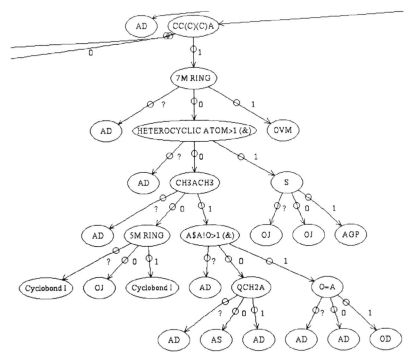

Fig. 4-17. Zoomed picture of the decision tree in the "nonaromatic" region. (0) indicates "no"; (1) indicates "yes". (A) for any atom except hydrogen. ($) indicates that the bond is part of a ring, and (!) bond is part of a chain. CC(C)(C)A for tBu.

Since this current study is restricted to the best enantioseparations ($\alpha > 1.8$), it is quite clear that the tree does not accurately reflect the full information contained in CHIRBASE.

However, it has provided some interesting results. At the top of the tree, the molecule population is first divided according to the presence or absence of the attribute "NH2" (primary amine). If the answer is "yes", the developed branches (on the right of the tree) mostly leads to the Crownpak CSP. The next attribute is "Aromatic". If the answer is "no", here the predominant CSP is Chiralpak AD. Aromatic compounds form the largest part of the tree and as expected the dominant CSP is Chiralcel OD which is disseminated in almost every region of the tree.

Some other interesting results are:

- The importance of steric and hydrogen bond interactions in chiral separations of nonaromatic samples
- Chiralpak OT(+) dominates the branches built under the "spiro" and "ARO-MATIC RING>1" molecular keys.
- CTA-I and Chiralcel OJ are found under the key "AROMATIC" and "8 member ring or larger". If sample also contains one amino group, then the tree leads to Chiralcel OJ.
- Whelk-O 1 well predominates the aromatic samples bearing an axial chirality created by a C–N bond.
- Chiralcel OB is associated with the separation of aromatic and nonaromatic sulfoxides and Chiralcel OD with aromatic alcohols.

Nonetheless, these results are partial and can be seen only as a test study, and clearly many improvements will be considered. For example, the decision at each node should not be restricted to the only use of molecular key attributes, but should also take into account the mobile phase constituents. Future works will also extend this approach to the full database and will probably lead to the introduction of knowledge rules in CHIRBASE. Knowledge rules will help the users not only in the choice of a wide range of columns but also in the selection of appropriate experimental conditions.

From these initial results we have seen that this approach has exciting practical issues. However, we have also found that it does not match the accuracy of a database structure search, and the latter will certainly continue to be the best approach for CSP prediction for separation of a particular structure.

4.8   Conclusions

Today, the use of CHIRBASE as a tool in aiding the chemist in the identification of appropriate CSPs has produced impressive and valuable results. Although recent developments diminish the need for domain expertise, today the user must possess a certain level of knowledge of analytical chemistry and chiral chromatography. Nevertheless, further refinements will notably reduce this required level of expertise. Part of this effort will include the design of an expert system which will provide rule sets for each CSP in a given sample search context. The expert system will also be able to query the user about the specific requisites for each sample (scale, solubility, etc.) and generate rules which will indicate a ranked list of CSPs as well their most suitable experimental conditions (mobile phase, temperature, pH, etc.).

Such an expert system can also be adapted for the evaluation of data in the published literature. However, this point raises a number of practical questions. A better exploitation of chromatographic data in this field would require an important effort to be made by analysts to constitute standards for quality control and interpretation of results. It is essential that authors report detailed and specific information on the techniques and experimental conditions used for chiral analysis. For instance, one simple means of improving the quality of data would be to report the exact value of the temperature and not "room temperature" or "ambient temperature", as is often found. An examination of CHIRBASE data completed by the authors shows that this expression can cover a large range of temperature from 15 °C to 30 °C. Reports should always include standardized chromatographic data information (k, α and resolution), as well as, if available, other important measurements such as elution order.

We have seen that the accumulation of data is now furnishing a variety of hypotheses that could be further verified in the laboratory. While predictive techniques can be designed without data, they cannot be evaluated in the absence of such data. Clearly, data are essential for study designs aimed at the investigation of CSP-solute chiral interactions. We are convinced that it is actually possible to attain a detailed structural knowledge of the mechanisms of chiral separations using statistical analysis techniques such as cluster analysis and other multivariate analysis methods combined with data mining, rather than only via the means of molecular model building. None of these concurrent studies can be performed today without the availability of large amounts of experimental data.

From this perspective, and with a continued lack of models and a better understanding of enantioseparation mechanisms, we can assume that the role of computers in this field will become increasingly determinant.

References

[1] (a) Okamoto, Y., Yashima, E., Angew. Chem. Int. Ed. Engl., 1998, 37, 1021–1043. (b) Francotte, E., Chem. Anal. Series, 1997, 142, 633–683. (c) Ahuja, S. (ed.) *Chiral Separations: Application and technology*, Washington DC: American Chemical Society, 1996. (d) Schreier, P., Bernreuther, A., Huffer, M. (eds.) *Analysis of Chiral Organic Molecules. Methodology and applications*, Berlin: Walter de Gruyter and Co., 1995. (e) Subramanian, G. (ed.), *A Practical Approach to Chiral Separations by Liquid Chromatography*, Weinheim, New York: VCH, 1994.
[2] (a) Roussel, C., Piras, P., Spectra 2000, Chromatographie Suppl., 1991, 16. (b) Koppenhoefer, B., Nothdurft, A., Pierrot-Sanders, J., Piras, P., Popescu, C., Roussel, C., Stiebler, M., Trettin, U., Chirality, 1993, 5, 213– 219.
[3] MDL Information Systems, Inc., 14600 Catalina Street, San Leandro, CA 94577
[4] (a) Roussel, C., Piras, P., Pure Appl. Chem., 1993, 65, 235–244. (b) Koppenhoeffer, B., Graf, R., Holzschuh, H., Nothdurft, A., Trettin, U., Piras, P., Roussel, C., J. Chromatogr. A, 1994, 666, 557–563. (c) Roussel, C., Piras, P., in: *Proceedings CHIRAL'94 USA*, 1994. (d) Roussel, C., Piras, P. in: *Data and Knowledge in a Changing World: Modeling Complex Data for Creating Information: Real and Virtual Objects*, J. E. Dubois, Gershon, N. (eds.) Springer Verlag, 1996.
[5] (a) Ash, J. E.; Warr, W. A.; Willett, P. (eds.), *Chemical Structure Systems*, Chichester: Ellis Horwood:, 1991. (b) Warr, W. A., Wilkins, M. P., Online, 1992, 16(1), 48–55. (c) Suhr, C.; Warr, W. A., *Chemical Information Management*, Weinheim: VCH, 1992.
[6] CHIRBASE Internet Home Page, http://chirbase.u-3mrs.fr.
[7] (a) Armstrong, D.W., Tang, Y., Chen, S., Zhou, Y., Bagwill, C., Chen, J.-R., Anal. Chem., 1994, 66, 1473–1484. (b) Berthod, A., Nair, U. B., Bagwill, C., Armstrong, D. W., Talanta, 1996, 43, 1767–1782. (c) Ekborg-Ott, K. H., Kullman, J. P., Wang, X., Gahm, K., He, L., Armstrong, D. W., Chirality, 1998, 10, 627–660.

[8] (a) Popescu, A., Hörnfeldt, A.-B., Gronowitz, S., Nucleosides & Nucleotides, 1995, 14(6), 1233–1249. (b) Levin, S., Sterin, M., Magora, A., Popescu, A., J. Chromatogr. A, 1996, 752, 131–146.

[9] Di Marco, M. P., Evans, C. A., Dixit, D. M., Brown, W. L., Siddiqui, M. A., Tse, H. L. A., Jin, H., Nguyen-Ba, N., Mansour, T. S., J. Chromatogr., 1993, 645, 107–114.

[10] Scypinski, S., Ross, A.J., J. Pharm. Biomed. Anal., 1994, 12(10), 1271–1276.

[11] Jin, H., Tse, H. L. A., Evans, C. A., Mansour, T. S., Beels, C. M., Ravenscroft, P., Humber, D. C., Jones, M. F., Payne, J. J., Ramsay, M. V. J., Tetrahedron Asymmetry, 1993, 4(2), 211–214.

[12] Csuk, R., von Scholz, Y., Tetrahedron, 1996, 52(18), 6383–6396.

[13] Nicklaus, M. C., Neamati, N., Hong, H., Mazumder, A., Sunder, S., J. Med. Chem., 1997, 40(6), 920–929.

[14] Neamati, N., Hong, H., Mazumder, A., Wang, S., Sunder, S., J. Med. Chem., 1997, 40, 942–951.

[15] Hong, H., Neamati, N., Winslow, H. E., Christensen, J. L., Orr, A., Pommier, Y., Milne, G. W., Antivir. Chem. Chemother, 1998, 9(6), 461–72.

[16] Sadowski, J., Gasteiger, J., Klebe, G., J. Chem. Inf. Comput. Sci., 1994, 34, 1000.

[17] Oxford Molecular Ltd., the Medawar Centre, Oxford Science Park, Sandford-on-Thames, Oxford OX4 4GA, England.

[18] Moock, T. E., Henry, D. R., Ozkabak, A. G., Alamgir, M., J. Chem. Inf. Comput. Sci., 1994, 34, 184.

[19] (a) Allenmark, S., Nielsen, L., Pirkle, W. H., Acta Chem. Scand., 1983, B37, 325–328. (b) Pirkle, W. H., Mc Cune, J. E., J. Chromatogr., 1989, 469, 67–75, 1989. (c) Azzolina, O., Collina, S., J. Liq. Chromatogr., 1995, 18, 81–92. (d) Pirkle, W. H., Koscho, M. E., J. Chromatogr. A, 1997, 761, 65–70.

[20] Good, A. C., Mason, J. S., in: *Reviews in Computational Chemistry*, Lipkowitz, K. B., Boyd, D. B. (eds.) New York: VCH Inc., 1996; Chapter 2.

[21] Marriott, D. P., Dougall, I. G., Meghani, P., Liu, Y. J., Flower, D. R., J. Med. Chem., 1999, 42, 3210–3216.

[22] Pirkle, W. H., Welch, C. J., J. Liq. Chromatogr., 1992, 15, 1947–1955.

[23] Dean, P. M. (ed.), *Molecular similarity in Drug Design*, Glasgow: Blackie Academic and Professional, Chapman & Hall, 1995.

[24] Downs, G. M., Willett, P., in: *Reviews in Computational Chemistry*, Lipkowitz, K. B., Boyd, D. B. (eds.) New York: VCH Inc., 1996; Chapter 1.

[25] Matter, H., J. Med. Chem., 1997, 40, 1219–1229.

[26] Brown, R. D., Martin, Y. C., J. Chem. Inf. Comput. Sci., 1997, 37, 1–9.

[27] InovaGIS Project makes available to the general public a wide variety of free geographic information software. Contact: Pedro Pereira Gonçalves, Departamento de Ciências e Engenharia do Ambiente, Faculdade de Ciências e Tecnologia / Universidade Nova de Lisboa, 2825 Monte de Caparica, Portugal.

[28] Available from Daicel Chemical Industries, Ltd. Chiral Chemicals Division, 2-5, Kasumigaseki 3-chome, Chiyoda-ku 100-607 7, Tokyo, Japan.

[29] Available from Regis Technologies, Inc., 8210 Austin Av, Morton Grove, IL 60053, USA.

[30] Roussel, C., Piras, P., Heitmann, I., Biomed. Chromatogr., 1997, 11, 311–316.

[31] Ehlert, G., Wuensch, G., J. Prakt. Chem./Chem.-Ztg., 1994, 336, 458–464.

[32] Meisel, W., Jolley, M., Heller, S. R., Milne, G. W. A., Anal. Chim. Acta, 1979, 112, 407–416.

[33] Ghuloum, A. M., Sage, C. R., Jain, A. N., J. Med. Chem., 1999, 42, 1739–1748.

[34] Koevesdi, I., Dancso, A., Hegedues, M., Jakoczy, I., Blasko, G., Arch. Mod. Chem., 1997, 134, 141–150.

[35] Bolis G., Di Pace, L., Fabrocini, F., J. Comput. Aided Mol. Des., 1991, 5, 617–628.

[36] Stauffer, S. T., Dessy, R. E., J. Chromatogr. Sci., 1994, 32, 228–235.

[37] Bryant, C. H., Adam, A. E., Taylor, D. R., Rowe, R. C., Chemometrics and Intelligent Laboratory Systems, 1996, 34, 21–40.

[38] (a) Quinlan, J. R., *Discovering rules by induction from large numbers of examples: a case study*, in: Michie, D. (Ed.), *Expert Systems in the Micro-Electronic Age*. Edinburgh, Scotland: Edinburgh University Press, 1979. (b) Quinlan, J. R., *Generating production rules from decision trees*, in: *Proceedings of the Tenth International Joint Conference on Artificial Intelligence*, Milan, Italy: Morgan Kaufmann, 1987, 304–307. (c) Quinlan, J. R., C4.5: Programs for machine learning, Los Altos, California: Morgan Kaufmann Publishers, Inc., 1993.

[39] MLC++ was initially developed at Stanford University (Kohavi, R., Sommerfield, D. and Dougherty, J.) and was public domain. The new version 2.0 is freely distributed by Silicon Graphics, Inc.

[40] ChromTech AB, Box 6056, 129 06 Hägersten, Sweden.

[41] Macherey-Nagel GmbH & Co. KG, Postfach 10 13 52, D-52313 Düren, Germany.

[42] Merck KGaA, Frankfurter Str. 250, D-64293 Darmstadt, Germany.

[43] Alltech Associates, Inc., 2051 Waukegan Road, Deerfield, IL 60015, USA.

[44] Shinwa Chemical Industries Ltd., Kyoto, Japan.

5 Membranes in Chiral Separations

Maartje F. Kemmere and Jos T. F. Keurentjes

5.1 Introduction

At present moment, no generally feasible method exists for the large-scale production of optically pure products. Although for the separation of virtually every racemic mixture an analytical method is available (gas chromatography, liquid chromatography or capillary electrophoresis), this is not the case for the separation of racemic mixtures on an industrial scale. The most widely applied method for the separation of racemic mixtures is diastereomeric salt crystallization [1]. However, this usually requires many steps, making the process complicated and inducing considerable losses of valuable product. In order to avoid the problems associated with diastereomeric salt crystallization, membrane-based processes may be considered as a viable alternative.

During the past decades, the range of conventional separation techniques has been extended by a wide range of membrane separation processes. The first applications of membranes are found in biomedical applications such as hemodialysis and plasmaphoresis [2]. The first industrial membrane application has been the desalination of water streams, mainly for potable water production [3]. A significant effort in membrane materials development has led to many industrial applications of a variety of processes, including gas separations, pervaporation, pertraction, electrodialysis and various filtration processes (reverse osmosis, RO; nano filtration, NF; ultrafiltration, UF; micro filtration, MF) [4]. Membrane separations often provide opportunities as a cost-efficient alternative to separations that are troublesome or even impossible, using classical methods. Additionally, since most membrane processes are performed at ambient temperature, they can offer clear advantages compared to other separation processes, e.g. reducing the formation of by-products.

For the separation of racemic mixtures, two basic types of membrane processes can be distinguished: a direct separation using an enantioselective membrane, or separation in which a nonselective membrane assists an enantioselective process [5]. The most direct method is to apply enantioselective membranes, thus allowing selective transport of one of the enantiomers of a racemic mixture. These membranes can either be a dense polymer or a liquid. In the latter case, the membrane liquid can be chiral, or may contain a chiral additive (carrier). Nonselective membranes can also provide essential nonchiral separation characteristics in combination with chirality outside the membrane. The required enantioselectivity can stem either from a selective physical interaction, or from a selective (bio)conversion.

In this chapter we will provide an overview of the application of membrane separations for chiral resolutions. As we will focus on physical separations, the use of membranes in kinetic (bio)resolutions will not be discussed. This chapter is intended to provide an impression, though not exhaustive, of the status of the development of membrane processes for chiral separations. The different options will be discussed on the basis of their applicability on a large scale.

5.2 Chiral Membranes

5.2.1 Liquid Membranes

In general, a liquid membrane for chiral separation contains an enantiospecific carrier which selectively forms a complex with one of the enantiomers of a racemic mixture at the feed side, and transports it across the membrane, where it is released into the receptor phase (Fig. 5-1).

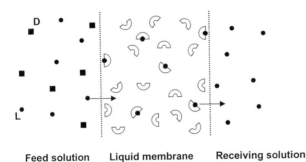

Feed solution   Liquid membrane   Receiving solution

Fig. 5-1. Schematic representation of a liquid membrane for chiral separation.

The carrier should not dissolve in the feed liquid or receptor phase in order to avoid leakage from the liquid membrane. In order to achieve sufficient selectivity, minimization of nonselective transport through the bulk of the membrane liquid is required. Liquid membranes can be divided into three basic types [6]: emulsion; supported; and bulk liquid membranes, respectively (Fig. 5-2).

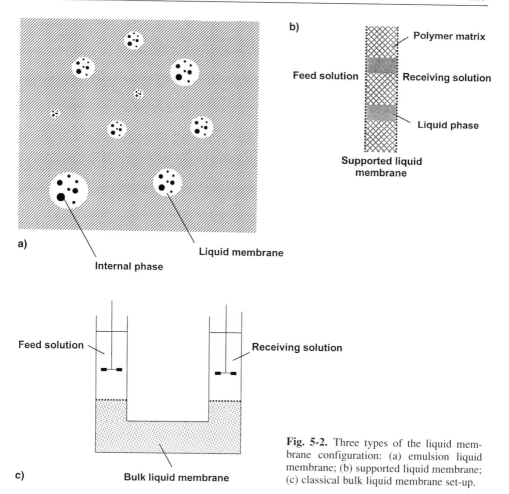

b)

Polymer matrix

Feed solution

Receiving solution

Liquid phase

Supported liquid
membrane

a)

Liquid membrane

Internal phase

Feed solution

Receiving solution

Fig. 5-2. Three types of the liquid membrane configuration: (a) emulsion liquid membrane; (b) supported liquid membrane; (c) classical bulk liquid membrane set-up.

c)

Bulk liquid membrane

5.2.1.1 Emulsion Liquid Membranes

The application of an emulsion liquid membrane system involves three consecutive steps [7]. First, two immiscible phases are stirred with a surfactant to generate an emulsion. Subsequently, the emulsion is mixed with another liquid containing the material to be extracted. The phases are then separated, and the emulsifying agents are recovered in a de-emulsification step. Two examples of the use of the liquid emulsion membrane configuration are the selective extraction of phenylalanine enantiomers, using copper(II)*N*-decyl-1-hydroxyproline (**1**) as the chiral selector [8], and the permeation of dipeptides and derivatives, using Rokwin 60 (a commercial nonionic surfactant, consisting of a mixture of esters of higher fatty acids with D-sorbitol), as reported by Skrzypinski et al.[9].

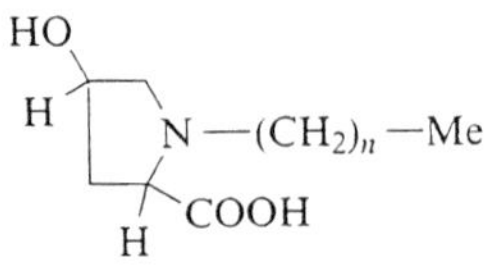

$$\underset{H}{\overset{HO}{\diagdown}}\boxed{}\underset{H}{\overset{}{\diagup}}\overset{}{\underset{COOH}{N-(CH_2)_n-Me}}$$

1

5.2.1.2 Supported Liquid Membranes

In supported liquid membranes, a chiral liquid is immobilized in the pores of a membrane by capillary and interfacial tension forces. The immobilized film can keep apart two miscible liquids that do not wet the porous membrane. Vaidya et al. [10] reported the effects of membrane type (structure and wettability) on the stability of solvents in the pores of the membrane. Examples of chiral separation by a supported liquid membrane are extraction of chiral ammonium cations by a supported (microporous polypropylene film) membrane [11] and the enantiomeric separation of propranolol (**2**) and bupranolol (**3**) by a nitrate membrane with a *N*-hexadecyl-L-hydroxy proline carrier [12].

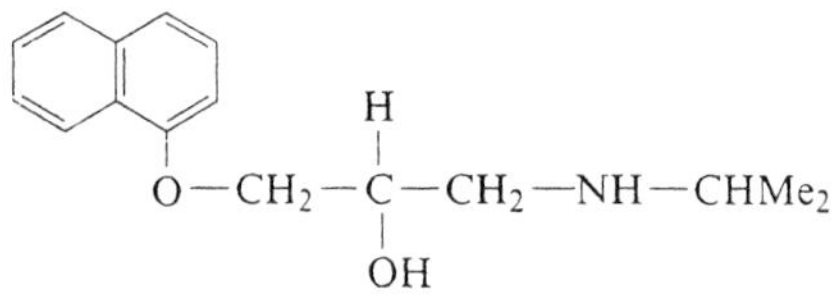

$$O-CH_2-\overset{\overset{\displaystyle H}{|}}{\underset{\underset{\displaystyle OH}{|}}{C}}-CH_2-NH-CHMe_2$$

2

$$O-CH_2-\overset{\overset{\displaystyle H}{|}}{\underset{\underset{\displaystyle OH}{|}}{C}}-CH_2-NH-CMe_3$$

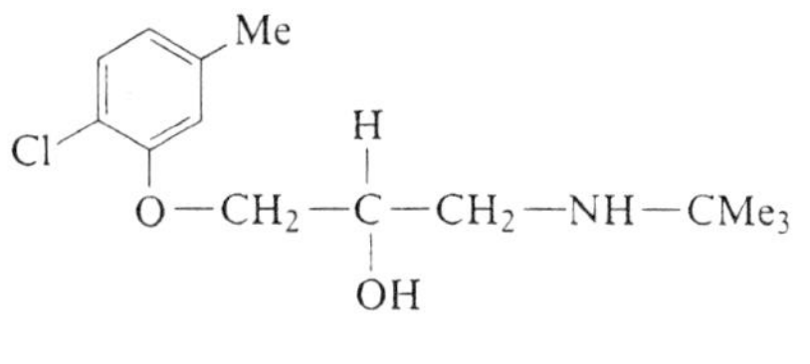

3

5.2.1.3 Bulk Liquid Membranes

In the classical set-up of bulk liquid membranes, the membrane phase is a well-mixed bulk phase instead of an immobilized phase within a pore or film. The principle comprises enantioselective extraction from the feed phase to the carrier phase, and subsequently the carrier releases the enantiomer into the receiving phase. As formation and dissociation of the chiral complex occur at different locations, suitable conditions for absorption and desorption can be established. In order to allow for effective mass transport between the different liquid phases involved, hollow fiber membranes can be used (Fig. 5-3). Scimin et al. [13] used a chiral lipophilic ligand as a carrier for the copper (II)-mediated, selective transport of α-amino acids across a bulk liquid membrane. Pirkle and Doherty [14] reported a bulk liquid membrane system, used for the selective transport of *N*-(3,5-dinitrobenzoyl)leucine.

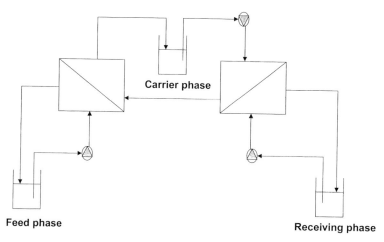

Fig. 5-3. Bulk liquid membrane set-up by Pirkle and Doherty [14].

With regard to the enantioselective transport through the membrane, one advantage of liquid membrane separation is the fact that the diffusion coefficient of a solute in a liquid is orders of magnitude higher as compared to the diffusion coefficient in a solid. The flux through the membrane depends linearly on the diffusion coefficient and concentration of the solute, and inversely on the thickness of the membrane [7].

Addition of a chiral carrier can improve the enantioselective transport through the membrane by preferentially forming a complex with one enantiomer. Typically, chiral selectors such as cyclodextrins (e.g. (**4**)) and crown ethers (e.g. (**5**) [21]) are applied. Due to the apolar character of the inner surface and the hydrophilic external surface of cyclodextrins, these molecules are able to transport apolar compounds through an aqueous phase to an organic phase, whereas the opposite mechanism is valid for crown ethers.

Armstrong and Jin [15] reported the separation of several hydrophobic isomers (including (1-ferrocenylethyl)thiophenol, 1'-benzylnornicotine, mephenytoin and disopyramide) by cyclodextrins as chiral selectors. A wide variety of crown ethers have been synthesized for application in enantioselective liquid membrane separation, such as binaphthyl-, biphenanthryl-, helicene-, tetrahydrofuran and cyclohexanediol-based crown ethers [16–20]. Brice and Pirkle [7] give a comprehensive overview of the characteristics and performance of the various crown ethers used as chiral selectors in liquid membrane separation.

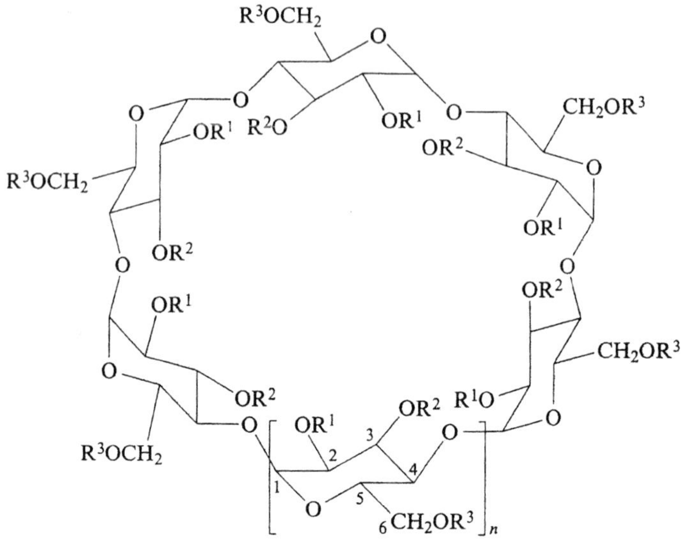

4

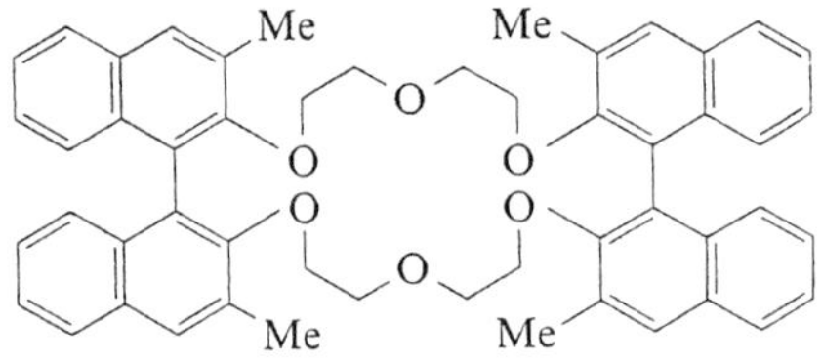

5

In general, high selectivities can be obtained in liquid membrane systems. However, one disadvantage of this technique is that the enantiomer ratio in the permeate decreases rapidly when the feed stream is depleted in one enantiomer. Racemization of the feed would be an approach to tackle this problem or, alternatively, using a system containing the two opposite selectors, so that the feed stream remains virtually racemic [21]. Another potential drawback of supported enantioselective liquid membranes is the application on an industrial scale. Often a complex multistage process is required in order to achieve the desired purity of the product. This leads to a relatively complicated flow scheme and expensive process equipment for large-scale separations.

5.2.2 Polymer Membranes

As the main disadvantage of liquid membrane systems is the instability over a longer period of time, another approach would be to perform separation through a solid membrane [22]. Enantioselective polymer membranes typically consist of a nonselective porous support coated with a thin layer of an enantioselective polymer. This type of polymer membrane requires a high specific surface, low mass transfer resistance, good mechanical strength and enantio recognition ability [23]. The separation mechanism involves enantiospecific interactions (solution and diffusion) between the isomers to be separated and the top layer polymer matrix. In case the required optical purity cannot be obtained in one single step, a cascade of membrane units can easily be applied to achieve the desired purity (see Section 5.2.4). Both the permeability P (flux) and enantioselectivity (α) determine the performance of an enantioselective membrane, for which α is defined as the ratio of the permeabilities of the L- and D-enantiomers, respectively:

$$\alpha = \frac{P_L}{P_D} \tag{1}$$

Enantiospecific polymers commonly used as stationary phases in chromatography are potentially applicable for chiral membrane separation, e.g. polysaccharides, acrylic polymers, poly(α-amino acids) and polyacetylene-derived polymers [24]. Additionally, chiral separations have been reported at high resolution and high rate by a bovine serum albumin (BSA)-multilayer-adsorbed porous hollow-fiber membrane as stationary phase [25]. In general, interest has been mainly focused on the separation of racemic amino acids by enantioselective polymer membranes [26–35]. Although almost complete resolution can be obtained in a dialysis configuration, the flux through this type of membranes is extremely low. Aoki and co-workers [36–41] reported the preparation of membranes completely made of chiral polymer in order to improve the permeability. The materials comprise derivatized poly-L-glutamate membranes and PDPSP (**6**) membranes.

6

A different approach is the use of an ultrafiltration membrane with an immobilized chiral component [31]. The transport mechanism for the separation of D,L-phenylalanine by an enantioselective ultrafiltration membrane is shown schematically in Fig. 5-4a. Depending on the trans-membrane pressure, selectivities were found to be between 1.25 and 4.1, at permeabilities between 10^{-6} and 10^{-7} m s^{-1}, respectively (Fig. 5-4b).

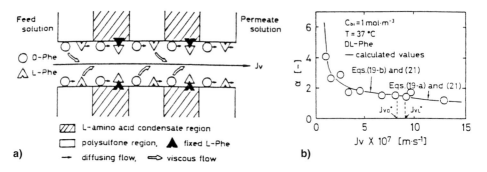

Fig. 5-4. (a) Separation of d,l-phenylalanine by an amino acid immobilized in the pores of a polysulfone ultrafiltration membrane. (b) Effect of volume flux on the separation factor, Jv = volume flux, T = 37 °C [32].

5.2.3 Molecular Imprinted Polymers

In addition to polymers typically used for chromatographic purposes, molecular imprinted polymers, i.e. polymers having enantiospecific cavities in the bulk phase, can be used as a basis for chiral membranes. In the past two decades, a novel technique of introducing molecular recognition sites into polymeric materials by molecular imprinting has been developed [42, 43]. Molecularly imprinted polymers (MIPs) find among others application as stationary phases in chromatography, sensors, membranes, and catalysts. The preparation procedure of a MIP involves the formation of a template– monomer complex. During the subsequent polymerization with crosslinking agents, the geometry of the self-assembled template–monomer complex is captured in the polymer matrix. After removal of the templates, the cavities of the MIP possess a shape and an arrangement corresponding to the functional groups of the template (Fig. 5-5).

Recently, an in-depth review on molecular imprinted membranes has been published by Piletsky et al. [4]. Four preparation strategies for MIP membranes can be distinguished: (i) in-situ polymerization by bulk crosslinking; (ii) preparation by dry phase inversion with a casting/solvent evaporation process [45–51]; (iii) preparation by wet phase inversion with a casting/immersion precipitation [52–54]; and (iv) surface imprinting.

With regard to MIP membranes, the polymer morphology directly affects permeability and selectivity of the membrane. The interactions of the template with polymer domains may cause conformational reorganization of the MIP network structure. It is suggested [44] that chiral separation by MIP membranes occurs by sieving as well as by selective transport through the template-specific cavities of the polymer barrier (Fig. 5-6). The former mechanism is provided by the polymer structure micropores, which are formed around template molecules during polymerization. The latter mechanism results from the ligand-functional groups interactions inside the MIP.

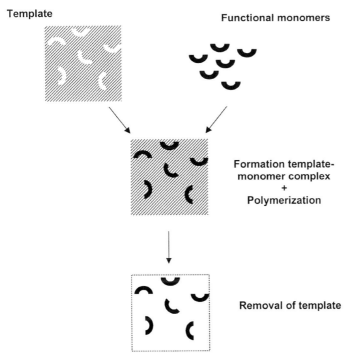

Fig. 5-5. Schematic representation of the preparation procedure of molecular imprinted polymers (MIP).

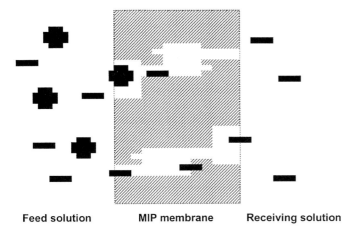

Fig. 5-6. Chiral separation by MIP membranes: a combination of sieving and selective transport [44].

Several selective interactions by MIP membrane systems have been reported. For example, an L-phenylalanine imprinted membrane prepared by in-situ crosslinking polymerization showed different fluxes for various amino acids [44]. Yoshikawa et al. [51] have prepared molecular imprinted membranes from a membrane material which bears a tetrapeptide residue (DIDE resin (7)), using the dry phase inversion procedure. It was found that a membrane which contains an oligopeptide residue from an L-amino acid and is imprinted with an L-amino acid derivative, recognizes the L-isomer in preference to the corresponding D-isomer, and vice versa. Exceptional difference in sorption selectivity between theophylline and caffeine was observed for poly(acrylonitrile-co-acrylic acid) blend membranes prepared by the wet phase inversion technique [53].

7

Possible applications of MIP membranes are in the field of sensor systems and separation technology. With respect to MIP membrane-based sensors, selective ligand binding to the membrane or selective permeation through the membrane can be used for the generation of a specific signal. Practical chiral separation by MIP membranes still faces reproducibility problems in the preparation methods, as well as mass transfer limitations inside the membrane. To overcome mass transfer limitations, MIP nanoparticles embedded in liquid membranes could be an alternative approach to develop chiral membrane separation by molecular imprinting [44].

5.2.4 Cascades of Enantioselective Membranes

Considering the limited enantioselectivities commonly found for chiral membranes, these membranes are not capable of separating a racemic mixture in one single step. For this reason a cascade of membrane steps must be used (Fig. 5-7). A description of multistage membrane separations is derived from the graphical description of other multistage separation processes (e.g. distillation) using the McCabe–Thiele diagram. The "equilibrium" curve is now obtained by plotting the retentate concentration versus the permeate concentration [55]. Since it is impossible to consider a membrane separation as an equilibrium separation, the term "selectivity curve" is more appropriate than equilibrium curve. A McCabe–Thiele diagram is shown in Fig. 5-8, in which the curved line represents the selectivity curve. The tangent of line AB in this diagram can be chosen freely, and equals the ratio of the permeate and retentate streams. Using this approach the required membrane surface area can be calculated for a given separation. The required number of stages and the total membrane surface area are plotted versus the enantioselectivity of the membrane in Fig. 5-9. From this graph it will be obvious that at the enantioselectivities commonly

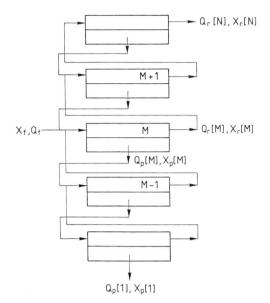

Fig. 5-7. Cascade configuration for the separation of a racemic mixture into a top and bottom concentration of X[N] and X[1], respectively [55].

found, a large number of stages is required, leading to a substantial membrane area. The possibility to make membranes of opposite chirality, e.g. by using the MIP technique, allows for the use of membranes of alternating chirality. According to the analysis described above, the required membrane surface area can then effectively be lowered by approximately 25 %. Generally, it can be concluded that the large number of (independent) stages required, resulting from low selectivities, in combination with low permeabilities leading to a large membrane area, are currently prohibitive for application of enantioselective membranes on an industrial scale. Obviously, there is a need for increased selectivities at higher permeabilities, which clearly will not be an easy task to perform.

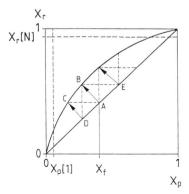

Fig. 5-8. McCabe–Thiele diagram for $\alpha < 1$. The curved line is the selectivity curve.

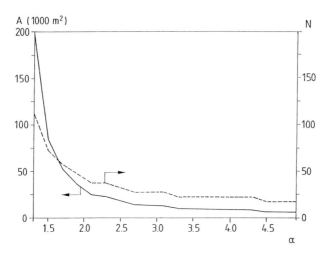

Fig. 5-9. Total number of stages and total membrane surface area versus membrane selectivity for the separation of 1 kg s^{-1} of a racemic mixture at a membrane permeability of 1.6×10^{-2} kg m^{-2}.s, yielding both enantiomers at 95 % purity [55].

5.3   Membrane-Assisted Chiral Separations

One of the major advantages of membrane processes for the separation of racemic mixtures lies in the ease of scale-up. As scale-up procedures are well established, this enables the implementation of membrane processes for separations on a multi-kilogram scale. From the foregoing sections, it will be obvious that enantioselective membranes are still in an early stage of development, and large-scale applications are not expected in the short term [5]. Nevertheless, enantioselective membranes do have considerable potential for large-scale separations because of the existing experience with techniques such as reverse osmosis and ultrafiltration. When looking at membrane processes for chiral separations which have found full-scale implementation, or which are close to this, the range is clearly limited to membrane-assisted processes.

Nonselective membranes can assist enantioselective processes, providing essential nonchiral separation characteristics and thus making a chiral separation based on enantioselectivity outside the membrane technically and economically feasible. For this purpose several configurations can be applied: (i) liquid–liquid extraction based on hollow-fiber membrane fractionation; (ii) liquid– membrane fractionation; and (iii) micellar-enhanced ultrafiltration (MEUF).

5.3.1 Liquid–Liquid Extraction

In principle, the same selectors can be used for chiral separation in conventional liquid–liquid extraction as are applied in supported liquid membranes. Generally, enantioselectivity is obtained by addition of a chiral selector to just one of the two phases. Although relatively high separation factors can be obtained in these types of systems, a multistage configuration is required to achieve high optical purities. In order to reach a large number of stages, good countercurrent flow is a prerequisite. The countercurrent extraction described by Takeuchi et al. [56] meets these specifications, though scale-up of the proposed apparatus is expected to be difficult. The performance of conventional extraction is often limited by back-mixing and flooding. This can be avoided by immobilizing the liquid–liquid interface in the pores of a membrane (Fig. 5-10), so-called "hollow-fiber membrane extraction". Advantages of hollow-fiber membrane extraction as compared to conventional extraction are the high surface area per unit volume, as well as the fact that flow ratios can be chosen independently [57–61].

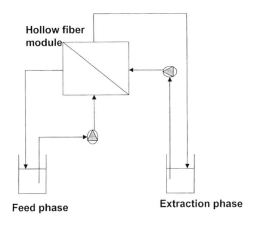

Hollow fiber module

Feed phase   **Extraction phase**

Fig. 5-10. Schematic representation of hollow-fiber membrane extraction.

For the separation of D,L-leucine, Ding et al. [62] used poly(vinyl alcohol) gel-coated microporous polypropylene hollow fibers (Fig. 5-11). An octanol phase containing the chiral selector (*N-n*-dodecyl-L-hydroxyproline) is flowing countercurrently with an aqueous phase. The gel in the pores of the membrane permits diffusion of the leucine molecules, but prevents convection of the aqueous and octanol phase. At a proper selection of the flow ratios it is possible to achieve almost complete resolution of the D,L-leucine (Fig. 5-12).

Important numbers in the design of (membrane-based) liquid– liquid extractions are the extraction factors for both enantiomers, E_L and E_D, defined as

$$E_L = \frac{m_L \cdot F_e}{F_r} \qquad (2)$$

$$E_D = \frac{m_D \cdot F_e}{F_r} \qquad (3)$$

in which m_L and m_D are the distribution coefficients of the L and D enantiomers over both phases, and F_e and F_r are the extract and raffinate flow, respectively [63]. In order to achieve complete resolution, E_L and E_D should be opposite from unity. Due to the fact that most racemic mixtures dissolve preferentially in either the aqueous or the organic phase, this often results in extreme flow ratios. In the example of D,L-leucine separation, the flow ratio is chosen as 4, yielding extraction factors of 1.32 and 0.64, respectively.

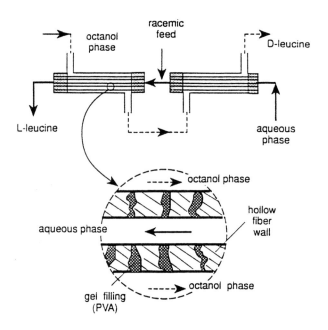

Fig. 5-11. Fractional extraction with gel-coated hollow fibers [57].

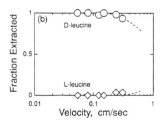

Fig. 5-12. Separation of d,l-leucine in hollow-fiber membrane extraction using a N-n-dodecyl-1-hydroxyproline solution in octanol as the enantioselective extraction liquid. The modules used were 32 cm long and contained 96 Celgard X-20 polypropylene fibers [57].

5.3.2 Liquid–Membrane Fractionation

As described above, the application of classical liquid– liquid extractions often results in extreme flow ratios. To avoid this, a completely symmetrical system has been developed at Akzo Nobel in the early 1990s [64, 65]. In this system, a supported liquid–membrane separates two miscible chiral liquids containing opposite chiral selectors (Fig. 5-13). When the two liquids flow countercurrently, any desired degree of separation can be achieved. As a result of the system being symmetrical, the racemic mixture to be separated must be added in the middle. Due to the fact that enantioselectivity usually is more pronounced in a nonaqueous environment, organic liquids are used as the chiral liquids and the membrane liquid is aqueous. In this case the chiral selector molecules are lipophilic in order to avoid transport across the liquid membrane.

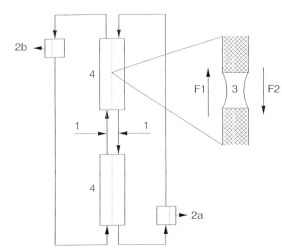

Fig. 5-13. Schematic representation of the Akzo Nobel enantiomer separation process. Two liquids containing the opposing enantiomers of the chiral selector (F1 and F2) are flowing countercurrently through the column (4) and are kept separated by the liquid membrane (3). The racemic mixture to be separated is added to the middle of the system (1), and the separated enantiomers are recovered from the outflows of the column (2a and 2b) [64].

The enantioselectivity α is defined as the distribution ratio of one single enantiomer over the two chiral phases and has been determined experimentally for a variety of compounds (Table 5-1). It has been known from work by Prelog [66, 67] that tartaric acid derivatives show selectivities towards α-hydroxyamines and amino acids. However, from Table 5-1 it is obvious that tartaric acid derivatives show selectivity for many other compounds, including various amino bases (e.g. mirtazapine (**10**)) and acids (e.g. ibuprofen (**11**)). The use of other chiral selectors (e.g. PLA)

8

9

10

Me

Table 5-1. Enantioselectivities determined for several drugs. All experiments were performed at room temperature, except those marked with *, which were performed at 4 °C. In some cases a lipophilic anion was used to facilitate the solubilization of the drug in the organic phases (PF_6^- = hexafluorophosphate; BPh_4^- = tetraphenyl borate). DHT = dihexyl tartrate; DBT = dibenzoyl tartrate; PLA = poly (lactic acid).

Compound	Selector	Solvent	Anion	α
Norephedrine [8]	0.25 M DHT	Heptane	–	1.19
	0.25 M DHT	Heptane	–	1.50*
	10 wt % PLA	Chloroform	–	1.07
	10 wt % PLA	Chloroform	–	1.10*
Ephedrine [9]	0.10 M DHT	Heptane	0.06 M PF_6^-	1.30
Mirtazapine [10]	0.10 M DHT	Heptane	–	1.05
	0.25 M DHT	Heptane	–	1.06
	0.50 M DHT	Heptane	–	1.08
	0.25 M DBT	Decanol	–	1.06
	10 wt % PLA	Chloroform	–	1.04
	0.10 M DHT	Heptane	0.01 M PF_6^-	1.07
	0.50 M DHT	Heptane	0.03 M PF_6^-	1.16
Phenylglycine	0.25 M DHT	Heptane	–	1.06
Salbutamol [11]	0.25 M DHT	Heptane	0.018 M BPh_4^-	1.06
	0.25 M DHT	Cyclohexane	0.0035 M BPh_4^-	1.04
Terbutaline [12]	0.43 M DHT	Dichloroethane	0.5 M PF_6^-	1.14
	2.15 M DHT	Heptane	0.0045 M BPh_4^-	1.05
Propranolol [2]	0.25 M DHT	Heptane	–	1.03
Ibuprofen [13]	0.10 M DHT	Heptane	–	1.10

11   **12**   **13**

also results in acceptable selectivities. Although most experiments were performed at room temperature, the effect of temperature can be significant, as indicated for norephedrine (**8**). When the temperature is lowered from 20 C to 4 °C, α increases from 1.19 to 1.50 when DHT in heptane is used.

The degree of separation achieved in this system can be calculated according to

$$\frac{R}{S} = \frac{\dfrac{\alpha_S}{\alpha_S - 1}\left(\exp\left(\dfrac{\alpha_S - 1}{2\alpha_S} NTU\right) - \dfrac{1}{\alpha_S}\right)}{\dfrac{\alpha_R}{\alpha_R - 1}\left(\exp\left(\dfrac{\alpha_R - 1}{2\alpha_R} NTU\right) - \dfrac{1}{\alpha_R}\right)} \tag{4}$$

In which R/S is the required product purity and α_R and α_S are the distribution ratios of the R- and S-enantiomers over the two chiral liquids, respectively (by definition $\alpha_R = 1/\alpha_S$). NTU is the number of transfer units required for the separation. From the data given in Table 5-1 it can be concluded that enantioselectivities typi-

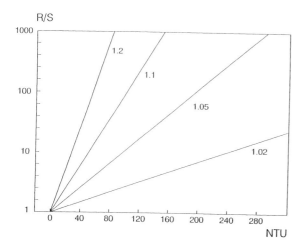

Fig. 5-14. The optical purity (*R/S*) of the outflows as a function of the number of transfer units (NTU) in the apparatus at equal flow rates. Different lines are given for different α values.

cally are in the range between 1.05 and 1.20. According to Equation (4), the number of transfer units can now be calculated. This is shown in Fig. 5-14. For a 99 % pure product (*R/S* = 100), about 190 NTU are required at an enantioselectivity of 1.05, a number decreasing to approximately 30 when α increases to a value of 1.20.

Once the required number of transfer units for a given degree of separation is known, the height of a transfer unit (HTU) must be determined in order to design the column. Membrane modules containing well-spaced Cuprophan regenerated cellulose hollow fibers with an internal diameter of 200 μm and a dry wall thickness of 8 μm were used. The walls of the fibers are filled with the aqueous "membrane liquid" and the two chiral liquids flow through the lumen and the shell side of the module, respectively. Typical values for HTU were found to be in range of 2 to 6 cm [65]. These extremely low values are mainly due to the high specific surface area that can be obtained in the hollow-fiber modules (typically in the order of 10 000 m² m⁻³) [62]. The total length of the membrane column used for the separation (L) is now given by

$$L = NTU \cdot HTU \tag{5}$$

Using the calculated and measured values for NTU and HTU, typical column lengths of 2–5 m are expected.

To demonstrate the potential of the process in obtaining both enantiomers at a high purity, experiments were performed using racemic norephedrine as the compound to be separated. Two columns of seven small membrane modules were used. The enantiomer ratios in the outflows during start-up are shown in Fig. 5-15. It can be concluded that the system reaches equilibrium within approximately 24 h, and that both enantiomers are recovered at 99.3–99.8 % purity.

To evaluate the economics of this process, a cost model has been developed to estimate the separation costs for a specific racemate [68, 69]. For this purpose, the sensitivity of the separation costs for several key process parameters have been established as compared to a base-case separation in which a purity of 99 % is

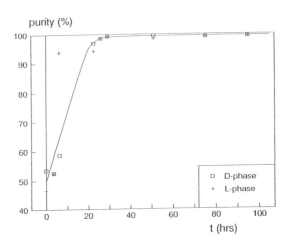

Fig. 5-15. Enantiomer ratio in the outflow versus time for the separation of racemic norephedrine.

required at an enantioselectivity of 1.15. The maximum solubility of the drug is set at 1 wt-%, and the membrane life time is taken 1.5 years. As can be seen from Fig. 5-16, the separation costs are not extremely sensitive to the required optical purity and the membrane life time. However, the achieved enantioselectivity and attainable concentration of the drug (in the center of the column) determine separation costs to a large extent. This is mainly due to the costs involved for product recovery from a large, relatively diluted stream. Clearly, the effectiveness of this separation process is predominantly determined by a proper selection and optimization of the selector–solvent combination.

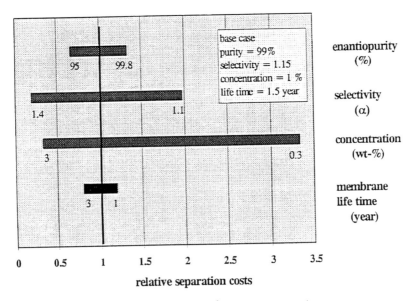

Fig. 5-16. Sensitivity of separation costs to key process parameters.

5.3.3 Micellar-Enhanced Ultrafiltration

Ultrafiltration of micellar solutions combines the high permeate flows commonly found in ultrafiltration systems with the possibility of removing molecules independent of their size, since micelles can specifically solubilize or bind low molecular weight components. Characteristics of this separation technique, known as micellar-enhanced ultrafiltration (MEUF), are that micelles bind specific compounds and subsequent ultrafiltration separates the surrounding aqueous phase from the micelles [70]. The pore size of the UF membrane must be chosen such, that the micelles are retained but the unbound components can pass the membrane freely. Alternatively, proteins such as BSA have been used in stead of micelles to obtain similar enantioselective aggregates [71].

For the separation of amino acids, the applicability of this principle has been explored. For the separation of racemic phenylalanine, an amphiphilic amino acid derivative, l-5-cholesteryl glutamate (**14**) has been used as a chiral co-surfactant in micelles of the nonionic surfactant Serdox NNP™10. Copper(II) ions are added for the formation of ternary complexes between phenylalanine and the amino acid co-surfactant. The basis for the separation is the difference in stability between the ternary complexes formed with d- or l-phenylalanine, respectively. The basic principle of this process is shown in Fig. 5-17 [72].

14

The intrinsic enantioselectivity of the micelles has been established based on single-component binding isotherms [73], resulting in a remarkably high value of 7.7.

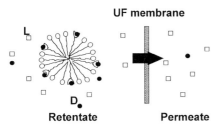

Fig. 5-17. Principle of micellar-enhanced ultrafiltration (MEUF). The d-enantiomer of a racemic mixture is preferentially bound to the micelles, which are retained by the membrane. The bulk containing the l-enantiomer is separated through the membrane [72].

However, when complexation experiments are performed with both D- and L-enantiomers (Fig. 5-18), this leads to selectivity values between 1.4 and 1.9. It was shown that complexation by enantioselective micelles can effectively be described using straightforward multicomponent Langmuir isotherms [74].

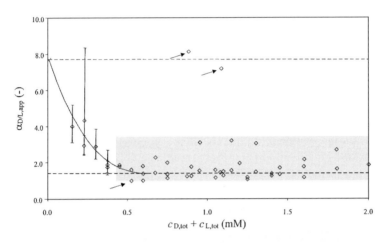

Fig. 5-18. Selectivity for D,L-phenylalanine, fitted with a multicomponent Langmuir isotherm [73].

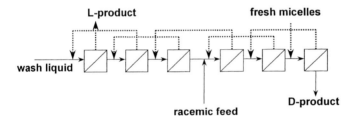

Fig. 5-19. Cascaded system for countercurrent 99 %+ separation of racemic mixtures [75].

From the foregoing it will be obvious that a single-stage MEUF system will not be capable of producing single enantiomers at high purity. Therefore, to achieve high optical purities, a multistage separation process is required (Fig. 5-19) [75]. This system is operated in a countercurrent mode, analogous to conventional extraction and distillation processes. Here, the enantioselective micellar phase flows in opposite direction of the water phase. In each stage a UF membrane separates the micellar phase from its surrounding aqueous phase. To regenerate the saturated micelles leaving the cascade, a simple decrease in pH leads to electrostatic repulsion between selector and bound enantiomer [76]. Using the experimental Langmuir constants, the process has been modeled in a cascade of 60 stages [77]. When the racemate is fed to stage 34, the flows leaving stages 1 and 60 have an enantiomeric excess of 99.1 % and 99.8 %, respectively (Fig. 5-20). An industrially important feature of the process comes from the fact that the feed is diluted by a factor of only 3.

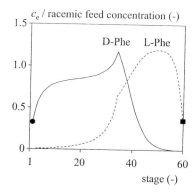

Fig. 5-20. Enantiomer separation in a cascaded ultrafiltration (UF) system of 60 stages [77]

Although the basic principles underlying this MEUF process for chiral separations have been established, no optimization has yet been performed. Based on the ability of UF to treat large streams relatively inexpensively, this process will potentially become an extension to the rather limited range of large-scale commercial separation processes. In addition, to facilitate further purification (e.g. by diastereomeric salt crystallization) a single-stage MEUF process can potentially serve as an effective low-grade separation method.

5.4 Concluding remarks

In this chapter, we have discussed the potential of membrane technology for chiral separations. As the focus of this chapter has been on large-scale separations, the various developments have been discussed from an engineering point of view. A significant effort is being put into the development of chiral polymer membranes by a number of research groups. Nevertheless, these membranes still are in an early stage of development, and for industrial use significant improvements on flux and/or enantioselectivity are required. Once these limitations have been overcome, the application of chiral membranes will be similar to reverse osmosis and ultrafiltration, which will allow a relatively rapid implementation. Similarly, a large number of chiral selectors have been developed for application in liquid membranes. A major issue for the application of liquid membranes on an industrial scale lies in the problem of staging. As a result of the many recycles in the process, selective liquid membranes will require a large number of storage vessels, which is undesirable from a process economics standpoint.

In the short term, we do not expect chiral membranes to find large-scale application. Therefore, membrane-assisted enantioselective processes are more likely to be applied. The two processes described in more detail (liquid–membrane fractionation and micellar-enhanced ultrafiltration) rely on established membrane processes and make use of chiral interactions outside the membrane. The major advantages of these processes are the ease of staging (this applies especially to the liquid–membrane fractionation) and the ease of treating large streams (which is a key advantage of the MEUF process). In the near future, we expect more of these "hybrid" technologies to be developed, which will certainly broaden the scope of membrane-based enantioseparations.

List of structures

1   *N*-alkyl-L-hydroxyproline
2   Propranolol
3   Bupranolol
4   Modified cyclodextrin
5   3,3'-dimethylbis(α,α'-dinaphthyl)-22-crown-6
6   Poly{1-[dimethyl(10-pinanyl)silyl]prop-1-yne} (PDPSP)
7   H-Asp(OcHex)-Leu-Asp(OcHex)-Glu(Obzl)-CH$_2$- (DIDE)
8   Norephedrine
9   Ephedrine
10 Mirtazipine
11 Salbutamol
12 Terbutaline
13 Ibuprofen
14 -5-cholesteryl glutamate

References

[1] Bayley, C. R., Vaidya, N. A., in *Chirality in Industry, The Commercial Manufacture and Applications of Optically Active Compounds*, Collins, A. N., Sheldrake, G. N., and Crosby, J. (eds.), Wiley, Chichester, 1992

[2] Dorson, W. J. and Pierson, J. S., *J. Membr. Sci.*, **44** (1989) 35

[3] Sourirajan, S., Matsuura, T., *Reverse Osmosis/Ultrafiltration, Process Principles*, National Research Council Canada, Ottawa, 1985

[4] Ho, W. S. W., and Sirkar, K. K., Membrane Handbook, Chapman & Hall, New York, 1992

[5] Keurentjes, J. T. F. and Voermans, F. J. M., in *Chirality in Industry II, Developments in the Manufacture and Applications of Optically Active Compounds*, Collins, A. N., Sheldrake, G. N., and Crosby, J. (eds.), Wiley, Chichester, 1997

[6] Pickering, P. J., Southern, C. R., *J. Chem. Biotechnol.*, **68** (1997) 417

[7] Brice, L. J., Pirkle, W. H., *Chiral separations: Applications and Technology*, Ahuja, Satinder (eds.) American Chemical Society, Washington D.C., 1997

[8] Pickering, P. J., Chaudhuri, J. B., *Chirality*, **9** (1997) 261

[9] Skrzypinski, W., Sierleczko, E., Plucinski, P., Lejczak, B., Pafarski, P., *J. Chem. Soc. Perkin Trans.*, **2** (1990) 689

[10] Vaidya, A. M., Bell, G., Halling, P. J., *J. Membr. Sci.*, **97** (1994) 13

[11] Victoiria Martinez-Diaz, M., de Mendoza, J., Torres, T., *Tetrahedron Lett.*, **35** (1994) 7669

[12] Heard, C. M., Hadgraft, J., Brain, K. R., *Bioseparations*, **4** (1994) 111

[13] Scrimin, P., Tonellato, U., Zanta, N., *Tetrahedron Lett.*, **29** (1988) 4967

[14] Pirkle, W.H., Doherty, E. M., *J. Am. Chem. Soc.*, **111** (1989) 4113

[15] Armstrong, D. W., Jin, H. L., *Anal. Chem.*, **59** (1987) 2237

[16] Pietraszkiewicz, M., Kozbial, M., Pietraskiewicz, O., *J. Enantiomer*, **2** (1997) 319

[17] Pietraszkiewicz, M., Kozbial, M., Pietraskiewicz, O., *J. Membr. Sci.*, **138** (1998) 109
[18] Tsukube, H., Shinoda, S., Uenishi, J., Kanatani, T., Itoh, H., Shiode, M., Iwachido, T., Yonemitsu, O., *Inorg. Chem.*, **37** (1998) 1585
[19] Horvath, V., Takacs, T., Horvai, G., Huszthy, P., Bradshaw, J. S., Izatt, R. M., *Anal. Lett.*, **30** (1997) 1591
[20] Kozbial, M., Pietraszkiewicz, M., Pietraskiewicz, O., *J. Incl. Phen. Mol. Rec. Chem.*, **30** (1998) 69
[21] Newcomb, M., Toner, J. L., Helgeson, R. C., Cram, D. J., *J. Am. Chem. Soc.*, **101** (1979) 4941
[22] Ceynowa, J., *Chem. Anal.*, **43** (1998) 917
[23] Szabo, L. P., *Hung. J. Ind. Chem.*, **25** (1997) 209
[24] Aboul-Enein, H. Y., *Saudi Pharm. J.*, **4** (1996) 1
[25] Nakamura, M., Kiyohara, S., Saito, K., Sugita, K., Sugo, T., *Anal. Chem.*, **71** (1999) 1323
[26] Ogata, N., *Polym. Prepr. Am. Chem. Soc. Div. Polym. Chem.*, **34** (1993) 96
[27] Bryjak, M., Kozlowski, J., Wieczorek, P., Kafarski, P., *J. Membr. Sci.*, **85** (1993) 221
[28] Kakuchi, T., Harada, Y., Satoh, T., Yokota, K., Hashimoto, H., *Polymer*, **35** (1994) 204
[29] Scrimin, P., Tecilla, P., Tonellato, U., *Tetrahedron*, **51** (1995) 217
[30] Kataoka, H., Hanawa, T., Katagi, T., *Chem. Pharm. Bull.*, **40** (1992) 570
[31] Inoue, K., Miyahara, A., Itaya, T., *J. Am. Chem. Soc.*, **119** (1997) 6191
[32] Masawaki, T., Sasai, M., Tone, S., *J. Chem. Eng. Japan*, **25** (1992) 33
[33] Maruyama, A., Adachi, N., Takatsuki, T., Torii, M., Sanui, K., Ogata, N., *Macromolecules*, **23** (1990) 2748
[34] Masawaki, T., Matsumoto, S., Tone, S., *J. Chem. Eng. Japan* 27 (1994) 517
[35] Ogata, N., *Macromol. Symp.*, **98** (1995) 543
[36] Aoki, T., Shinohara, K., Oikawa, E., *Macromol. Rapid. Commun.*, **13** (1992) 565
[37] Aoki, T., Kokai, M., Shinohara, K., Oikawa, E., *Chem. Lett.*, **12** (1993) 2009
[38] Shinohara, K., Aoki, T., Oikawa, E., *Polymer*, **36** (1995) 2403
[39] Aoki, T., Tomizawa, S., Oikawa, E., *J. Membr. Sci.*, **99** (1995) 117
[40] Aoki, T., Shinohara, K., Kaneko, T., Oikawa, E, *Macromol.*, **29** (1996) 4192
[41] Aoki, T., Ohshima, M., Shinohara, K., Kaneko, T., Oikawa, E, *Polymer*, **38** (1997) 235
[42] Wulff, G., Sarhan, A., *Angew. Chem. Int. Ed. Engl.*, **11** (1972) 341
[43] Wulff, G., *Angew. Chem. Int. Ed. Engl.*, **34** (1995) 1812
[44] Piletsky, S. A., Panasyuk, T. L., Piletskaya, E. V., Nicholls, I. A., Ulbricht, M., *J. Membr. Sci.*, **157** (1999) 263
[45] Yoshikawa, M., Izumi, J., Kitao, T., *Chem. Lett.*, **8** (1996) 611
[46] Yoshikawa, M., Izumi, J., Kitao, T., Sakamoto, S., *Macromol.*, **29** (1996) 8197
[47] Yoshikawa, M., Izumi, J., Kitao, T., *Pol. J.*, **29** (1997) 205
[48] Yoshikawa, M., Izumi, J., Kitao, T., Sakamoto, S., *Macromol. Rapid. Commun.*, **18** (1997) 761
[49] Yoshikawa, M., Izumi, J., Ooi, T., Kitao, T., Guiver, M.D., Robertson, G.P., *Pol. Bull.*, **40** (1998) 517
[50] Yoshikawa, M., Fujisawa, T., Izumi, J., Kitao, T., Sakamoto, S., *Anal. Chem.*, **365** (1998) 59
[51] Yoshikawa, M., *ACS Symp. Ser.*, **703** (1998) 170
[52] Kobayashi, T., Wang, H.Y., Fuji, N., *Chem. Lett.*, (1995) 927
[53] Wang, H.Y., Kobayashi, T., Fuji, N., *Langmuir*, **12** (1996) 4850
[54] Kobayashi, T., Wang, H. Y., Fuji, N., *Anal. Chim. Acta.*, **365** (1998) 81
[55] Keurentjes, J. T. F., Linders, L. J. M., Beverloo, W. A. and Van 't Riet, K., *Chem. Engng. Sci.* **47** (1992) 1561
[56] Takeuchi, T., Horikawa, R., Tanimura, T., *Sep. Sci. Tech.*, **25** (1990) 941
[57] Dahuron, L., Cussler, E. L., *AIChE J.*, **34** (1988) 130
[58] D'Elia, N. A., Dahuron, L., Cussler, E. L., *J. Membr. Sci.*, **29** (1986) 309
[59] Yang, M. C., Cussler, E. L., *AIChE J.*, **32** (1986) 1910
[60] Prasad, R., Sirkar, K. K., *AIChE J.*, **34** (1988) 177
[61] Keurentjes, J. T. F., Sluijs, J. T. M., Franssen, R. J. H., van 't Riet, K., *Ind. Eng. Chem. Res.*, **31** (1992) 581
[62] Ding, H. B., Carr., P. W., Cussler, E. L., *AIChE J.*, **38** (1992) 1493
[63] Beek, W. J., Muttzall, K. M. K., *Transport Phenomena*, Wiley, London, 1975
[64] Keurentjes, J. T. F., *PCT Int. Pat. Appl.*, WO 94/07814, 1994
[65] Keurentjes, J. T. F., Nabuurs, L. J. W. M., Vegter, E. A., *J. Membr. Sci.*, 113 (1996) 351
[66] Prelog, V., Stojanac, Z., Kovacevic, *Helv. Chim. Acta*, **65** (1982) 377

[67] Prelog, V., Mutak, S., Kovacevic, *Helv. Chim. Acta*, **66** (1982) 2279
[68] Riegman, R. L. M., Dirix, C. A. M. C., Zsom, R. L. J., *Chiral USA '97 Symposium, Spring Innovations*, Stockport, 1997
[69] Keurentjes, J. T. F., Lammers, H., Schneider, K., *Proc. 8th European Congress on Biotechnology*, Budapest, Hungary (1997) 132
[70] Scamehorn, J. F., Christian, S. D., Ellington, R. T., in *Surfactant-based Separation Processes*, Scamehorn, J. F., Harwell, J. H. (eds.), Marcel Dekker, New York, 1989
[71] Poncet, S., Randon, J., Rocca, J. L., *Sep. Sci. Technol.* **32** (1997) 2029
[72] Creagh, A. L., Hasenack, B. B. E., Van der Padt, A., Sudhölter, E. J. R., Van 't Riet, K., *Biotechnol. Bioeng.* **44** (1994) 690
[73] Overdevest, P. E. M., Keurentjes, J. T. F., Van der Padt, A. and Van 't Riet, K., in *Surfactant-based Separations, Science and Technology*, Scamehorn, J. F. and Harwell, J. H. (eds.), ACS Symposium Series 740, American Chemical Society, Washington, 1999
[74] Overdevest, P. E. M., Van der Padt, A., Keurentjes, J. T. F., Van 't Riet, K., *Colloids and Surfaces A*, 163 (2000), 209
[75] Overdevest, P. E. M., De Bruin, T. J. M. and Van der Padt, A., *ChemTech*, 29, **12** (1999), 17
[76] Overdevest, P. E. M., De Bruin, T. J. M., Sudhölter, E. J. R., Keurentjes, J. T. F., Van 't Riet, K., Van der Padt, A., submitted
[77] Overdevest, P. E. M., *Proc.1998 Membrane Technology/Separations Planning Conference*, Newton, Mass. (1998) 140

6 Enantiomer Separations using Designed Imprinted Chiral Phases

Börje Sellergren

6.1 Introduction

The need for efficient high-throughput techniques in the production of enantiomerically pure compounds is growing in parallel to the increasing structural complexity of new drug compounds [1].

In the absence of synthetic methods allowing the drug to be synthesized in optically pure form, the resolution of racemates using characterized chiral selectors or auxiliaries is the first step in this process. These techniques have the additional advantage of providing both enantiomers in preparative amounts, which means that the requirements for biological testing of both enantiomers can be met. Conventionally, preparative optical resolution is performed by fractional crystallization, microbiological methods, kinetic enzymatic resolution and by chromatography. Of growing importance are methods allowing continuous production of pure enantiomers. In chromatography, these can be based on liquid–solid partitioning as in simulated moving bed (SMB) chromatography (see Chapter 10) or liquid–liquid partitioning as in countercurrent distribution [2, 3] or chromatography [4]. In the case of phases exhibiting particularly high enantioselectivities, batch- [5], membrane-, [6–8] or bubble- based [9] separation techniques may be more attractive.

In chromatography, polysaccharide-based phases (modified amylose or cellulose) are, due to their high site density and broad applicability, the most common phases used for preparative-scale separations [10]. A problem with these, as well as other common CSPs, is the limited predictability of elution orders and separability, making screening of stationary phase libraries a necessary step in the method development [10]. Polymers imprinted with chiral templates here promise to alleviate these problems offering a new generation of custom-made CSPs with predictable selectivities [11]. In view of the high selectivity often exhibited by these phases, preparative applications in the above-mentioned formats are being investigated. This review will summarize the present state of this research field.

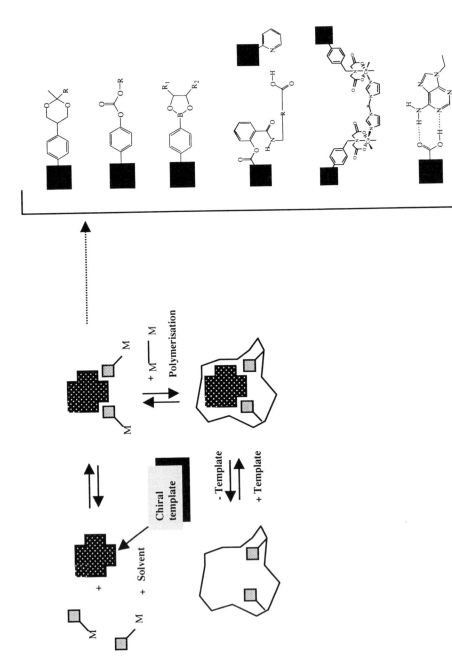

Fig. 6-1. Approaches to generate imprinted binding sites.

6.2   Molecular Imprinting Approaches

Molecularly imprinted polymers (MIPs) can be prepared according to a number of approaches that are different in the way the template is linked to the functional monomer and subsequently to the polymeric binding sites (Fig. 6-1). Thus, the template can be linked and subsequently recognized by virtually any combination of cleavable covalent bonds, metal ion co-ordination or noncovalent bonds. The first example of molecular imprinting of organic network polymers introduced by Wulff was based on a covalent attachment strategy i.e. covalent monomer–template, covalent polymer–template [12].

Currently, the most widely applied technique to generate molecularly imprinted binding sites is represented by the noncovalent route developed by the group of Mosbach [13]. This makes use of noncovalent self-assembly of the template with functional monomers prior to polymerization, free radical polymerization with a crosslinking monomer, and then template extraction followed by rebinding by noncovalent interactions. Although the preparation of a MIP by this method is technically simple, it relies on the success of stabilization of the relatively weak interactions between the template and the functional monomers. Stable monomer–template assemblies will in turn lead to a larger concentration of high affinity binding sites in the resulting polymer. The materials can be synthesized in any standard equipped laboratory in a relatively short time, and some of the MIPs exhibit binding affinities and selectivities in the order of those exhibited by antibodies towards their antigens. Nevertheless, in order to develop a protocol for the recognition of any given target, all of the alternative linkage strategies must be taken into account.

Most MIPs are synthesized by free radical polymerization of functional monounsaturated (vinylic, acrylic, methacrylic) monomers and an excess of crosslinking di- or tri- unsaturated (vinylic, acrylic, methacrylic) monomers, resulting in porous organic network materials. These polymerizations have the advantage of being relatively robust, allowing polymers to be prepared in high yield using different solvents (aqueous or organic) and at different temperatures [14]. This is necessary in view of the varying solubilities of the template molecules.

The most successful noncovalent imprinting systems are based on commodity acrylic or methacrylic monomers, such as methacrylic acid (MAA), crosslinked with ethyleneglycol dimethacrylate (EDMA). Initially, derivatives of amino acid enantiomers were used as templates for the preparation of imprinted stationary phases for chiral separations (MICSPs), but this system has proven generally applicable to the imprinting of templates allowing hydrogen bonding or electrostatic interactions to develop with MAA [15, 16]. The procedure applied to the imprinting with l-phenylalanine anilide (L-PA) is outlined in Fig. 6-2. In the first step, the template (L-PA), the functional monomer (MAA) and the crosslinking monomer (EDMA) are dissolved in a poorly hydrogen bonding solvent (porogen) of low to medium polarity. The free radical polymerization is then initiated with an azo initiator, commonly azo-*N,N*'-bis-isobutyronitrile (AIBN) either by photochemical homolysis below room temperature [16, 17] or thermochemically at 60 °C or higher [15]. Lower thermochemical initiation at temperatures down to 40 °C or 30 °C is also possible using less stable azoinitiators [18]. In the final step, the resultant polymer is crushed using a mortar and pestle or in a ball mill, extracted using a Soxhlet apparatus, and sieved to a particle size suitable for chromatographic (25–38 µm) or batch (150– 250 µm) applications [16]. The polymers are then evaluated as stationary phases in chromatography by comparing the retention time or capacity factor (k') [19] of the template with that of structurally related analogs (Fig. 6-3). We will refer to the system shown in Fig. 6-2 as the L-PA-model system.

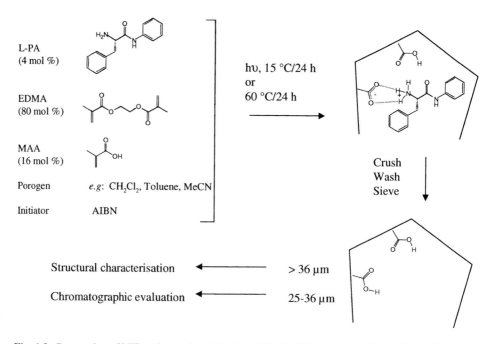

Fig. 6-2. Preparation of MIPs using L-phenylalanine anilide (L-PA) as template. The L-PA model system.

In the elucidation of retention mechanisms, an advantage of using enantiomers as templates is that nonspecific binding, which affects both enantiomers equally, cancels out. Therefore the separation factor (α) uniquely reflects the contribution to binding from the enantioselectively imprinted sites. As an additional comparison the retention on the imprinted phase is compared with the retention on a nonimprinted reference phase. The efficiency of the separations is routinely characterized by estimating a number of theoretical plates (N), a resolution factor (R_s) and a peak asymmetry factor (A_s) [19]. These quantities are affected by the quality of the packing and mass transfer limitations, as well as of the amount and distribution of the binding sites.

Some restrictions of this molecular imprinting technique are obvious. The template must be available in preparative amounts, it must be soluble in the monomer mixture, and it must be stable and unreactive under the conditions of the polymer-

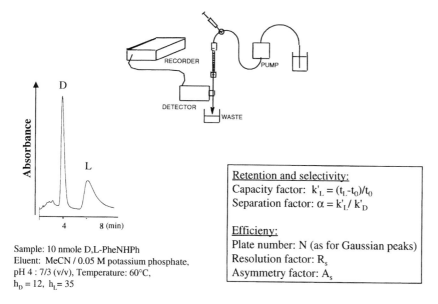

Retention and selectivity:
Capacity factor: $k'_L = (t_L-t_0)/t_0$
Separation factor: $\alpha = k'_L / k'_D$

Efficieny:
Plate number: N (as for Gaussian peaks)
Resolution factor: R_s
Asymmetry factor: A_s

Sample: 10 nmole D,L-PheNHPh
Eluent: MeCN / 0.05 M potassium phosphate,
pH 4 : 7/3 (v/v), Temperature: 60°C,
$h_D = 12$, $h_L = 35$

Fig. 6-3. Principle of the chromatographic evaluation of the recognition properties of MIPs.

ization. The solvent must be chosen considering the stability of the monomer– template assemblies and whether it results in the porous structure necessary for a rapid kinetics in the interaction of the template with the binding sites. However if these criteria are satisfied, a robust material capable of selectively rebinding the template can be easily prepared and evaluated in a short time.

Table 6-1. Examples of racemates successfully resolved on MIPs.

Racemate[a]	Separation factor[b] α	Resolution factor[b] R_s	Note[c]	Reference
Amino acids				
Phenylalanine	1.6	1.5	d,e	[104]
Phenylglycine		0.98		
Tyrosine		0.75		
Amino acid derivatives				
Phenylalanine ethyl ester	1.3	ND		[15]
Phenylalanine anilide (PA)	4.9	1.2		[105]
Phenylalanine ethyl amide	2.0	0.5		[15]
p-Aminophenylalanine ethyl ester	1.8	0.8		[15]
Arginine ethyl ester	1.5	ND		[105]
Tryptophan ethyl ester	1.8	0.5		[105]
p-Aminophenylalanine anilide	5.7	0.9		[78]
Phenylalanine-N-methyl-anilide (PMA)	2.0	n.d.		[25]
Leucine-β-naphthylamide	3.8	0.7		[26]

Racemate[a]	Separation factor[b] α	Resolution factor[b] R_s	Note[c]	Reference
N,N'-Dimethyl-phenylalanine anilide	3.7	1.4		[26]
Proline anilide	4.5	1.0		[26]
Pyridylmethyl-phenylalanine anilide	8.4	1.1		[106]
Pyridoxyl-phenylalanine anilide	2.7	0.4		[106]
Cbz-Glutamic acid	2.5	2.9		[27]
Cbz-Aspartic acid	2.2	1.7		[27]
Cbz-Phenylalanine	2.3	3.1		[73]
Cbz-Alanine	1.9	–	TRIM	[107]
Cbz-Tyrosine	4.3	1.9	VPy-MAA	[108]
Boc-Tryptophan	4.4	1.9	VPy-MAA	[108]
Boc-Phenylalanine	2.0	1.5	VPy-MAA	[108]
Dansyl-Phenylalanine	3.2	1.6	VPy-MAA	[108]
Boc-Proline-*N*-hydroxysuccinimide ester	1.3	0.8		[27]
Acetyl-Tryptophan methyl ester	3.9	2.2		[108]
Diethyl-2-amino-3-phenyl-propylphosphonate	2.3	ND		[109]
Peptides				
Phenylalanylglycine anilide	5.1	0.5		[26]
Cbz-Ala-Ala-OMe	3.2	4.5	TRIM	[110]
Cbz-Ala-Gly-Phe-OMe	3.6	4.2	TRIM	[110]
N-Ac-Phe-Trp-OMe	3.3	> 2		[31]
Cbz-Asp-Phe-OMe	2.5		VPY-MAA	[111]
Commercial drugs				
Propranolol	2.8	1.3		[112]
Timolol	2.9	2.0		[112]
Metoprolol	1.08	1.2	TRIM[d]	[113]
Ephedrine	3.4	1.6		[29]
Naproxen	1.7	0.8	VPY	[114]
Ropivacaine	7.7 / 5.7 [f]		[d]	[113]
Carboxylic acids				
R-(–)-Mandelic acid	1.5	–	VPy	[108]
R-Phenylsuccinic acid	3.6	2.0	VPy	[108]
2-Phenylpropionic acid	high	high	PYAA/DVBd	[96]
Amines				
N-(3,5-dinitrobenzoyl)-methylbenzylamine	1.9	–	MAA/DPGL	[115]
(*R*)-α-methylbenzylamine	>1.5	1.0		[116]

[a]) Each racemate was applied on a polymer (ca. 0.1 μmol per gram dry polymer) imprinted with one antipode of the racemate. The standard mobile phase, consisting of acetonitrile containing various amounts of acetic acid, was used in most cases. Cbz = Carbobenzyloxy, Boc = t-butyloxycarbonyl.

[b]) α was calculated as the ratio of the capacity factor (k') of the template enantiomer to the capacity factor of its antipode. R_s is the resolution factor.

[c]) The polymers were prepared using MAA as functional monomer and EDMA as crosslinking monomer if not otherwise noted. VPY= 2- or 4-vinylpyridine; TRIM = trimethylolpropane trimethacrylate; DPGL = (*R*)-N,O-dimethacryloylphenylglycinol; PYAA = 3-(4-pyridinyl)acrylic acid.

[d]) The polymer was evaluated in capillary electrophoresis.

[e]) The polymer was imprinted with L-PA.

[f]) Migration times of the two enantiomers.

6.3 Structure-Binding Relationships

A large number of racemates have been successfully resolved on tailor-made MIC-SPs (Table 6-1). Using MAA as functional monomer, good recognition is obtained for templates containing Brönsted-basic or hydrogen bonding functional groups close to the stereogenic center. On the other hand, templates containing acidic functional groups are better imprinted using a basic functional monomer such as vinylpyridine. This emphasizes the importance of functional group complementarity when designing the MICSPs. Furthermore, the separation factors are high and higher than those observed for many of the widely used commercial CSPs [20]. However, the columns are tailor-made and the number of racemates resolved equals nearly the number of stationary phases, i.e. each column can resolve only a limited number of racemates. Although the separation factors are high, the resolution factors are low, but the performance can often be enhanced by running the separations at higher temperatures [15] and by switching to an aqueous mobile phase (Fig. 6-3) [21], or by performing the imprinting in situ in fused silica capillaries for use in capillary electrochromatography [22, 23]. At low sample loads, the retention on the MICSPs is extremely sensitive to the amount of sample injected, indicating overloading of a small amount of high energy binding sites. [24] Moreover the peaks corresponding to the template are usually broad and asymmetric. This is ascribed to the mentioned site heterogeneity together with a slow mass transfer (see Section 6.3.1).

Table 6-2. Examples of highly selective recognition by MIPs.

Template	$k'_L(1)$	$\alpha(1)$	$k'_L(2)$	$\alpha(2)$
1 a	6.6	4.2	1.05	1.07
2 a	1.7	1.4	2.1	2.0
1 b	2.4	2.0	0.9	1.3
2 b	0.4	1.1	0.8	2.3

The polymers were prepared by the standard procedure using MAA as functional monomer (see Fig. 6-2) as described elsewhere [25]. [a] Mobile phase: acetonitrile/acetic acid: 90/10 (v/v). Sample: 0.2 µmol racemate g^{-1}.
[b] Mobile phase: acetonitrile/water/acetic acid: 96.3/1.2/2.5 (v/v).

6.3.1 High Selectivity

MICSPs are often highly selective for their respective template molecule. This was the case for polymer imprinted with L-phenylalanine anilide (L-PA) and L-phenylalanine-*N*-methylanilide respectively (comparing a secondary and tertiary amide as template) (Table 6-2) [25]. The racemate corresponding to the template was well resolved on the corresponding MICSP, whereas the analogue racemate was less retained and only poorly resolved. Similar results were obtained when comparing a polymer imprinted with L-phenylalanine ethyl ester and one with its phosphonate analogue (Table 6-2) and have also been observed in comparisons of a primary (**1**) and a tertiary (**2**) amine, different in two amino methyl groups, two diacids, N-protected aspartic (**3**) and glutamic (**4**) acid, which differed only in one methylene group

in the alkyl chain [26, 27]. Pronounced discrimination of minor structural differences have also been reported in the imprinting of N-protected amino acids as (**5**) and (**6**) [28], aminoalcohols such as ephedrine (**7**) and pseudoephedrine (**8**) [29], monosaccharides [30] and peptides such as **9–11** [31]. Since the polymers imprinted with templates containing bulky substituents discriminated against those containing

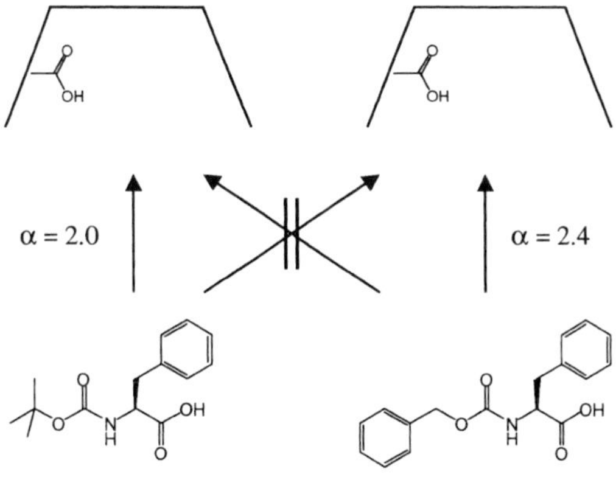

α = 2.0

5

α = 2.4

6

α = 3.2

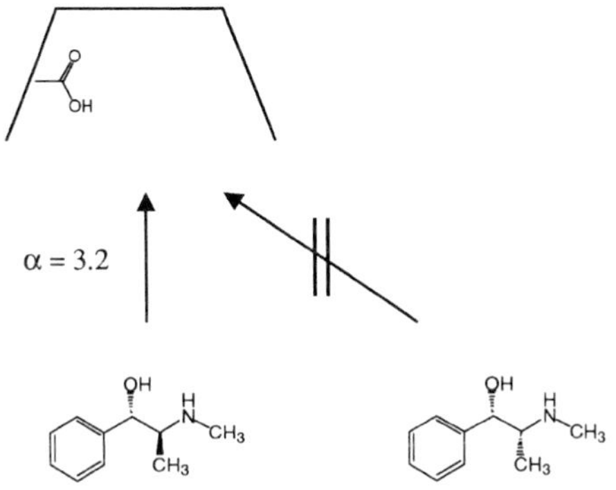

8

7

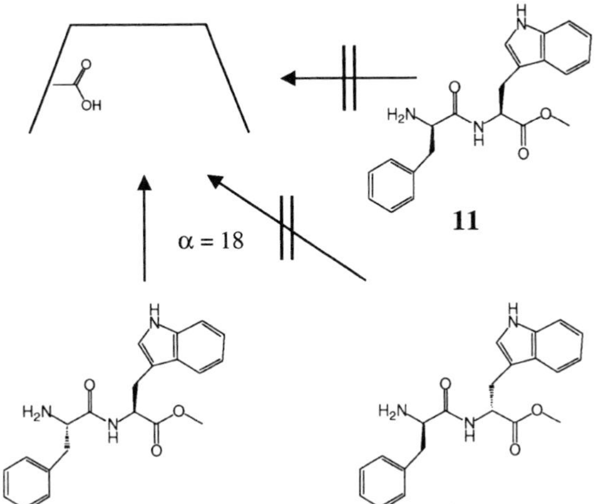

11

α = 18

9

10

smaller substituents, the recognition is not purely size exclusion but instead must be driven by shape complementarity between the site and the substrate, or conformational differences between the derivatives. It was concluded on the basis of ^{1}H-NMR nuclear Overhauser enhancement experiments and molecular mechanics calculations that L-PA and the *N*-methylanilide exhibit large conformational differences. Thus, the torsional angles between the anilide ring plane and the amide plane, as well as in the E-Z preference over the amide bond (Fig. 6-4) are different [25]. The low energy conformer of the anilide has the phenyl group in a *cis* conformation to the carbonyl oxygen with a torsional angle of about 30 °, whereas in the *N*-methylanilide the phenyl group is found in a *trans* conformation twisted almost 90 ° out of the amide plane. This will result in a different arrangement of the functional groups at the site. In this context it is interesting to note (Table 6-2) that the polymer imprinted with the *N*-methylanilide is less selective for its template, i.e. a lower separation factor is seen for the template compared to what is observed using the L-PA-imprinted polymer and furthermore, a significant separation of the enantiomers of D,L-PA is also observed. This can be explained considering the smaller space requirements of D,L-PA that thus can be forced into a conformation matching the site of the *N*-methylanilide.

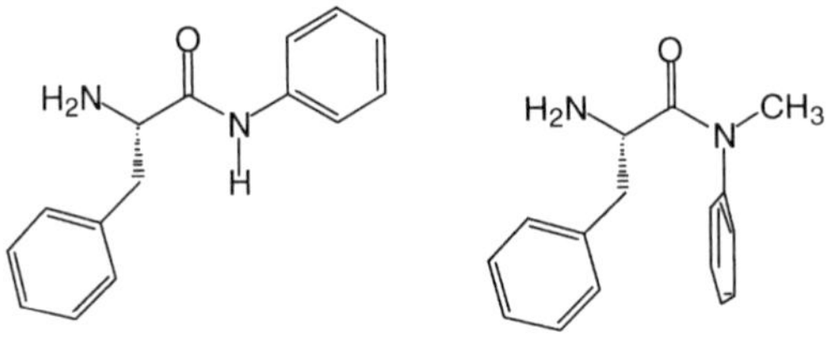

Fig. 6-4. Minimum energy conformations of L-PA and L-phenylalanine-*N*-methyl-anilide (L-PMA) based on molecular mechanics calculations and UV- and NMR-spectroscopic characterizations. (From Lepistö and Sellergren [25].)

Table 6-3. Resolution of amino acid derivatives on a MIP imprinted with L-phenylalanine anilide (L-PA).

Racemate	k'_L	α
Phenylalanine anilide	3.5	2.3
Tyrosine anilide	2.9	2.2
Tryptophan anilide	2.4	2.0
Phenylalanine *p*-nitroanilide	3.1	2.1
Leucine *p*-nitroanilide	2.1	1.6
Alanine *p*-nitroanilide	2.0	1.6

Data taken from reference [117].

6.3.2 Low Selectivity

Numerous examples of MICSPs that are capable of resolving more than the racemate corresponding to the template have been reported [17, 32]. In these cases some structural variations are tolerated without seriously compromising the efficiency of the separation. For instance, a polymer imprinted with L-phenylalanine anilide resolved amino acid derivatives with different side chains or amide substituents [17]. Anilides of all aromatic amino acids were here resolved as well as β-naphthylamides and *p*-nitroanilides of leucine and alanine (Table 6-3). Furthermore, in aqueous mobile phases, the free amino acid phenylalanine could also be base line resolved on an L-PA-imprinted polymer [32]. Apparently, substitution of groups that are not involved in potential binding interactions only leads to a small loss in enantioselectivity. Also it was noted that the dipeptide, D,L-phenylalanylglycine anilide was resolved, while glycyl-D,L-phenylalanine anilide was not. This observation emphasizes the importance of the spatial relationship between the functional groups at the sites, and indicates that substitutions made at some distance away from the center of chirality are allowed.

6.3.3 Studies of the Monomer–Template Solution Structures

To what extent do the solution complexes formed between the monomer and the template in solution reflect the architecture of the polymeric binding sites ? This question is important, since a thorough characterization of the monomer template assemblies may assist in deducing the structure of the binding sites in the polymer and thus have a predictive value. ^{1}H-NMR spectroscopy and chromatography were used to study the association between MAA and the template L-PA in solution as a mimic of the pre-polymerization mixture [15]. The ^{1}H-NMR chemical shifts of either the template or the monomer versus the amount of added MAA as well as the chromatographic retention of D,L-PA versus the amount of acid in the mobile phase, varied in accordance with the formation of multimolecular complexes between the template and the monomer in the mobile phase. A 1:2 template–monomer complex was proposed to exist prior to polymerization based on the modeled complex distribution curves. Based on these results, hydrogen bond theory, and the assumption that the solution structure was essentially fixed by the polymerization, a structure of the template bound to the site was proposed (Fig. 6-5). Since these initial studies, a number of other examples support this model, i.e. the recognition is due to functional group complementarity and a correct positioning of the functional groups in the sites as well as steric fit in the complementary cavity [33-36]. Rebinding to sites formed of residual nonextracted template have also been proposed as a contributing factor to the observed recognition [37]. In most imprinted systems however, rebinding selectivity or catalytic efficiency increase with increasing recovery of the template [38] and the Langmuir-type adsorption indicates a true receptor behavior [39].

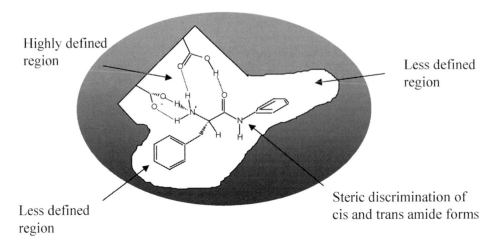

Highly defined region

Less defined region

Less defined region

Steric discrimination of cis and trans amide forms

Fig. 6-5. Model of the binding site for L-PA based on spectroscopic and chromatographic characterization of the prepolymerization monomer–template assemblies.

6.4   Adsorption Isotherms and Site Distribution

Adsorption isotherms can yield important information concerning binding energies, modes of binding and site distributions in the interaction of small molecule ligands with receptors [40]. In the case of MIPs, a soluble ligand interacts with binding sites in a solid adsorbent. The adsorption isotherms are then simply plots of equilibrium concentrations of bound ligand (adsorbate) versus concentration of free ligand. The isotherms can be fitted using various models where different assumptions are made. The most simple is the Langmuir-type adsorption isotherm (Equation (1)), where the adsorbent is assumed to contain only one type of site, where adsorbate–adsorbate interactions are assumed not to occur, and where the system is assumed ideal. This isotherm depends on two parameters: the saturation capacity (site density), q_s, and the adsorption energy, b [41, 42].

$$q = \frac{a_1 C}{1 + b_1 C} \tag{1}$$

$$q = \frac{a_1 C}{1 + b_1 C} + \frac{a_2 C}{1 + b_2 C} \tag{2}$$

$$q = aC^{1/n} \quad (a \text{ and } n = \text{numerical parameters}) \tag{3}$$

The bi-Langmuir model (Equation (2)) or tri-Langmuir model, the sum of two or three Langmuir isotherms, correspond to models that assume the adsorbent surface to be heterogeneous and composed of two or three different site classes, and finally the Freundlich isotherm model (Equation (3)) with no saturation capacity but instead a complete distribution of sites of different binding energies. Depending on the template-functional monomer system, the type of polymer, the conditions for its preparation and the concentration interval covered in the experiment the adsorption isotherms of MIPs have been well fitted with all the isotherm models [39, 43-45].

Thus, most MIPs suffer from a heterogeneous distribution of binding sites. In noncovalent imprinting, two effects contribute primarily to the binding site heterogeneity. Due to the amorphous nature of the polymer, the binding sites are not identical, but are somewhat similar to a polyclonal preparation of antibodies. The sites may for instance reside in domains with different crosslinking density and accessibility [46]. Secondly, this effect is reinforced by the incompleteness of the monomer–template association [15]. In most cases the major part of the functional monomer exists in a free or dimerized form, not associated with the template. As a consequence, only a part of the template added to the monomer mixture gives rise to selective binding sites. This contrasts with the situation in covalent imprinting [33, 45, 47] or stoichiometric noncovalent imprinting [48, 60, 90] where theoretically all of the template split from the polymer should be associated with a templated binding site. The poor yield of binding sites results in a strong dependence of selectivity and binding on sample load at least within the low sample load regime.

For determining the adsorption isotherm, the equilibrium concentrations of bound and free template must be reliably measured within a large concentration interval. Since the binding sites are part of a solid, this experiment is relatively simple and can be carried out in a batch equilibrium rebinding experiment or by frontal analysis.

One powerful technique for the study of the interactions between solutes and stationary phases and for the investigation of the parameters of these interactions is frontal analysis [49]. This method allows accurate determination of adsorption and kinetic data from simple breakthrough experiments, and the technique has proven its validity in a number of previous studies. This has also been used for estimating the adsorption energies and saturation capacities in the binding of templates to MIPs, but often the data have been modeled only at one temperature and graphically evaluated using a simple Langmuir mono site model which in most cases gives a poor fit of the data [50]. Furthermore, the breakthrough curves are interpreted assuming thermodynamic equilibrium, which is often an invalid assumption in view of the slow mass transfer in these systems. Rather, based on the mass balance equation and by assuming kinetic and isotherm values to best-fit isotherms and elution profiles obtained at different temperatures, a more accurate picture of the thermodynamics and mass transfer data can be obtained [49].

The isotherms for the two enantiomers of phenylalanine anilide were measured at 40, 50, 60 and 70 C, and the data fitted to each of the models given in Equations (1–3) [42]. The isotherms obtained by fitting the data to the Langmuir equation were of a quality inferior to the other two. Fittings of the data to the Freundlich and to the bi-Langmuir equations were both good. A comparison of the residuals revealed that the different isotherms of D-PA were best fitted to a bi-Langmuir model, while the isotherms for L-PA were slightly better fitted to a Freundlich isotherm model, particularly at low temperatures. However, at concentrations higher than 17 μm (4 × 10^{-3} g L^{-1}), the isotherm data of L-PA were equally well fitted to the Freundlich and to the bi-Langmuir isotherm models, suggesting the existence of binding sites with higher binding energies ($K > 50\,000$ M^{-1}). At 40 °C for L-PA, the binding constants and site densities are respectively 84 M^{-1} and ca. 90 μmol g^{-1} for the low-affinity sites, and 16 000 M^{-1} and 1 μmol g^{-1} for the high-affinity sites. For D-PA the respective values are 48 M^{-1} and 136 μmol g^{-1} for the low-affinity sites, and 5520 M^{-1} and 0.4 μmol g^{-1} for the high-affinity sites. These values agree well with those determined in previous studies [18]. In view of the small saturation capacities observed for D-PA on these sites at the other temperatures studied (50, 60, 70 °C) or after thermal annealing of the materials [24], the second site class appears to be specific for L-PA.

For preparative or semipreparative-scale enantiomer separations, the enantioselectivity and column saturation capacity are the critical factors determining the throughput of pure enantiomer that can be achieved. The above-described MICSPs are stable, they can be reproducibly synthesized, and they exhibit high selectivities – all of which are attractive features for such applications. However, most MICSPs have only moderate saturation capacities, and isocratic elution leads to excessive peak tailing which precludes many preparative applications. Nevertheless, with the L-PA MICSP described above, mobile phases can be chosen leading to acceptable resolution, saturation capacities and relatively short elution times also in the isocratic mode (Fig. 6-6).

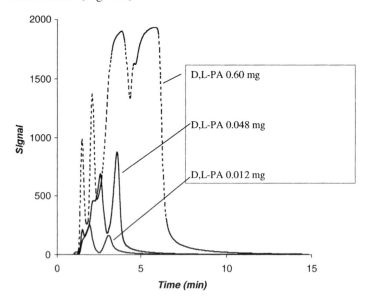

Fig. 6-6. Overload elution profiles of D,L-PA injected on a column (125 4 mm) packed with the L-PA imprinted stationary phase used in Fig. 6-5. Mobile phase: MeCN: TFA (0.01 %): H_2O (2.5 %). The tendency for fronting and the increase in retention with sample load is attributed in part to saturation of the mobile phase modifier.

6.5 Adsorption-Desorption Kinetics and Chromatographic Band Broadening

For most applications of specific molecular recognition elements, a rapid association dissociation kinetics in the ligand receptor binding is important. In chemical sensors the response time depends on the association rate between the sensor-bound receptor and the target analyte, whereas the dissociation rate determines if, and how quickly, the sensor can be regenerated [51]. The kinetics thus influences the sample throughput of the analysis, i.e. how many samples that can be analyzed in a certain time interval. Furthermore, in catalysis the binding kinetics will determine the maximum rate of the chemical transformation, and in chromatographic separations it will influence the spreading of the chromatographic peaks.

When a solute band passes a chromatographic column it is broadened continuously due to various dispersion processes [52]. These include processes that show little or no flow rate dependence, such as eddy-diffusion or extracolumn effects and flow rate-dependent processes such as axial diffusion, mass transfer processes including mobile phase, intraparticle and stationary phase diffusion and slow kinetic processes upon interaction with the stationary phase. Other factors such as nonlinear binding isotherms and slow desorption kinetics instead affect the shape of the peak [53]. Altogether, these processes counteract the separation of two compounds and lead to lower resolutions. An understanding of their origin is important in order to improve the separations as well as to gain insight into the kinetics and mechanism of solute retention.

The dependence of the chromatographic parameters on flow rate and sample load was studied in enantiomer separations of d- and l-phenylalanine anilide (D,L-PA) on L-PA-imprinted chiral stationary phases (CSPs) [54].Using a thermally annealed stationary phase, a strong dependence of the asymmetry factor (A_s) of the l-form on sample load and a weak dependence on flow rate suggested that column overloading contributed strongly to the peak asymmetry (Fig. 6-7). This is to be expected in view of the site heterogeneity discussed in the previous section. However, slow kinetic processes is another contributing factor to the pronounced band broadening in the chromatography using MIP-based columns. In view of the high binding constants observed for MIPs, the desorption rate at the high-energy binding sites should be much slower than that at the low-energy sites. The mass transfer rate coefficients, estimated using a MIP prepared in dichloromethane as diluent, were small and strongly dependent on the temperature and concentration, in particular the rate coefficients corresponding to the imprinted L-enantiomer [42]. Recent related studies of the retention mechanism of both enantiomers of dansyl-phenylalanine on a dansyl-L-phenylalanine MICSP led to similar conclusions [55], although these processes are strongly dependent on the system studied, i.e. template-monomer system, crosslinking monomer, porogen and method of polymerization.

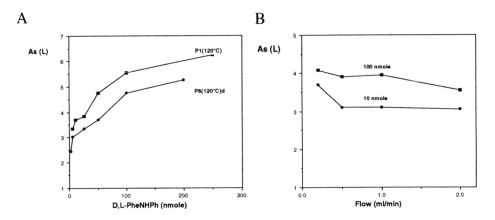

Fig. 6-7. Asymmetry factor (A_s) of the L-enantiomer versus sample load (A) and versus flow rate (B) on L-PA-imprinted polymers. Flow rate: 1.0 ml min^{-1}. Mobile phase: MeCN/[potassium phosphate 0.05 M, pH 7] (7/3, v/v).

6.6 Factors to Consider in the Synthesis of MICSPs

In spite of the fact that molecular imprinting allows materials to be prepared with high affinity and selectivity for a given target molecule, a number of limitations of the materials prevent their use in real applications. The main limitations are:

1 Binding site heterogeneity
2 Extensive nonspecific binding
3 Slow mass transfer
4 Bleeding of template
5 Low sample load capacity
6 Unpractical manufacturing procedure
7 Poor recognition in aqueous systems
8 Swelling–shrinkage: may prevent solvent changes
9 Lack of recognition of a number of important compound classes
10 Preparative amounts of template required

It is clear that improvements aiming at increasing the yield of high-energy binding sites or modifying the site distribution in other ways will have a large impact on the performance of the materials (affecting limitations 1, 2, 4 and 5). The strategies adopted to achieve this have been focusing either on prepolymerization measures, aimed at stabilization of the monomer template assemblies prior to polymerization, or postpolymerization measures aimed at modifying the distribution of binding sites by either chemical or physical means. The most important of these factors will now be discussed, together with techniques allowing their optimization.

6.6.1 Factors Related to the Monomer-Template Assemblies

It is of obvious importance that the functional monomers interact strongly with the template prior to polymerization, since the solution structure of the resulting assemblies presumably defines the subsequently formed binding sites. By stabilizing the monomer–template assemblies, it is possible to achieve a large number of imprinted sites. At the same time, the number of nonspecific binding sites will be minimized, since free functional monomer not associated with the template is likely to be accessible for binding. Considering one particular binding site, the following factors have been identified that are likely to affect the recognition properties of the site (Fig. 6-8).

• Choice of the functional monomer

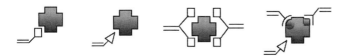

• Stabilization of monomer-template assemblies

• Template size and shape

• Monomer-template conformational rigidity

Fig. 6-8. Factors affecting the recognition properties of MIPs related to the monomer template assemblies.

The strength and positioning of the monomer–template interactions are of importance for materials with good molecular recognition properties to be obtained. The broad applicability of MAA as a functional monomer is related to the fact that the carboxylic acid group serves well as a hydrogen bond and proton donor and as a hydrogen bond acceptor [56]. In aprotic solvents such as in acetonitrile carboxylic acids and amine bases form contact hydrogen-bonded assemblies where the association strength for a given acid increases with the basicity of the base [57]. Thus, templates containing Brönsted-basic or hydrogen-bonding functional groups are potentially suitable templates for the MAA/EDMA system [15]. Furthermore, more stable cyclic hydrogen bonds can form with templates containing acid [27], amide[26] or functionalized nitrogen heterocycles [39, 44]. The potential for a given monomer template pair to produce templated sites can be predicted by measuring the stability constants, e.g. by spectroscopic techniques, in a homogeneous solution mimicking the monomer mixture prior to polymerization [15]. This can ultimately be used as a preliminary screening procedure to search for suitable functional monomers. Thus, estimated solution association constants can be correlated with the heterogeneous binding constants determined for the polymer (Table 6-4). For the prepolymerization complexes discussed thus far, the electrostatic interactions are sensitive to the presence of polar protic solvents. One exception is the complex formed between carboxylic acids and guanines or amidines [58, 59]. Here, cyclic hydrogen-bonded ion-pairs are formed with stability constants that are order of magnitude higher than those previously discussed (Table 6-4). This allows amidines such as pentamidine (**12**) to be imprinted using *iso*-propanol–water as a porogenic solvent mixture, resulting in polymers that bind pentamidine strongly in aqueous media [60, 90].

12

Table 6-4. Association constants for complexes between carboxylic acids and nitrogen bases in aprotic solvents and corresponding association constants and site densities for binding of the base to a molecularly imprinted polymer.

Acid	Base	Solvent	K_a (m^{-1})	n (μmol g^{-1})	Reference
Acetic acid	Atrazine	CCl$_4$	210	–	[80]
Butyric acid	9-Ethyladenine	CDCl$_3$	(1) 114	–	[79]
			(2) 41		
4-Methylbenzoic acid	(**13**)	CDCl$_3$	>10^6	–	[59]
PMAA	Atrazine	CHCl$_3$	(1) 8.3 × 10^4	20	[118]
			(2) 1.0 × 10^4	40	
PMAA	9-Ethyladenine	CHCl$_3$	(1) 7.7 × 10^4	20	[39]
			(2) 2.4 × 10^3	86	

PMAA refers to polymers imprinted with respective base using MAA as functional monomer.

Apart from the successful imprinting discussed above, the recognition for many templates is far from that is required for the particular application, even after careful optimization of the other factors affecting the molecular recognition properties. Often, a large excess of MAA in the synthesis step is required for recognition to be observed and then only in solvents of low to medium polarity and hydrogen bond capacity [61]. In fact, in these cases the optimum rebinding solvent is often the solvent used as porogen.[62] Thus, the polymer exhibits memory for the template as well as the porogen. Moreover, the excess of functional monomer results in a portion of the functional monomer not being associated with imprinted sites. These sites interact nonselectively with solutes binding to carboxylic acids and limit the degree of separation that can be achieved. Hence MAA is not a universal monomer. Instead, for the recognition of any given target molecule access to functional monomers targeted towards structural features, specific for particular compounds or classes of compounds are required.

Based on the structural features of the templates that generate good sites, an interesting possibility would be to incorporate these structures in new functional monomers for the recognition of carboxylic acids. This concept is somewhat similar to the reciprocity concept in the design of chiral stationary phases [63]. Thus, Wulff et al. synthesized *N,N'*-substituted *p*-vinylbenzamidines (**13**) and showed that these monomers could be used to generate high-fidelity sites for the molecular recognition of chiral carboxylic acids [59]. The binding is here strong enough to provide efficient recognition also in aqueous media. Furthermore, due to strong binding the functional monomer is quantitatively associated with the template, thus minimizing the nonspecific binding. Functional group complementarity is thus the basis for the choice of functional monomer. The search for the optimal structural motif to complement the template functionality is preferentially guided by results from the area of host–guest chemistry and ligand– receptor chemistry. Thus cyclodextrins have been used to template binding sites for cholesterol [64] or to enhance the selectivity in the imprinting of enantiomers of amino acids [65]. Based on metal ion co-ordination of amino acids and *N*-(4-vinylbenzyl)iminodiacetic acid (**14**), imprinting and subsequent chiral separation of free amino acids in aqueous solutions has also been possible [66].

13

14

15

Based on chiral functional monomers such as (**15**), MICSPs can be prepared using a racemic template. Thus, using racemic *N*-(3,5-dinitrobenzoyl)-a-methylbenzylamine (**16**) as template, a polymer capable of racemic resolution of the template was obtained [67]. Another chiral monomer based on L-valine (**17**), was used to prepare MIPs for the separation of dipeptide diastereomers [68]. In these cases the configurational chirality inherent in the pendant groups of the polymer are to some extent themselves chiral selectors, and the effect of imprinting is merely to enhance the selectivity. A good example of this was shown in the imprinting of *N*-benzyl-L-valine as a bidentate ligand to a styrene-based chiral cobalt complex (**18**) [69]. The strong enantioselectivity of the imprinted polymer should here be viewed with respect to the enantioselectivity of the control polymer.

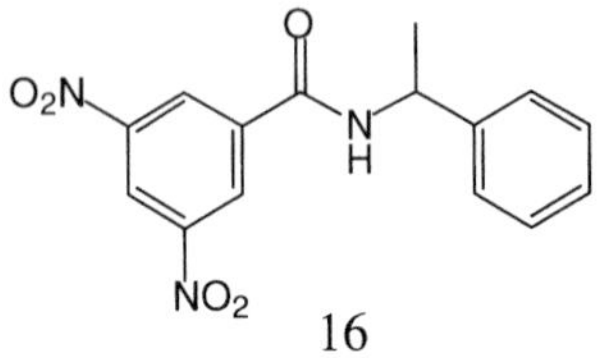

16

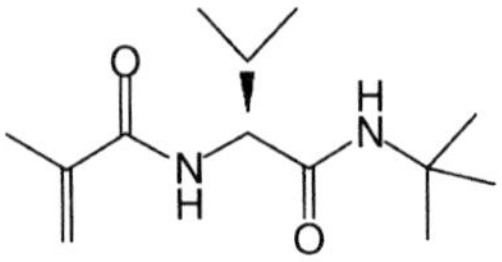

17

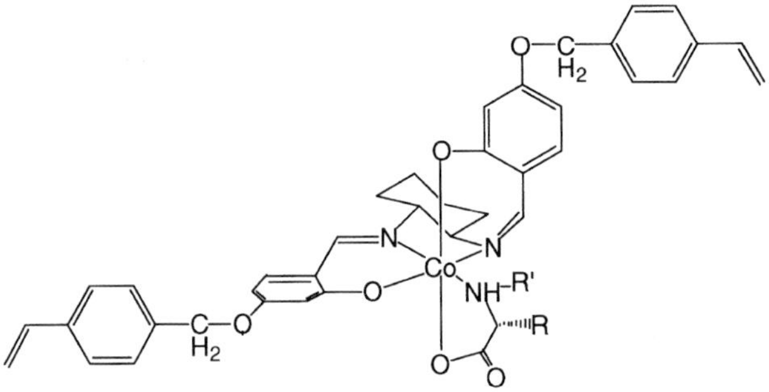

18

Thus, enhanced separations can be obtained using chiral selectors with configurational chirality in combination with molecular imprinting. What about selectors with conformational chirality ? Can chirality be induced by molecular imprinting ? This concept was elegantly demonstrated by Welch using a brush-type stationary phases containing a slowly interconverting (in the order of a day) racemic atropisomer (**19**) as imprintable selector (Fig. 6-9) [70]. Leaving the selector in contact with an enantiomerically pure template molecule (**20**) for more than 2 weeks led to induction of the most stable selector selectand complex. After washing out the selectand, the selector could be used to separate the racemate of **20** with similar separation factors as obtained using the reciprocal phase. However due to interconversion, the CSP racemized over a period of 2–3 days, a period that possibly can be extended by storing the CSP at low temperatures. Also mentioned was the interesting possibility of using a selection of slowly interconverting selectors to achieve a broadly applicable system for atropisomer-based imprinting.

Alternative approaches to imprint peptides via strong monomer template association have recently been reported, although no results of the chromatographic application of these phases have been shown. Strong complexation inducing a β-sheet conformation was possible using a designed functional monomer (**21**) [71]. Peptides

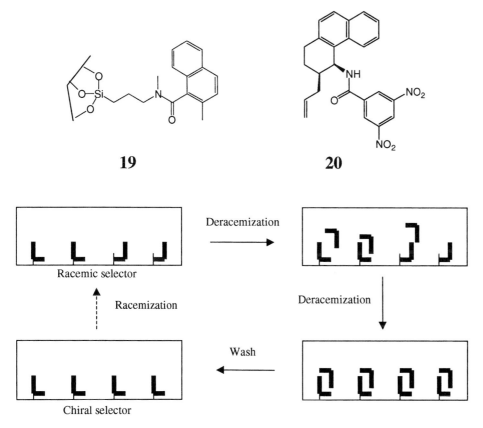

19   **20**

Fig. 6-9. Imprintable brush-type selectors. (From Welch [70].)

21

can also be imprinted via a sacrificial spacer approach which potentially will result in a high yield of templated sites exhibiting pronounced selectivity towards the target peptide (**22**) [72].

Considering functional group complementarity, other commodity monomers may also be used. Thus for templates containing acid groups, basic functional monomers are preferably chosen. The 2- or 4-vinylpyridines (VPY) are particularly well-suited for the imprinting of carboxylic acid templates and provide selectivities of the same order as those obtained using MAA for basic templates [73, 74]. These polymers are, however, susceptible to oxidative degradation and require special handling.

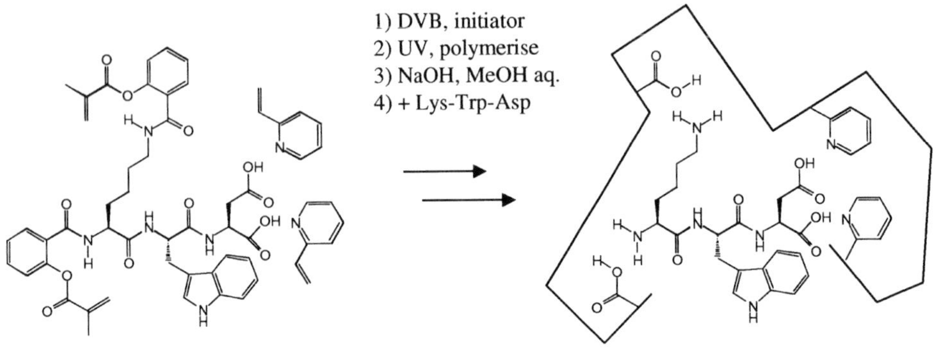

1) DVB, initiator
2) UV, polymerise
3) NaOH, MeOH aq.
4) + Lys-Trp-Asp

22

In the imprinting of carboxylic acids and amides, high selectivities are also seen using acrylamide (AAM) as functional monomer [28]. Furthermore, combinations of two or more functional monomers, giving terpolymers or higher polymers, have in a number of cases resulted in better recognition ability than the recognition observed from the corresponding co-polymers [67, 73-75]. These systems are particularly complex when the monomers constitute a donor–acceptor pair, since monomer–monomer association will compete strongly with template–monomer association if neither of the monomers has a particular preference for the template. In a recent series of papers by the group of Liangmo, careful optimization showed that a combination of acrylamide and 2-vinylpyridine gave significantly higher enantioselectivities in the imprinting of N-protected amino acids than the combination of 2-vinylpyridine with MAA (**23**) [76]. Furthermore, better results were obtained using acetonitrile as the porogen, in contrast to other systems where solvents of lower polarity (e.g. toluene, CH_2Cl_2, $CHCl_3$) give the best results. These results show that adequate performance can only be achieved after careful optimization where the related factors are systematically varied.

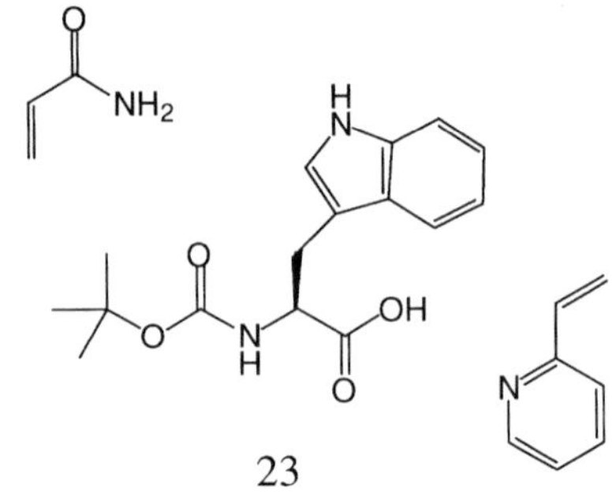

23

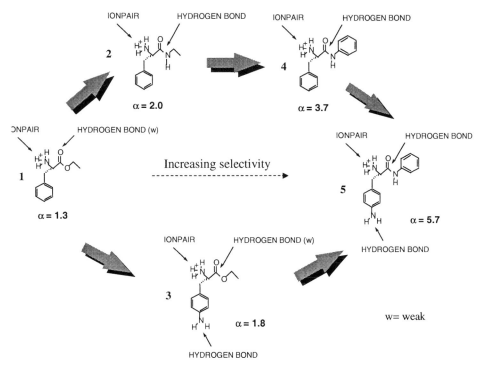

Fig. 6-10. Influence of the number of basic interaction sites of the template versus the separation factor measured in chromatography for the corresponding racemate. The templates were imprinted using MAA as functional monomer by thermochemical initiation at 60/90/120 °C (24 h at each temperature) and using acetonitrile as porogen. (From Sellergren et al. [15].)

6.6.2 Influence of the Number of Template Interaction Sites

Molecular recognition in the biological machinery takes place by the combination of several complementary weak interactions between a biological binding site and the molecule to be bound [77]. A larger number of complementary interactions will increase the strength and fidelity in the recognition. Thus, templates offering multiple site of interaction for the functional monomer are likely to yield binding sites of higher specificity and affinity for the template [12]. One example of this effect was observed in a study of the molecular imprinting of enantiomers of phenylalanine derivatives (Fig. 6-10) [15, 78]. Starting with L-phenylalanine ethyl ester (**1**) as the template, interactions with carboxylic acids in acetonitrile should consist of the ammonium carboxylate ion pair, as well as a weak ester– carboxylic acid hydrogen bond (indicated by arrows). By replacing either the ester group with the stronger hydrogen bonding amide group in (**2**), or by introducing an aromatic amino group as in (**3**) – which allows an additional hydrogen bond interaction with another carboxylic acid group – the enantiomeric selectivity increased. In L-PA (**4**), where the ethyl amide substituent has been replaced by an anilide group, an additional increase in selectivity is seen. Combining the structural modifications in one molecule, *p*-amino-phenylalanine-anilide (**5**), the highest separation factor was obtained. Similar observations have been made in the imprinting of a number of different classes of compounds and thermodynamic evidence for the existence of multiple additive interactions in the sites have been provided [35]. In the search for optimal synthetic conditions for MIPs, useful start-up information can be obtained from the vast literature existing on solution studies of molecular interactions and molecular recognition [For example see: 79-81].

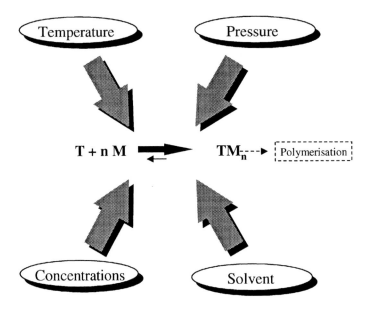

Fig. 6-11. Stabilization of monomer template assemblies by thermodynamic considerations.

6.6.3 Thermodynamic Factors

An important part of the optimization process is the stabilization of the monomer–template assemblies by thermodynamic considerations (Fig. 6-11). The enthalpic and entropic contributions to the association will determine how the association will respond to changes in the polymerization temperature [18]. The change in free volume of interaction will determine how the association will respond to changes in polymerization pressure [82]. Finally, the solvent's interaction with the monomer–template assemblies relative to the free species indicates how well it will stabilize the monomer–template assemblies in solution [16]. Here each system must be optimized individually. Another option is simply to increase the concentration of the monomer or the template. In the former case, a problem is that the crosslinking as well as the potentially nonselective binding will increase simultaneously. In the latter case, the site integrity will be compromised. The above factors have been studied for theL-PA model system. In aprotic media of low polarity, MAA and templates containing polar functional groups are only weakly solvated, and the interactions holding the monomer template assemblies together are mainly electrostatic in nature [77]. In such cases the association of the monomer and template is associated with a loss of one set of rotational and translational degrees of freedom which leads to a net decrease in entropy [83]. From this follows that the interaction is weakened at increasing temperature. On the other hand, when the monomer and the template is more strongly solvated, the association may lead to release of part of the solvent shell, leading in turn to a net increase in rotational and translational entropy. In this case the interaction will be favored by increasing the temperature.

6.6.4 Factors Related to Polymer Structure and Morphology

For the formation of defined recognition sites, the structural integrity of the monomer–template assemblies must be preserved during polymerization to allow the functional groups to be confined in space in a stable arrangement complementary to the template. This is achieved by the use of a high level of crosslinking, usually >80 % [18]. The role of the polymer matrix, however, is to contain the binding sites not only in a stable form but also in an accessible form (Fig. 6-12). Porosity is achieved by carrying out the polymerization in presence of a porogen. Most of the crosslinked network polymers used for molecular imprinting have a wide distribution of pore sizes associated with various degrees of diffusional mass transfer limitations and a different degree of swelling. Based on the above criteria, i.e. site accessibility, integrity, and stability, the sites can be classified according to different types. The sites associated with meso- and macro-pores (>20 Å) (sites A and B in Fig. 6-12) are expected to be easily accessible compared to sites located in the smaller micropores (<20 Å) (sites C) where the diffusion is slow. The number of the latter may be higher since the surface area, for a given pore volume, of micropores are higher than that of macropores. One undesirable effect of adding an excess of template is the loss of site integrity due to coalescence of the binding sites, which is related to the extent of template selfassociation. The optimum amount of template is usually about 5 % of the total amount of monomer, but can be higher when trivinyl monomers such as TRIM (**24**) are used as crosslinkers, where a larger fraction of functional monomer is used [84]. In this case higher sample load capacities have been observed. The amount of template is of course also limited by the solubility and availability of the template, although recycling is possible.

24

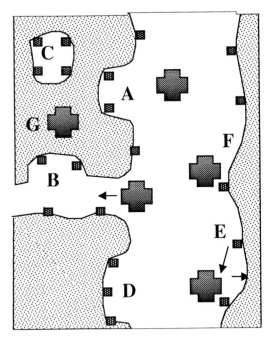

Fig. 6-12. Different types of binding sites in polymers containing micro- (site B), meso- and macropores (site A); C) Embedded site, D) Site complementary to dimer or multimer, E) Induced binding site, F) Nonselective site, G) Residual template.

Often the materials swell to different extents depending on the type of diluent. The swelling is here normally high in solvents and low in nonsolvents for the polymer. Unfortunately, this may lead to large changes in the accessibility and density of the binding sites when the solvent is changed [16].

6.7   Methods for Combinatorial Synthesis and Screening of Large Numbers of MIPs

For a complete optimization of all factors, the above-described procedure is not practical. In order to perform this rapidly, parallel synthesis and screening techniques must be developed. These can consist of a scaled-down version of the MIPs in vials that can be handled automatically and analyzed in situ (Fig. 6-13) [85, 86].

The principle was demonstrated using triazine herbicides as templates and by varying the type of functional monomer and the monomer composition. With a final batch size of ca. 40 mg of monomer, the consumption of monomers and template is significantly reduced and the synthesis and evaluation can take place in standard high-performance liquid chromatography (HPLC) autosample vials. After synthesis,

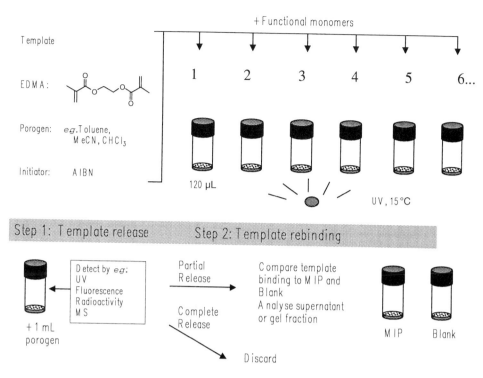

Fig. 6-13. Combinatorial imprinting technique suitable for automation.

the primary assessment is based on quantitative HPLC or UV-absorbance analysis of the amount of template released from the polymer in the porogenic solvent. Thus in the case of a rapid and quantitative release the resulting polymer cannot be expected to rebind a significant amount of the template, and may thus be discarded. After having established useful functional monomers, a secondary screening for selectivity is performed. Here, the rebinding of the template to the MIPs was investigated in parallel to the rebinding to a corresponding control nonimprinted MIP [86]. Alternatively, an internal standard, structurally related to the template, may be added and the differential binding investigated [85]. An important question is whether the equilibrium rebinding results reflect the selectivity observed when investigating an upscaled batch in the chromatographic mode [87]. This was shown in the case of the triazines, but for other systems suffering from particularly slow mass transfer this may not be the case. Here, chiral resolution is observed only at low flow rates.

6.8 New Polymerization Techniques

As indicated above, MIPs have so far been prepared in the form of continuous blocks that need to be crushed and sieved before use. This results in a low yield of irregular particles, a high consumption of template, and a material exhibiting low chromatographic efficiency. There is therefore a need for MI-materials that can be prepared in high yield in the form of regularly shaped particles with low size dispersity and a controlled porosity. These are expected to be superior in terms of mass transfer characteristics and sample load capacity compared to the materials obtained from the monolith approach. However, the results obtained so far using alternative approaches, although showing some improvements, have been disappointing.

Bead-sized MIPs have been previously prepared through suspension polymerization techniques either using fluorocarbons (Fig. 6-14) [88] or water [89] as continuous phase, dispersion polymerization or precipitation polymerization [90, 91]. This resulted in spherical particles of a narrow size distribution. These procedures have the limitation of being sensitive to small changes in the manufacturing conditions and the type of solvents and polymerization conditions that can be applied, but once appropriate conditions have been found they should offer an economic alternative for up-scaling. An alternative to this procedure is the coating of preformed support materials [92-94]. MIPs have been prepared as grafted coatings on metal oxide supports [92, 93] on organic polymer supports [94] and on the walls of fused silica capillaries [95-97]. These techniques however involve many steps and are thus associated with larger batch-to-batch variations. In addition, problems appear in achieving homogeneous coatings and to suppress secondary interactions with the support surface.

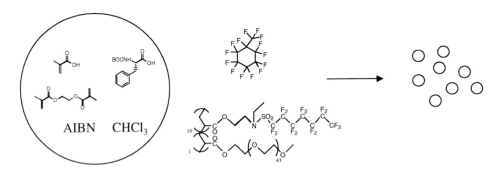

Fig. 6-14. Suspension polymerization technique for noncovalent imprinting.

Much effort has been devoted to the development of a multi-step swelling polymerization technique using water as suspension medium [98]. This has resulted in polymers showing similar selectivities but slightly improved mass transfer characteristics compared with the corresponding monolithic polymers. Of particular rele-

vance for bioanalytical applications was the functionalization of the outer surface of
a polymer imprinted with (S)-naproxen with a hydrophilic polymer layer (Fig. 6-15).
This led to a slight decrease in the separation efficiency, but allowed on the other
hand direct injection of plasma samples on the columns.

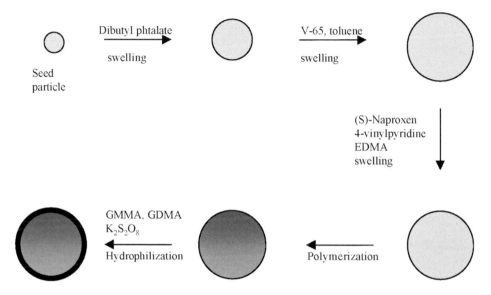

Fig. 6-15. Synthetic scheme of surface-modified MIP for (S)-naproxen. V65 = 2,2'-azobis(2,4-dimethylvaleronitrile); GMMA = glycerolmonomethacrylate; GDMA = glyceroldimethacrylate.

6.9 Other Separation Formats

As mentioned in the introduction, due to the high enantioselectivities exhibited by
the imprinted chiral phases, applications in batch-, SMB-, bubble- or membrane-
based separation processes may become attractive. The concept of applying MICSPs
for bubble fractionation of enantiomers was demonstrated recently [99]. This separation principle can be useful for separations of large amounts of material at very
low costs, and is an important technique for concentrating sulfide ores. For this process to be practical a high enrichment factor is needed and the chiral collector should
be easy to recycle. This is the case of solid collectors such as imprinted polymers
which also have the benefit of high robustness. Thus L-PA-imprinted polymer particles of less than 20 μm adhered to air bubbles and were effectively transported to the
top of the bubble column (Fig. 6-16). The particles were first pre-equilibrated with
a solution of the racemate, and then added to the separator. Here, they separated after
bubble flotation to the top of the column. Enantiomerically enriched compound was
then obtained by washing of the particles that in turn could be recycled. By using

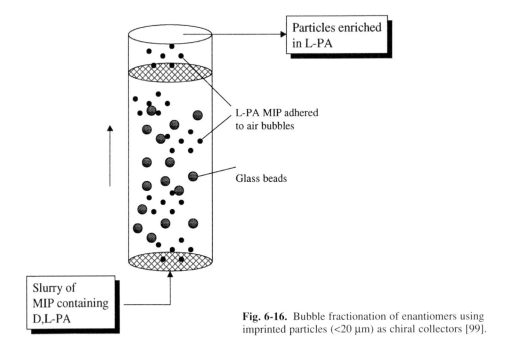

Particles enriched in L-PA

L-PA MIP adhered to air bubbles

Glass beads

Slurry of MIP containing D,L-PA

Fig. 6-16. Bubble fractionation of enantiomers using imprinted particles (<20 μm) as chiral collectors [99].

fine floating particles selective for one enantiomer, and large sinking particles selective for the opposite enantiomer, the efficiency of this process can most likely be enhanced.

A number of studies have recently been devoted to membrane applications [8, 100-102]. Yoshikawa and co-workers developed an imprinting technique by casting membranes from a mixture of a Merrifield resin containing a grafted tetrapeptide and of linear co-polymers of acrylonitrile and styrene in the presence of amino acid derivatives as templates [103]. The membranes were cast from a tetrahydrofuran (THF) solution and the template, usually N-protected d- or l-tryptophan, removed by washing in more polar nonsolvents for the polymer (Fig. 6-17). Membrane applications using free amino acids revealed that only the imprinted membranes showed detectable permeation. Enantioselective electrodialysis with a maximum selectivity factor of ca. 7 could be reached, although this factor depended inversely on the flux rate [7]. Also, the transport mechanism in imprinted membranes is still poorly understood.

In summary, the present limitations in saturation capacities and selectivity of imprinted polymers preclude their applications in the above-mentioned preparative separation formats.

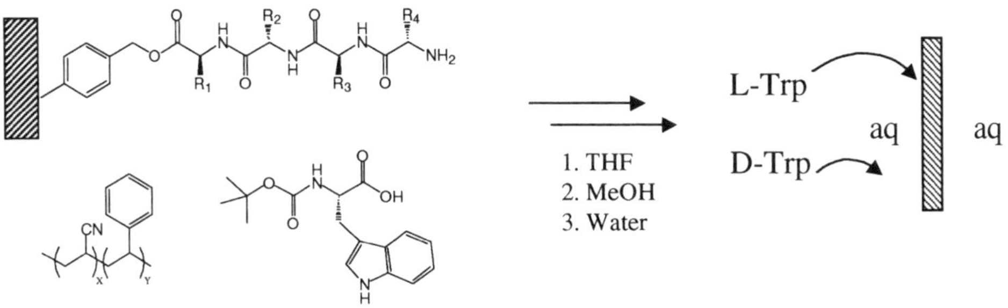

Fig. 6-17. Cast-imprinted membranes. (From Yoshikawa et al. [103].)

6.10 Conclusions

A number of conditions will directly influence the development of a new MICSP. The availability of the template in preparative amounts will determine whether it will have to be recycled, or a template analogue must be used. The latter alternative should also be considered in cases where the template is unstable or poorly soluble in the monomer mixture. Depending on the format of the separation, the polymer must meet certain requirements. If the material is to be used as a HPLC stationary phase, then monodisperse spherical particles are desirable and rapid adsorption–desorption of the template to the sites is necessary for high-performance separations. However, broad and asymmetric band shapes and low saturation capacities due to the heterogeneous distribution of binding sites and slow mass transfer processes are important problems that strongly limit the possible applications of these phases in analytical and preparative chromatography. The use of imprinted polymers in a foam flotation apparatus or in membrane separations have been demonstrated, although probably also here no viable application can be expected in the near future. Nevertheless, with designed functional monomers, new polymerization techniques and combinatorial synthesis and screening techniques, MICSPs that meet the above-mentioned requirements may soon be a reality.

Acknowledgment

The author is grateful to Dr. Francesca Lanza for assistance in preparing the manuscript.

References

[1] R. McCague, G. Casy, in G. P. Ellis, D. K. Luscombe (eds.): *Progress in Medicinal Chemistry, Vol. 34*, Elsevier, Amsterdam 1997, p. 204–261.

[2] B. Sellergren, B. Ekberg, P.-Å. Albertsson, K. Mosbach, *J. Chromatogr. 450* (1988) 277–280.

[3] B. Ekberg, B. Sellergren, P.-Å. Albertsson, *J. Chromatogr. 333* (1985) 211–214.

[4] D. W. Armstrong, R. Menges, I. Wainer, *J. Liq. Chromatogr. 13* (1990) 3571–3581.

[5] G. Cao, M. E. Garcia, M. Alcalá, L. F. Burgess, T. E. Mallouk, *J. Am. Chem. Soc. 114* (1992) 7574–7575.

[6] W. H. Pirkle, E. M. Doherty, *J. Am. Chem. Soc. 111* (1989) 4114–4116.

[7] M. Yoshikawa, T. Fujisawa, J. Izumi, *Macromol. Chem. Phys. 200* (1999) 1458–1465.

[8] A. Dzgoev, K. Haupt, *Chirality 11* (1999) 465–469.

[9] D. W. Armstrong, J. M. Schneiderheinze, Y.-S. Hwang, B. Sellergren, *Anal. Chem. 70* (1998) 3304–3314.

[10] L. Miller, C. Orihuela, R. Fronek, D. Honda, O. Dapremont, *J. Chromatogr. 849* (1999) 309–317.

[11] R. A. Bartsch, M. Maeda, Molecular and ionic recognition with imprinted polymers, *ACS Symposium Series 703*, Oxford University Press, Washington 1998.

[12] G. Wulff, *Angew. Chem., Int. Ed. Engl. 34* (1995) 1812–1832.

[13] K. Mosbach, *Trends Biochem. Sci. 19* (1994) 9–14.

[14] J. M. G. Cowie, *Polymers: Chemistry & Physics of modern materials*, Blackie and Son Ltd., Glasgow 1991.

[15] B. Sellergren, M. Lepistö, K. Mosbach, *J. Am. Chem. Soc. 110* (1988) 5853–5860.

[16] B. Sellergren, K. J. Shea, *J. Chromatogr. 635* (1993) 31.

[17] D. J. O'Shannessy, B. Ekberg, L. I. Andersson, K. Mosbach, *J. Chromatogr. 470* (1989) 391–399.

[18] B. Sellergren, *Makromol. Chem. 190* (1989) 2703–2711.

[19] L. R. Snyder, J. J. Kirkland, *Introduction to Modern Liquid Chromatography*, Wiley, US 1979.

[20] G. Subramanian, A practical approach to chiral separations by liquid chrom. VCH, Weinheim 1994.

[21] B. Sellergren, K. J. Shea, *J. Chromatogr. A 654* (1993) 17–28.

[22] K. Nilsson, J. Lindell, O. Norrlöw, B. Sellergren, *J. Chromatogr. 680* (1994) 57.

[23] S. Nilsson, L. Schweitz, M. Petersson, *Electrophoresis 18* (1997) 884–890.

[24] Y. Chen, M. Kele, P. Sajonz, B. Sellergren, G. Guiochon, *Anal. Chem. 71* (1999) 928–938.

[25] M. Lepistö, B. Sellergren, *J. Org. Chem. 54* (1989) 6010–6012.

[26] L. I. Andersson, D. J. O'Shannessy, K. Mosbach, *J. Chromatogr 513* (1990) 167–179.

[27] L. I. Andersson, K. Mosbach, *J. Chromatogr. 516* (1990) 313–322.

[28] C. Yu, K. Mosbach, *J. Org. Chem. 62* (1997) 4057–4064.

[29] O. Ramström, C. Yu, K. Mosbach, *J. Mol. Recognit. 9* (1996) 691–696.

[30] A. G. Mayes, L. I. Andersson, K. Mosbach, *Anal. Biochem. 222* (1994) 483–488.

[31] O. Ramström, I. A. Nicholls, K. Mosbach, *Tetrahedron: Asymmetry 5* (1994) 649–656.

[32] J.-M. Lin, T. Nakagama, T. Uchiyama, T. Hobo, *J. Pharm. Biomed. Anal. 15* (1997) 1351–1358.

[33] K. J. Shea, T. K. Dougherty, *J. Am. Chem. Soc. 108* (1986) 1091–1093.

[34] K. J. Shea, D. Y. Sasaki, *J. Am. Chem. Soc. 111* (1989) 3442–3444.

[35] H. S. Andersson, A.-C. Koch-Schmidt, S. Ohlson, K. Mosbach, *J. Mol. Recognit. 9* (1996) 675–682.

[36] G. Wulff, S. Schauhoff, *J. Org. Chem. 56* (1991) 395–400.

[37] J. L. Morrison, M. Worsley, D. R. Shaw, G. W. Hodgson, *Can. J. Chem. 37* (1959) 1986–1995.

[38] B. Sellergren, K. J. Shea, *Tetrahedron Asymmetry 5* (1994) 1403.

[39] K. J. Shea, D. A. Spivak, B. Sellergren, *J. Am. Chem. Soc. 115* (1993) 3368–3369.

[40] K. A. Connors, *Binding constants. The measurement of molecular complex stability.*, John Wiley & Sons, New York 1987.

[41] A. M. Katti, M. Diack, M. Z. El Fallah, S. Golshan-Shirazi, S. C. Jacobson, A. Seidel-Morgenstern, G. Guiochon, *Acc. Chem. Res. 25* (1992) 366–374.

[42] P. Sajonz, M. Kele, G. Zhong, B. Sellergren, G. Guiochon, *J. Chromatogr. 810* (1998) 1–17.

[43] G. Wulff, R. Grobe-Einsler, W. Vesper, A. Sarhan, *Makromol. Chem. 178* (1977) 2817–2825.

[44] G. Vlatakis, L. I. Andersson, R. Müller, K. Mosbach, *Nature 361(6413)* (1993) 645–647.

[45] M. J. Whitcombe, M. E. Rodriguez, E. N. Vulfson, *Spec. Publ. – R. Soc. Chem. 158* (1994) 565–571.

[46] K. J. Shea, D. Y. Sasaki, *J. Am. Chem. Soc. 113* (1991) 4109– 4120.

[47] G. Wulff, W. Vesper, R. Grobe-Einsler, A. Sarhan, *Makromol. Chem. 178* (1977) 2799–2816.

[48] G. Wulff, T. Gross, R. Schönfeld, *Angew. Chem. Int. Ed. Engl. 36* (1997) 1962–1964.

[49] G. Guiochon, S. Golshan-Shirazi, A. Katti, *Fundamentals of preparative and nonlinear chromatography.*, Academic Press, New York 1994.

[50] M. Kempe, K. Mosbach, *Anal. Lett. 24* (1991) 1137–1145.

[51] T. E. Mallouk, D. J. Harrison, Interfacial design and chemical sensing *ACS Symposium Series 561*, American Chemical Society, Washington DC 1994.

[52] J. Å. Jönsson, Chromatographic theory and basic principles. Marcel Dekker Inc., New York 1987.

[53] J. C. Giddings, *Anal. Chem. 13* (1963) 1999.

[54] B. Sellergren, K. J. Shea, *J. Chromatogr. A 690* (1995) 29–39.

[55] T. P. O'Brien, N. H. Snow, N. Grinberg, L. Crocker, *J. Liq. Chromatogr. & Rel. Technol. 22* (1999) 183–204.

[56] M. H. Abraham, P. P. Duce, D. V. Prior, D. G. Barrat, J. J. Morris, P. J. Taylor, *J. Chem. Soc. Perkin Trans. II* (1989) 1355– 1375.

[57] G. Albrecht, G. Zundel, *Z. Naturforsch. 39a* (1984) 986–992.

[58] E. Fan, S. A. Van Arman, S. Kincaid, A. D. Hamilton, *J. Am. Chem. Soc. 115* (1993) 369–370.

[59] G. Wulff, R. Schönfeld, *Adv. Mater. 10* (1998) 957–959.

[60] B. Sellergren, *Anal. Chem. 66* (1994) 1578.

[61] S. H. Cheong, S. McNiven, A. Rachkov, R. Levi, K. Yano, I. Karube, *Macromolecules 30* (1997) 1317–1322.

[62] D. Spivak, M. A. Gilmore, K. J. Shea, *J. Am. Chem. Soc. 119* (1997) 4388–4393.

[63] W. H. Pirkle, T. C. Pochapsky, *J. Am. Chem. Soc. 108* (1986) 352–354.

[64] H. Asanuma, M. Kakazu, M. Shibata, T. Hishiya, M. Komiyama, *Chem. Commun.* (1997) 1971–1972.

[65] S. A. Piletsky, H. S. Andersson, I. A. Nicholls, *Macromolecules 32* (1999) 633–636.

[66] S. Vidyasankar, M. Ru, F. H. Arnold, *J. Chromatogr., A 775* (1997) 51–63.

[67] K. Hosoya, Y. Shirasu, K. Kimata, N. Tanaka, *Anal. Chem. 70* (1998) 943–945.

[68] K. Yano, T. Nakagiri, T. Takeuchi, J. Matsui, K. Ikebukuro, I. Karube, *Anal. Chim. Acta. 357* (1997) 91.

[69] Y. Fujii, K. Matsutani, K. Kikuchi, *J. Chem. Soc. Chem. Commun.* (1985) 415–417.

[70] C. J. Welch, *J. Chromatogr. A 689* (1995) 189–193.

[71] C. Kirsten, T. Schrader, *J. Am. Chem. Soc. 119* (1997) 12061.

[72] J. U. Klein, M. J. Whitcombe, F. Mulholland, E. N. Vulfson, *Angew. Chem. Int. Ed. 38* (1999) 2057–2060.

[73] M. Kempe, L. Fischer, K. Mosbach, *J. Mol. Recognit. 6* (1993) 25–29.

[74] O. Ramström, L. I. Andersson, K. Mosbach, *J. Org. Chem. 58* (1993) 7562–7564.

[75] J. Matsui, Y. Miyoshi, T. Takeuchi, *Chem. Lett.* (1995) 1007– 1008.

[76] M. Zihui, Z. Liangmo, W. Jinfang, W. Quinghai, Z. Daoquian, *Biomed. Chromatogr. 13* (1999) 1–5.

[77] A. Fersht, *Enzyme structure and mechanism*, W. H. Freeman and Company, New York 1985.

[78] B. Sellergren, K. G. I. Nilsson, *Methods Mol. Cell. Biol. 1* (1989) 59–62.

[79] G. Lancelot, *J. Am. Chem. Soc. 99* (1977) 7037.

[80] G. J. Welhouse, W. F. Bleam, *Environ. Sci. Technol. 27* (1993) 500–505.

[81] J. Rebek, *Angew. Chem. Int. Ed. Engl. 29* (1990) 245–255.

[82] B. Sellergren, C. Dauwe, T. Schneider, *Macromolecules 30* (1997) 2454–2459.

[83] I. A. Nicholls, *Chem. Lett.* (1995) 1035–1036.

[84] M. Kempe, *Anal. Chem. 68* (1996) 1948–1953.

[85] T. Takeuchi, D. Fukuma, J. Matsui, *Anal. Chem. 71* (1999) 285.

[86] F. Lanza, B. Sellergren, *Anal. Chem. 71* (1999) 2092–2096.

[87] E. Tobler, M. Lämmerhofer, W. R. Oberleitner, N. M. Maier, W. Lindner, *Chromatographia 51* (2000) 65–70.

[88] A. G. Mayes, K. Mosbach, *Anal. Chem. 68* (1996) 3769–3774.

[89] J. Matsui, M. Okada, M. Tsuruoka, T. Takeuchi, *Anal. Commun. 34* (1997) 85–87.

[90] B. Sellergren, *J. Chromatogr. A 673* (1994) 133–141.

[91] Y. Lei, P. A. G. Cormack, K. Mosbach, *Anal. Commun. 36* (1999) 35–38.

[92] G. Wulff, D. Oberkobusch, M. Minarik, *React. Polym., Ion Exch., Sorbents 3* (1985) 261–275.

[93] F. H. Arnold, S. Plunkett, P. K. Dhal, S. Vidyasankar, *Polym. Prepr. 36(1)* (1995) 97–98.

[94] M. Glad, P. Reinholdsson, K. Mosbach, *React. Polym. 25* (1995) 47–54.

[95] L. Schweitz, L. I. Andersson, S. Nilsson, *Anal. Chem. 69* (1997) 1179–1183.

[96] O. Brüggemann, R. Freitag, M. J. Whitcombe, E. N. Vulfson, *J. Chromatogr. 781* (1997) 43–53.

[97] J. M. Lin, T. Nakagama, X. Z. Wu, K. Uchiyama, T. Hobo, *Fresenius' J. Anal. Chem. 357* (1997) 130–132.

[98] J. Haginaka, H. Takehira, K. Hosoya, N. Tanaka, *J. Chromatogr. 849* (1999) 331–339.

[99] D. W. Armstrong, J. M. Schneiderheinze, Y. S. Hwang, B. Sellergren, *Anal. Chem. 70* (1998) 3717–3719.

[100] J. Mathew-Krotz, K. J. Shea, *J. Am. Chem. Soc. 118* (1996) 8154–8155.

[101] M. Yoshikawa, J.-i. Izumi, T. Kitao, S. Koya, S. Sakamoto, *J. Membr. Sci. 108* (1995) 171–175.

[102] T. Kobayashi, H. Y. Wang, N. Fujii, *Anal. Chim. Acta 365* (1998) 81–88.

[103] M. Yoshikawa, T. Fujisawa, J.-i. Izumi, T. Kitao, S. Sakamoto, *Anal. Chim. Acta 365* (1998) 59–67.

[104] J.-M. Lin, T. Nakagama, K. Uchiyama, T. Hobo, *J. Liq. Chromatogr. Relat. Technol. 20* (1997) 1489–1506.

[105] B. Sellergren, *Chirality 1* (1989) 63–68.

[106] L. I. Andersson, K. Mosbach, *Makromol. Chem., Rapid Commun. 10* (1989) 491–495.

[107] M. Kempe, K. Mosbach, *J. Chromatogr., A 694* (1995) 3–13.

[108] O. Ramström, L. I. Andersson, K. Mosbach, *J. Org. Chem. 58* (1994) 7562–7564.

[109] B. Sellergren, in G. Subramanian (Ed.): *A practical approach to chiral separation by liquid chromatography*, VCH, Weinheim 1994, p. 69–93.

[110] M. Kempe, K. Mosbach, *J. Chromatogr., A 691* (1995) 317– 323.

[111] L. Ye, O. Ramström, K. Mosbach, *Anal. Chem. 70* (1998) 2789–2795.

[112] L. Fischer, R. Müller, B. Ekberg, K. Mosbach, *J. Am. Chem. Soc. 113* (1991) 9358–9360.

[113] L. Schweitz, L. I. Andersson, S. Nilsson, *J. Chromatogr. A 792* (1997) 401–409.

[114] M. Kempe, K. Mosbach, *J. Chromatogr., A 664* (1994) 276– 279.

[115] K. Hosoya, Y. Shirasu, K. Kimata, T. Araki, N. Tanaka, *Kuromatogurafi 17* (1996) 312–313.

[116] Z. H. Meng, L. M. Zhou, Q. H. Wang, D. Q. Zhu, *Chin. Chem. Lett. 8* (1997) 345–346.

[117] D. J. O'Shannessy, L. I. Andersson, K. Mosbach, *J. Mol. Recognit. 2* (1989) 1–5.

[118] J. Matsui, O. Doblhoff-Dier, T. Takeuchi, *Chem. Lett.* (1995) 489.

7 Chiral Derivatization Chromatography

Michael Schulte

7.1 Introduction

Discrimination between the enantiomers of a racemic mixture is a complex task in analytical sciences. Because enantiomers differ only in their structural orientation, and not in their physico-chemical properties, separation can only be achieved within an environment which is unichiral. Unichiral means that a counterpart of the racemate to be separated consists of a pure enantiomeric form, or shows at least enrichment in one isomeric form. Discrimination or separation can be performed by a wide variety of adsorption techniques, e.g. chromatography in different modes and electrophoresis. As explained above, the enantioseparation of a racemate requires a non-racemic counterpart, and this can be presented in three different ways:

1. As a unichiral template which is used as a stationary phase itself, or which is bonded to a solid support (silica particles or fused silica capillaries).
2. As a unichiral additive which is mixed with the racemate of interest to form non-covalent diastereomeric complexes which can be distinguished by achiral techniques.
3. As a unichiral group which reacts with the racemate to form diastereomeric molecules which can be separated by achiral adsorption processes.

This chapter will focus on topic 3, which is normally regarded to be chiral derivatization chromatography, but will also cover other topics that might be considered when applying derivatization techniques. The goal for the separation of the racemates may be their analysis or their preparation. Both topics will be covered in this chapter.

7.2 Different Approaches for Derivatization Chromatography

The educt, a racemate, is derivatized before the separation with an agent which might be achiral or unichiral (Fig. 7-1), and afterwards is passed through a chromatographic system which is equipped with a stationary phase. This stationary phase may also be achiral or unichiral in nature.

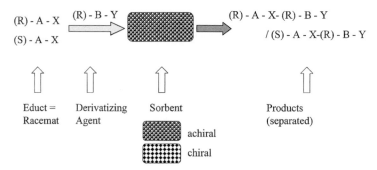

(R) - A - X   (R) - B - Y   →   (R) - A - X- (R) - B - Y
(S) - A - X   /(S) - A - X-(R) - B - Y

⇑   ⇑   ⇑   ⇑

Educt =   Derivatizing   Sorbent   Products
Racemat   Agent   (separated)

▓ achiral

▓ chiral

Fig. 7-1. General reaction scheme of chiral derivatization chromatography.

Derivatization techniques are divided into pre-column and post-column techniques. Post-column derivatization is especially useful to enhance the detection of compounds, whilst pre-column derivatization is the method of choice for enantioseparations via derivatization.

Pre-column derivatization offers some general advantages:

● every racemate with a functional group can be derivatized and separated;
● nonreacted derivatization reagent can be removed before the separation;
● the reaction can be used as a pre-purification step, if the reaction is selective for one compound;
● the derivatized compounds may show better chromatographic properties, e.g. derivatization of free amino groups can reduce tailing, or the introduction of derivatizing agents decrease detection limits; and
● it has a high success rate compared to direct methods, which cannot guarantee a separation.

Several strategies can be distinguished to achieve certain solutions, and these are summarized in Table 7-1. The strategies each have their benefits in special cases. The different approaches will be shown in the following examples.

Table 7-1. Different approaches for derivatization chromatography

Type	Educt	Derivatizing Agent number	type	Stationary phase	type of binding or reaction
I	one	one	unichiral	achiral	covalent
II	multiple	one	achiral or unichiral	achiral or unichiral	covalent
III	one	one	achiral	unichiral	covalent
IV	one	multiple	achiral	unichiral	covalent
V	one	one	achiral	achiral	enzymatic

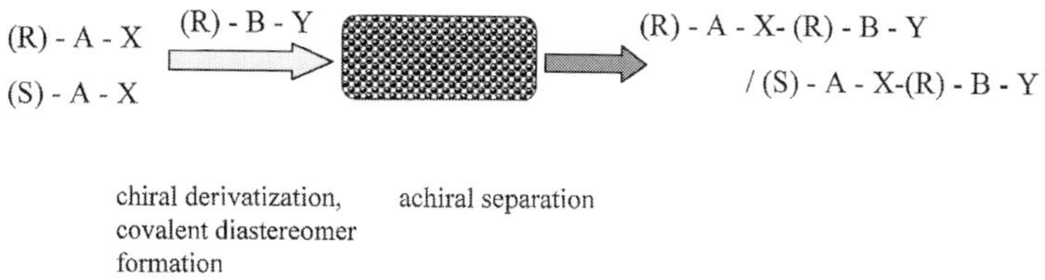

Fig. 7-2. Type I: Covalent derivatization with a unichiral reagent.

7.2.1 Type I: Covalent Derivatization with a Unichiral Derivatizing Agent

This strategy is the one most commonly used for the analytical determination of ena-tiopurity. A given racemate is reacted with a unichiral derivatizing agent, and the resulting pair of diastereomers is separated on an achiral stationary phase, in most of the cases on a reversed-phase type (Fig. 7-2).

Some prerequisites are essential in the use of these methods:

- the reaction must be rapid and reproducible, and not cause racemization of the product to be analyzed;
- the reaction must be complete;
- the derivatizing agent must be easily removable if the product is to be recovered; and
- the enantiopurity of the derivatizing agent must be high.

This final point will be highlighted by an example (Fig. 7-3).

Let us assume that a given compound has a purity of 98 % ee, and that this compound is reacted with a derivatizing agent which has also a purity of 98 % ee. The two major compounds plus the minor impurities in the compound to be analyzed and the derivatizing agent will create a set of four diastereomers. Two pairs of diastereomers (+)-A(+)B and (−)-A(−)-B as well as (−)-A(+)-B and (+)-A(−)-B are enantiomeric pairs, and thus elute together on an achiral column. Therefore, a peak area of 98.011 % will be detected for (+)-A(+)-B, which leads to a purity of 96.03 % ee for (+)-A. This is a quite significant deviation from the true value for (+)-A.

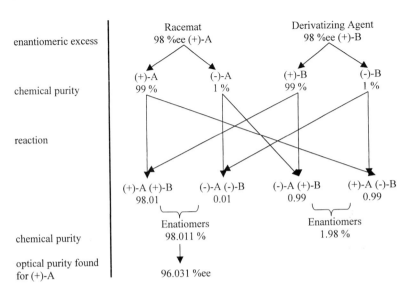

enantiomeric excess

chemical purity

reaction

chemical purity

optical purity found
for (+)-A

Fig. 7-3. Influence of reagent purity on quantification results.

7.2.1.1 Types of Modifications for Different Groups

One prerequisite of applying a chiral derivatization reaction to a racemate is the presence of a derivatizable functional group. In most of the cases these are amino-, acido-, alcohol- and carboxy-groups. Less common – but also derivatizable – are epoxides, olefins and thiols. There are legions of different derivatizing agents which differ in their reaction type and their most important features, e.g. speed and selectivity of reaction or sensitivity of the derivatizing agent. The different types of reaction for the different functional groups are illustrated in Figs. 7-4 to 7-7.

The most important group of derivatives for the amino function (Fig. 7-4) is the carbamate group, which can be formed by reactions with acids, acid chlorides or acid anhydrides. A series of chlorides as 2-chloroisovalerylchloride [1], chrysanthemoylchloride [2] and especially chloride compounds of terpene derivatives (camphanic acid chloride [3], camphor-10-sulfonyl chloride [4]) are used. The α-methoxy-α-trifluoromethylphenylacetic acid or the corresponding acid chloride introduced by Mosher in the 1970s are very useful reagents for the derivatization of amines and alcohols [5].

By using chloroformates instead of acid chlorides, the resultant urethanes are useful and stable derivatives. The chloroformate derivatives most commonly used are menthylchloroformate [6] and 1-(9-fluorenyl)ethylchloroformate (FLEC) [7], which exhibits excellent properties for fluorimetric detection.

The reaction of *ortho*-phthalaldehyde and a thiol compound with an amino acid to form an isoindole derivative can be used to enhance the detection sensitivity for the normally only weakly UV-detectable amino acid compounds, and to introduce an unichiral center which renders the amino acid derivatives separable on a reversed-phase column. An example of this technique will be given later.

A series of reactions was developed to transfer amines to ureido- and thioureido-derivatives for separation. The reaction of ureido-derivatives is widely used by the reaction with 1-phenylethyl isocyanate (PEIC) [8] or the naphthyl-analogue 1-(1-naphthyl)ethyl isocyanate (NEIC) [9]. Both reactions can be used not only for chiral amines but also for alcohols and thiols.

Thioisocyanates as derivatizing reagents are often based on unichiral carbohydrate compounds. One very frequently used reagent in the analysis of amino acids is 2,3,4,6-tetra-*O*-acetyl-β-D-glucopyranosyl isothiocyanate (TAGIT or GITC) [10]. Other derivatizing reagents of the same type are based on galactose or arabinose as unichiral molecules (Fig. 7-4).

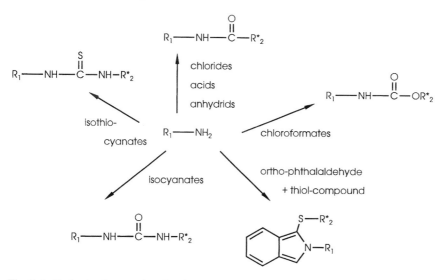

Fig. 7-4. Derivatization reactions for the amine function.

Most of the reactions applied to amines can also be transferred to alcohols (Fig. 7-5). One large group of chiral alcohols are the β-adrenoreceptor blockers, for which a variety of derivatization agents was developed. One highly versatile reagent for the separation of β-blockers is *N*-[(2-isothiocyanato)cyclohexyl]3,5-dinitrobenzoylamide (DDITC) [11]. Alternatively, unichiral drugs such as β-blockers or (*S*)-naproxen [12] may be used in a reciprocal approach to derivatize racemic amine compounds.

Methyl-substituted primary alcohols can be separated after derivatization with [6-methoxy-2,5,7,8-tetramethylchromane-2-carboxylic acid] (Trolox™ methyl ether) [13] while *sec.*- and *tert.*-alcohols are derivatized with 2-dimethylamino-1,3-dimethyl-octahydro-1*H*-1,3,2-benzodiazaphosphole [14] (Fig. 7-5).

While amino compounds can be derivatized with acids and acid chlorides, it is possible to separate racemic acids (vice versa) with unichiral amino compounds

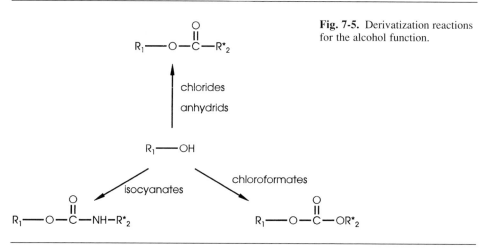

Fig. 7-5. Derivatization reactions for the alcohol function.

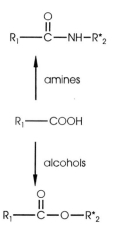

Fig. 7-6. Derivatization reactions for the acid function.

(Fig. 7-6). Two unichiral amides which have been known capable of this reaction are 1-phenylethylamine [15] and 1-(1-naphthyl)ethylamine [16]. Marfey's reagent [N-α-(2,4-dinitro-5-fluorophenyl)-L-alaninamide] was introduced as a reagent to derivatize amino acids with cyclopentane, tetrahydroisoquinoline or tetraline structures [17]. Simple chiral alcohols such as 2-octanol can also be used to derivatize acids such as 2-chloro-3-phenylmethoxypropionic acid [18].

Several alternative routes can be used in order to derivatize the carboxy function (Fig. 7-7). Ketones can be transferred by hydrazines and diols to the corresponding hydrazines or acetals. 2,2,2-Trifluoro-1-phenylethylhydrazine [19] is an example of the first group, while 2,3-butanediol or 1,4-dimethoxy-2,3-butanediol can be used to form diastereomeric acetals.

O-(–)-Menthylhydroxylamine may be used to form hydroxylamine derivatives for the determination of carbohydrates [20].

Although most of the reactions are performed in liquid state – which is the simplest reaction medium – a combination of solid-phase extraction (SPE) and reaction

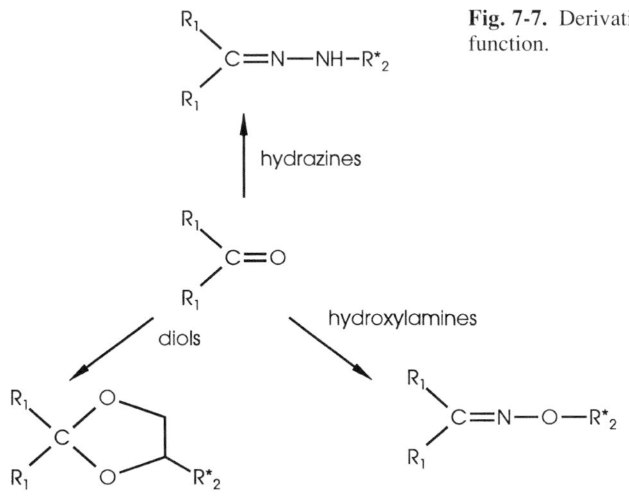

Fig. 7-7. Derivatization reactions for the carboxy function.

is being increasingly applied [21]. The most striking advantage of solid-phase reactions is the combined reaction and sample clean-up that occurs in one step, eliminating the need to remove excess of reagents.

7.2.1.2 Separation of Amino Acid Enantiomers after Derivatization with *Ortho*-Phthaldialdehyde (OPA) and a Unichiral Thiol Compound

One of the most useful applications of chiral derivatization chromatography is the quantification of free amino acid enantiomers. Using this indirect method, it is possible to quantify very small amounts of enantiomeric amino acids in parallel and in highly complex natural matrices. While direct determination of free amino acids is in itself not trivial, direct methods often fail completely when the enantiomeric ratio of amino acid from protein hydrolysis must be monitored in complex matrices.

One method that combines the good chromatographic properties with improved limit of detection is the separation of isoindole derivatives of amino acids that may be detected fluorimetrically. This method may be applied to protein hydrolysates, and used in automated format in routine analyses [22].

The mixture of free amino acids is reacted with OPA (Fig. 7-8) and a thiol compound. When an achiral thiol compound is used, a racemic isoindole derivative results. These derivatives from different amino acids can be used to enhance the sensitivity of fluorescence detection. Figure 7-9 shows the separation of 15 amino acids after derivatization with OPA and mercaptothiol; the racemic amino acids may be separated on a reversed-phase column. If the thiol compound is unichiral, the amino acid enantiomers may be separated as the resultant diastereomeric isoindole compound in the same system. Figure 7-10 shows the separation of the same set of amino acids after derivatization with the unichiral thiol compound *N*-isobutyryl-L-cysteine (IBLC).

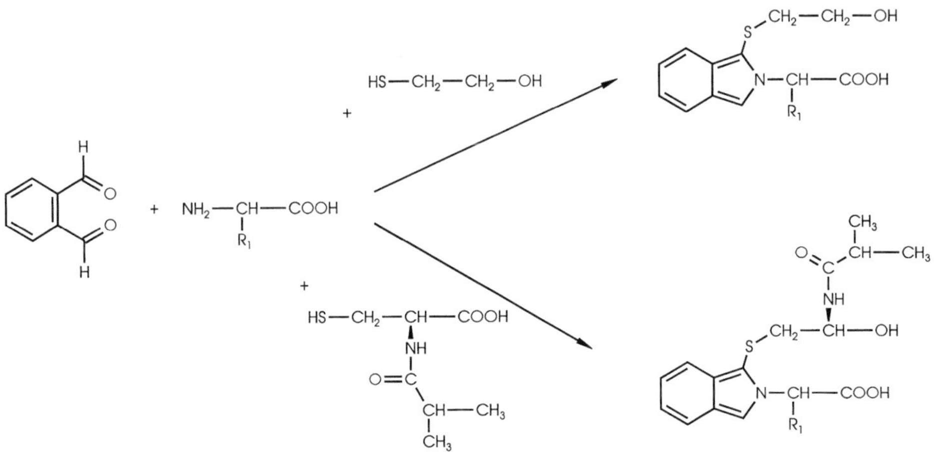

Fig. 7-8. Derivatization of amino acids with OPA and a thiol compound.

The advantages of this method are a short reaction time and the nonfluorescence of the OPA reagent. Therefore, excess reagent must not be removed before the chromatography stage. Using this method, it is possible to measure tryptophan, but not secondary amino acids such as proline or hydroxyproline. Cysteine and cystine can be measured, but because of the low fluorescence of their derivatives, they must be detected using an UV system, or alternatively oxidized to cysteic acid before reaction.

Detection is carried out using a fluorescence detector, with an extinction wavelength of 340 nm and an emission wavelength of 445 nm. With this method it is possible to detect amino acid at concentrations of 5 pmol ml^{-1} in the sample, which corresponds to 450 fmol per amino acid injected. The method may be applied to samples containing between 5 and 400 pmol ml^{-1} per amino acid.

The relative standard deviations for repetitive injections vary between 2 % and 7 % for the different amino acids.

The method described above is applicable to a wide range of samples for the determination of amino acids in different matrices. For example, the amino acid composition and distribution of single enantiomers has been determined in protein hydrolysates, orange juice (Fig. 7-11), yogurt and seawater [23].

CH . 1 C .S 5 .00 ATT 8 OFFS 0

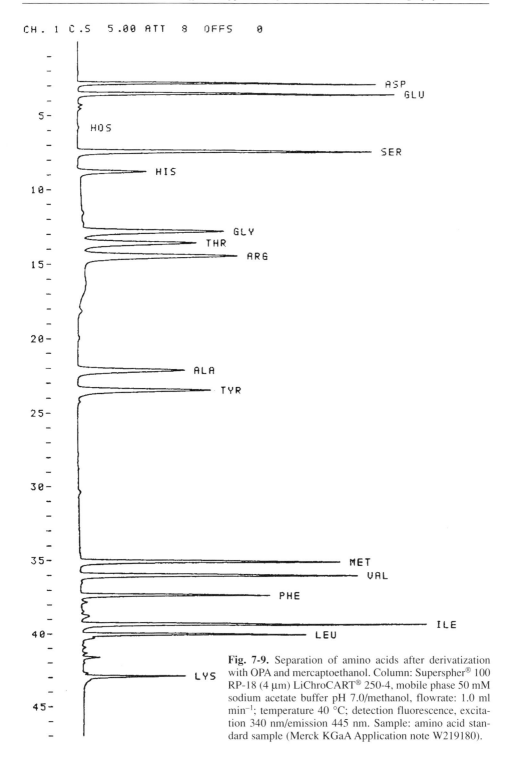

Fig. 7-9. Separation of amino acids after derivatization with OPA and mercaptoethanol. Column: Superspher® 100 RP-18 (4 μm) LiChroCART® 250-4, mobile phase 50 mM sodium acetate buffer pH 7.0/methanol, flowrate: 1.0 ml min^{-1}; temperature 40 °C; detection fluorescence, excitation 340 nm/emission 445 nm. Sample: amino acid standard sample (Merck KGaA Application note W219180).

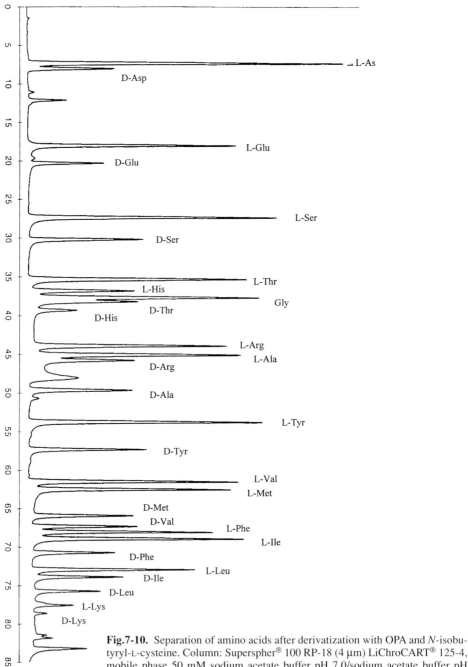

Fig.7-10. Separation of amino acids after derivatization with OPA and *N*-isobutyryl-L-cysteine. Column: Superspher® 100 RP-18 (4 μm) LiChroCART® 125-4, mobile phase 50 mM sodium acetate buffer pH 7.0/sodium acetate buffer pH 5.3/methanol, flowrate: 1.0 ml min⁻¹; temperature 25 °C; detection fluorescence, excitation 340 nm/emission 445 nm. Sample: amino acid standard mixture. (Merck KGaA Application note W219189; reproduced with permission from H. P. Fitznar, Alfred-Wegener-Institute for Polar and Marine Research.)

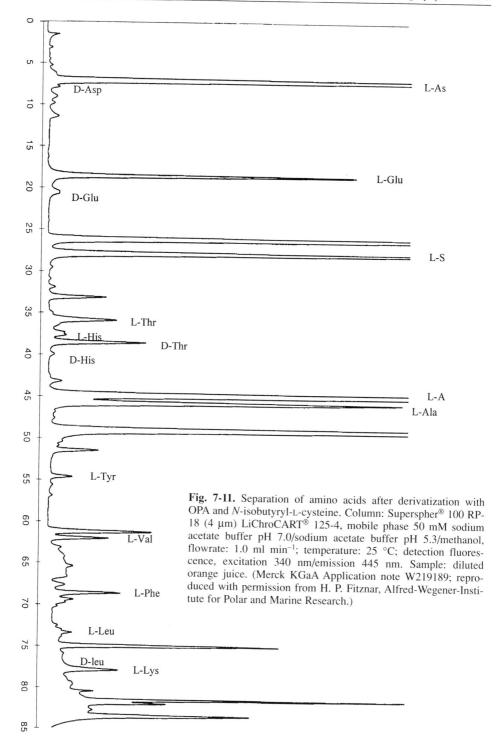

Fig. 7-11. Separation of amino acids after derivatization with OPA and *N*-isobutyryl-L-cysteine. Column: Superspher® 100 RP-18 (4 μm) LiChroCART® 125-4, mobile phase 50 mM sodium acetate buffer pH 7.0/sodium acetate buffer pH 5.3/methanol, flowrate: 1.0 ml min⁻¹; temperature: 25 °C; detection fluorescence, excitation 340 nm/emission 445 nm. Sample: diluted orange juice. (Merck KGaA Application note W219189; reproduced with permission from H. P. Fitznar, Alfred-Wegener-Institute for Polar and Marine Research.)

7.2.2 Type II: Selective Derivatization of One Compound

Despite the use of unichiral derivatizing agents, achiral derivatization also offers certain advantages for chromatographic enantioseparation. The newly introduced group can enhance the sensitivity for UV or fluorimetric detection. In addition, the chromatographic properties can be positively influenced with, on occasion, only the derivatization step making it possible to separate the enantiomers on a CSP. A further benefit of the derivatization step may be its selectivity for a single compound; in this way, derivatization may be used to measure one compound selectively, in a complex mixture (Fig. 7-12).

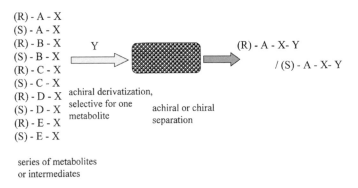

Fig. 7-12. Type II: Selective derivatization of one compound.

The latter approach is used in the enantioselective determination of a Phase I metabolite of the antihistaminic drug, terfenadine. Terfenadine is metabolized to several Phase I compounds (Fig. 7-13), among which the carboxylic acid MDL 16.455 is an active metabolite for which plasma concentrations must often be determined. Although terfenadine can be separated directly on Chiralpak AD® – an amylose-based CSP – the adsorption of the metabolite MDL 16.455 is too high to permit adequate resolution. By derivatizing the plasma sample with diazomethane, the carboxylic acid is converted selectively to the methyl ester, which can be separated in the presence of all other plasma compounds on the above-mentioned CSP Chiralpak AD® [24] (Fig. 7-14). Recently, MDL 16.455 has been introduced as a new antihistaminic drug, fexofenadine.

The method described is of particular value in the determination in complex matrices of metabolites or impurities which differ in the presence of functional groups. Notable examples of this are assays of compounds with free amino or acidic groups.

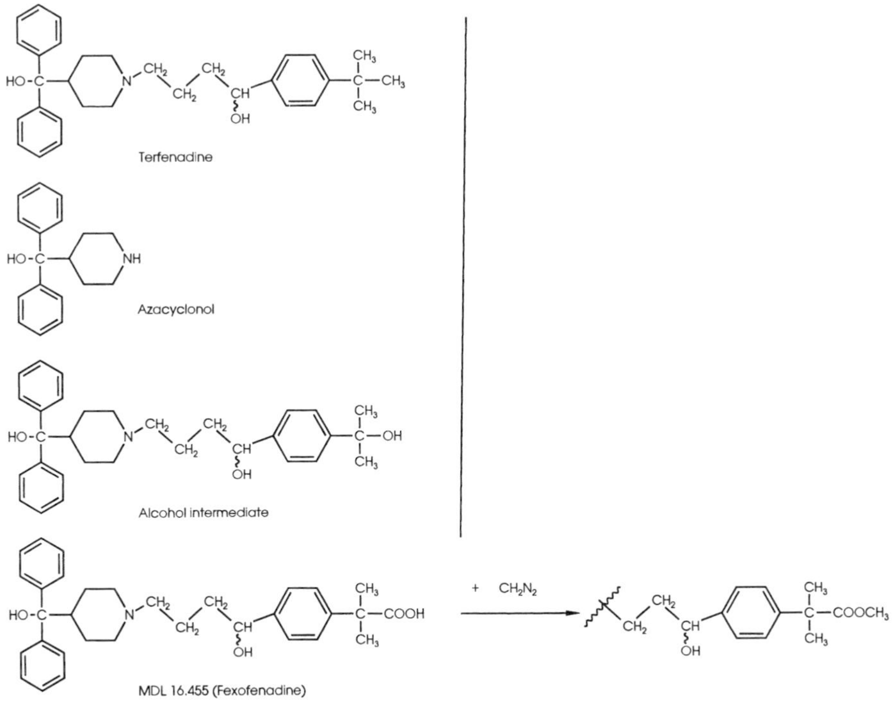

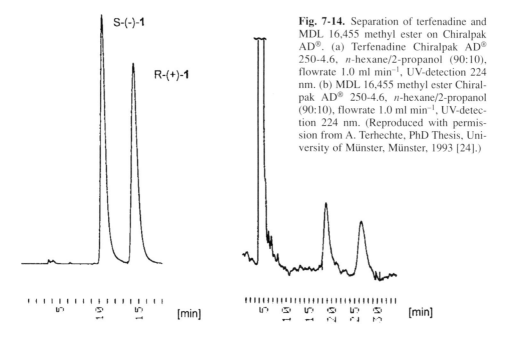

Fig. 7-13. Metabolic pathway of terfenadine and derivatization of MDL 16,455.

Fig. 7-14. Separation of terfenadine and MDL 16,455 methyl ester on Chiralpak AD®. (a) Terfenadine Chiralpak AD® 250-4.6, *n*-hexane/2-propanol (90:10), flowrate 1.0 ml min⁻¹, UV-detection 224 nm. (b) MDL 16,455 methyl ester Chiralpak AD® 250-4.6, *n*-hexane/2-propanol (90:10), flowrate 1.0 ml min⁻¹, UV-detection 224 nm. (Reproduced with permission from A. Terhechte, PhD Thesis, University of Münster, Münster, 1993 [24].)

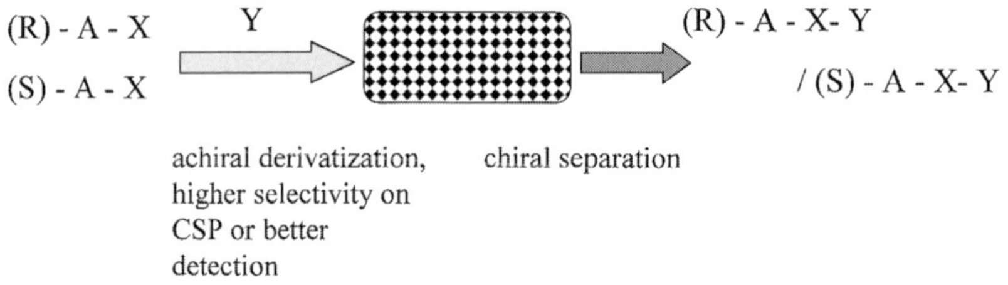

Fig. 7-15. Type III: Increase in selectivity.

7.2.3 Type III: Increase in Selectivity

Derivatization of a racemic compound with an achiral group may play an important role in the analysis of a chiral compound (Fig. 7-15). In the case of substances with low or no UV-activity, the compounds can be rendered detectable by introducing an UV-absorbing or fluorescent group. If the racemate itself shows selectivity on a chiral stationary phase (CSP), this method can be applied to reduce the limit of detection. Examples have been reported in the literature, especially for the derivatization of amino acids which are difficult to detect using UV detection. Different derivatization strategies can be applied (Fig. 7-16).

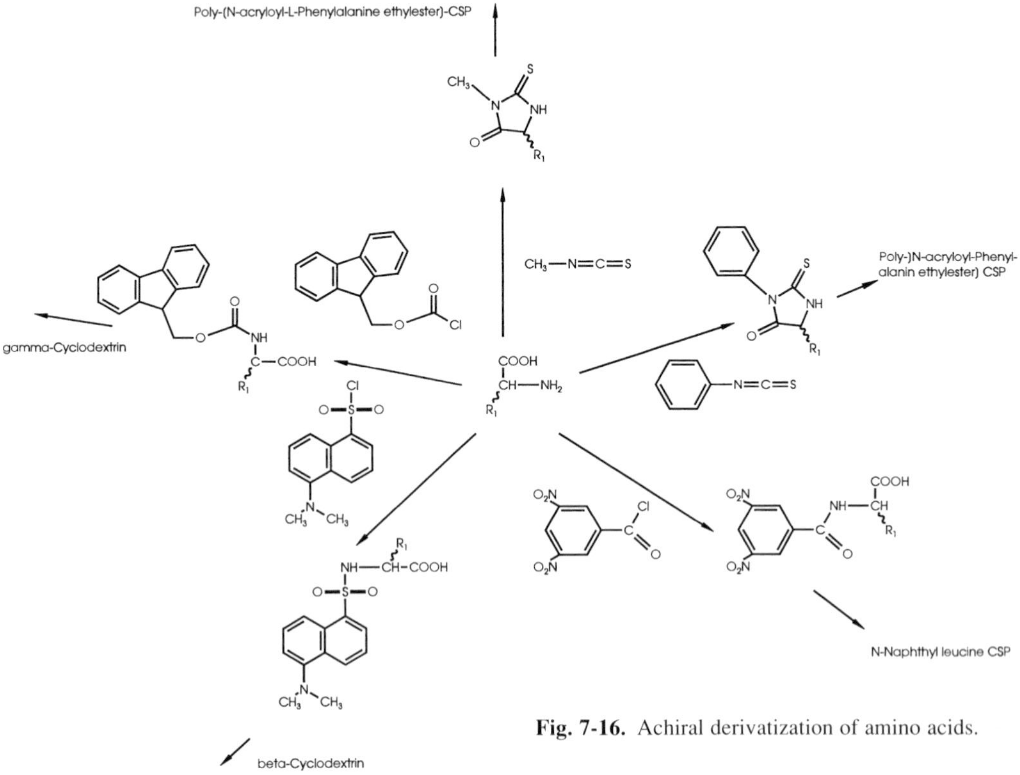

Fig. 7-16. Achiral derivatization of amino acids.

- Free amino acids can be derivatized with isothiocyanates to phenyl- or methyl-thiohydantoin derivatives. The thiohydantoins can be separated on a CSP with poly-[*N*-acryloyl-L-phenylalanine ethylester] (Chiraspher®) as a chiral selector [25]. This CSP offers a known selectivity for many five-membered heterocyclic rings.
- A derivatization with acid chlorides is also possible. Amino acids can be derivatized with 9-fluorenylmethyl chloroformate (FMOC) and separated on a CSP with χ-cyclodextrin (ChiraDex gamma®), a cyclic oligosaccharide which consists of eight glucose units.
- A CSP with a smaller β-cyclodextrin moiety (seven glucose units) immobilized on silica gel (ChiraDex®) is able to separate the dansyl-derivatives [5-(dimethylamino)-naphthalin-1-sulfonylchloride] of amino acids [26].

The separations described above can be used because of the known selectivities with certain groups which, in the case of derivatized compounds, are mostly the derivatizing groups. This has led towards the concept of rational CSP design. One of the most successful methods was the introduction of the brush-type CSP by Pirkle. The CSPs are designed to offer distinct attractive groups in a fixed stereochemical orientation towards an analyte. These interactions can be polar functions, H-donor and H-acceptor functions and π–π-interactions. When a CSP is designed which shows recognition for a certain derivatizing group, this derivatizing agent may be used for the separation of a whole range of compounds.

An example for this approach is the immobilization of (*S*)-(−)-α-*N*-(2-naphthyl)leucine, a π-donating group on silica. This chiral selector exhibits excellent recognition for 3,5-dinitrobenzoyl (DNB)- and 3,5-dintroanilido (DNAn)-derivatives. Amines and alcohols can be derivatized with DNB- or DNAn-chloride to the esters or carbamates and separated on the CSP, as shown by Pirkle for a wide variety of compounds [27].

7.2.4 Type IV: Derivative with best Selectivity

It is important not only that a multiplicity of compounds in the sample mixture may be selectively derivatized – as was shown for Type III reactions – but also that one racemate may be derivatized with a multiplicity of derivatizing agents (Fig. 7-17). Although this approach can be used to optimize the analogues of a compound [28, 29], it is of special interest when a compound is required to be separated on a preparative scale.

A major interest in the field of preparative enantioseparation has begun to emerge from the pharmaceutical industry, in which the preparation of unichiral drugs on a large scale is especially important. Since the introduction of simulated moving bed (SMB) technology (which has been used in the petrochemical and sugar industries for more than 30 years) to the field of chromatographic enantioseparation, the production of several tons of pure enantiomer per year has become possible, and indeed is currently performed on a regular basis [30, 31]. It is clear that for such a large-

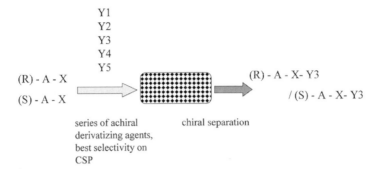

Fig. 7-17. Type IV: Derivative with best selectivity.

scale separation, only optimized conditions must be used, as small increases in selectivity, resolution and loading capacity will yield major increases in productivity. Several intermediates in the synthetic route of an enantiopure compound must be thoroughly examined. If an intermediate can be derivatized to a compound which has a known selectivity on a CSP, this may lead to a tremendous increase in productivity.

An example of this approach was described by Francotte, in which racemates bearing the benzoyl group could be separated on the CSP cellulose-tribenzoate (CTB). Using this CSP, differently substituted benzoyl-derivatives showed different selectivities. By testing the *ortho-*, *meta-* and *para*-methoxy-benzoyl-derivatives of different racemic alcohols it was possible to identify for every drug substance the optimum derivative [32]. Unfortunately, the same benzoyl-isomer was not the best for all substances to be separated, indicating that chiral recognition and therefore separation is dependent not only on the derivatizing group but also on the structure of the derivatized compound.

7.2.5 Type V: Reactive Separation

The final type of reaction covers not only chromatographic resolution but also conversion of a derivatized racemate to an unichiral product. Derivatives of chiral compounds can be converted by enzymes with a high enantiospecificity. Again, this approach offers the highest value when applied to the preparative separation of racemates. A production strategy for an unichiral compound can be set up when the target compound is derivatized to a molecule which is stereochemically specifically converted by an enzyme. The resulting two compounds (the target compound and the derivate of the other enantiomer of the target compound) can be separated on a normal silica or reversed-phase sorbent (Fig. 7-18).

This strategy, for the production of (*S*)-ibuprofen, is illustrated in Fig. 7-19. Ibuprofen is derivatized to the corresponding sulphonmethyl ester, but only one enantiomer of this compound is converted by a protease to (*S*)-ibuprofen [33]. The resulting (*S*)-ibuprofen and the unreacted ibuprofen sulphonmethyl ester can be sep-

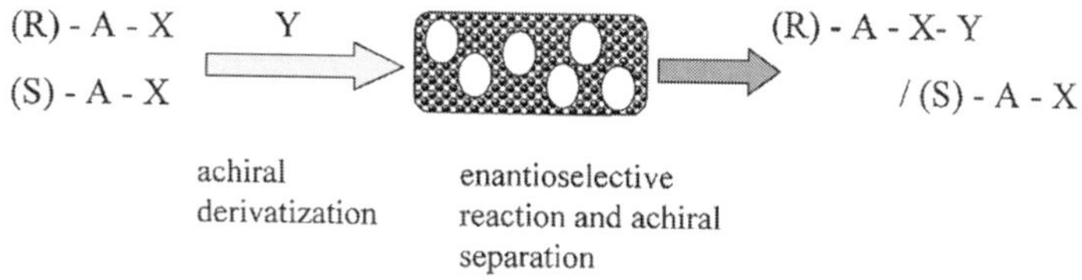

(R) - A - X
(S) - A - X
Y
(R) - A - X - Y
/ (S) - A - X

achiral
derivatization

enantioselective
reaction and achiral
separation

Fig. 7-18. Type V: Reactive separation.

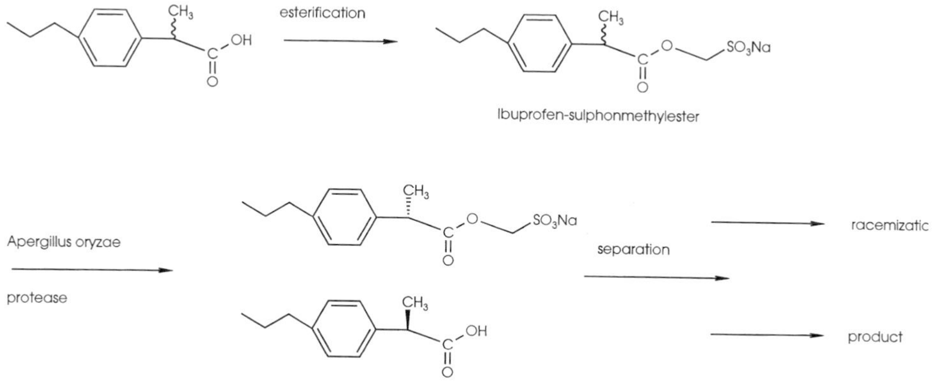

esterification

Ibuprofen-sulphonmethylester

Apergillus oryzae

protease

separation

racemizatic

product

Fig. 7-19. Enzymatic resolution and separation of ibuprofen sulphonmethyl ester.

arated chromatographically. Integration of the reaction and the separation may be performed in a chromatographic reactor and, by applying the principle of SMB chromatography, a continuous reaction/separation system can be set up [34]. Today, a wide variety of enzymes is known which can convert different derivatives specifically, with sufficient yield and in short time, so that they may be used in a chromatographic reactor.

7.3 Conclusions

Although indirect chiral separation seems to be a less elegant method than direct separation on CSPs, the importance of the flexibility and broad applicability of these derivatization methods cannot be overestimated. While direct methods will either work or not – and are subject to only minimal influence by the operator – derivatization methods offer a wide range of different methodologies, with a guaranteed success. On the basis of this general approach – which may be applied to entire groups of compounds – rules may be deduced from the literature. Therefore, this arsenal of indirect separation methods should be borne in mind and investigated thoroughly in order to determine the optimum solution for any given problem involving separation.

References

[1] Blackburn, I., *CRC Handbook of chromatography*, Boca Raton, Florida, CRC Press, **1983**; 205

[2] Allenmark, S., *Chromatographic enantioseparation*, Chichester, Ellis Horwood Ltd., **1988**; 55

[3] Olsen, L., Bronnum-Hansen, K., Helboe, P., Joergensen, G. H., Kryger, S., *J. Chromatogr.*, **1993**, *636 (2)*, 231

[4] Aberhart, D. J., Cotting, J. A., Lin, H. J., *Anal. Biochem*, **1985**, *151 (1)*, 88

[5] Dale, J. A., Dull, D. L., Mosher, H. S., *J. Org. Chem.*, **1969**, *34*, 2543

[6] Christensen, E. B., Hansen, S. H., Rasmussen, S. N., *J. Chromatogr. B*, **1995**, *670 (2)*, 243

[7] Vogt, C., Georgi, A., Werner, G., *Chromatographia*, **1995**, *40 (5–6)*, 287

[8] Pirkle, W. H., Hauske, J. R., *J. Org. Chem.*, **1977**, *42*, 183,9

[9] Terhechte, A., Blaschke, G., *J. Chromatogr.*, **1995**, *694 (1)*, 219

[10] Péter, A., Töth, G., Török, G., Tourné, D., *J. Chromatogr.*, **1996**, *728*, 455

[11] Kleidernigg, O. P., Posch, K., Lindner, W., *J. Chromatogr.*, **1996**, *729 (1)*, 33

[12] Spahn, H., Henke, W., Langguth, P., Schloos, J., Mutschler, E., *Arch. Pharm*, **1990**, *323 (8)*, 465

[13] Almquist, S. R., Petersson, P., Walther, W., Markides, K. E., *J. Chromatogr.* **1994**, *679 (1)*, 139

[14] Alexakis, A., Mutti, S., Mangeney, P., *J. Org. Chem*, **1992**, *57 (4)*, 1224

[15] Tamura, S., Kuzuna, S., Kawai, K., Kishimato, S., *J. Pharm. Pharmacol.*, **1981**, *33*, 701

[16] Vecci, M., Müller, R. K., *J. High Resol. Chromatogr.*, **1979**, *2*, 195

[17] Peter, A., Töth, G., *Anal. Chim. Acta*, **1997**, *352 (1–3)*, 335

[18] Anelli, P. L., Tomba, C., Uggeri, F., *J. Chromatogr.* **1992**, *589 (1-2)*, 346

[19] Pereira, W. E., Solomon, M., Halpern, B., *Aust. J. Chem.*, **1971**, *24*, 1103

[20] Schweer, H., *J. Chromatogr.*, **1982**, *243*, 149

[21] Bourque, A. J., Krull, I. S., *J. Chromatogr.*, **1991**, *537 (1– 2)*, 123

[22] Merck KGaA Application Note w219189, **1997**

[23] Fitznar, H. P., Lobbes, J. M., Kattner, G., *J. Chromatogr. A*, **1999**, *832 (1–2)*, 123

[24] Terhechte, A., Ph.D-thesis, University Münster, **1993**

[25] Kinkel, J. N., *GIT Supplement Chromatographie*, **1988**, *3*, 29

[26] Cabrera, K., Lubda, D., Bohne, S., Schäfer, M., *Poster presentation 3rd* International Symposium on Chiral Discrimination, Tübingen, **1992**

[27] Pirkle, W. H., Bowen, W. E., *J. High Resol. Chromatogr.*, **1994**, *17 (9)*, 629

[28] Dyas, A. M., Robinson, M. L. Fell, A. F., *Chromatographia*, **1990**, *30 (1–2)*, 73

[29] van Overbecke, A., Baegens, W., Dewaele, C., *Anal. Chim. Acta*, **1996**, *321 (2-3)*, 245

[30] Ditz, R., Schulte, M., Strube, J., *Drug Discov. Develop.*, **1998**, *1 (3)*, 264

[31] Juza, M., Mazzotti, M., Morbidelli, M., *Trends Biotechnol.*, **2000**, *18 (3)*, 108

[32] Francotte, E. R., *Chirality*, **1998**, *10 (5)*, 492

[33] West, S., Godfrey, T., *Industrial Enzymology*, New York, Stockton Press, *1996,* 166

[34] Meurer, M., Altenhöhner, U., Strube, J., Schmidt-Traub, H., *J. Chromatogr.*, **1997**, *769*, 71

8 Nonchromatographic Solid-Phase Purification of Enantiomers

Neil E. Izatt, Ronald L. Bruening, Krzysztof E. Krakowiak,
Reed M. Izatt and Jerald S. Bradshaw

8.1 Introduction

In recent years, there has been increasing interest in the preparation of enantiomerically pure compounds [1, 2]. This interest was intensified by a statement issued in 1990 by the U.S. Food and Drug Administration (FDA) concerning development of chiral drugs as single isomers or racemates [3]. Even though the FDA did not mandate development of single isomers (racemates may be appropriate in certain cases), pharmaceutical companies took this announcement as an indication of things to come, and began careful study of both isomers in potential drugs. The logic behind this development is clear. It has long been recognized that isomers, including enantiomers, can have quite different properties. Apart from the fact that the dosage needed is increased by having the "inert" isomer present, the "inert" isomer may have properties that range from benign to beneficial to fatal. An example of an "inert" isomer that produced devastating results was one of the enantiomers of thalidomide [4], which caused severe malformations in children born to pregnant women who took the drug by prescription. Only the unwanted isomer had this effect. As a result of the factors described above, there has been a tremendous increase in the production of chiral compounds. Sales of chiral drugs for antibiotic, cardiovascular, hormone, central nervous system, cancer, antiviral, hematology, respiratory, and gastrointestinal uses were nearly US $ 100 billion in 1998 [3]. In addition, single enantiomer applications are found in the preparation of pesticides, biochemicals, flavors, and aromas. The requirement for pure enantiomers in these applications requires a critical assessment of the most cost-effective way to accomplish their analysis and preparation.

Among the methods used to separate enantiomers are crystallization using a chiral auxiliary, chiral synthesis, and large-scale chromatography [3]. The choice of these methods usually involves trade-offs, particularly in large-scale separations. In the first and third cases, the similar chemical properties of enantiomers result in small separation factors (α values), making it necessary for multiple separations stages to be used in order to achieve satisfactory separations of 98 % or better. In addition, each of these technologies requires the use of large amounts of solvent, and the chiral throughput per sized separation material in the case of chromatography is relatively low. An alternative to isomer separation is the synthesis of the pure isomer. In many cases, this alternative is not practical because of the many and/or costly synthetic steps involved in the preparation.

For chiral separations, it is desirable to increase the product throughput, reduce the number of total separation stages, and minimize the use of operating chemicals. An approach used by us to accomplish these aims is to design molecules that can differentiate between enantiomers based on the preferential fit of the molecule to one of the enantiomers. There has been an extensive body of work on the design and performance of such molecules in single-phase homogeneous systems. These studies have been particularly successful for the separation of enantiomeric primary amine guests by chiral macrocyclic hosts. In these systems, $\Delta \log K$ values (Equation (1)) for the interaction of chiral hosts with enantiomeric guests exceeded 0.6, which corresponds to an α value (Equation (2)),

$$K = \frac{[\text{host} - \text{guest complex}]}{[\text{host}][\text{guest}]} \tag{1}$$

$$\alpha = \frac{K \text{ enantiomer } A}{K \text{ enantiomer } B} \tag{2}$$

$$\log \alpha = \Delta \log K$$

of 4. Reviews of this work have been published [5, 6]. The symbol K, as used in this paper, refers to an apparent equilibrium constant for single-phase homogeneous solutions and for interactions between supported ligand hosts and guests.

A significant advance toward the use of these host–guest systems for separations was the attachment of the host to a solid support, thereby making desired separations possible using few stages [7–9]. This approach has several advantages. First, synthetic methods of modern supramolecular chemistry can be used to design and construct appropriate hosts capable of selective interaction with desired enantiomeric guests. Second, attachment of the chiral host to a solid support allows the separations system to be used many times before replacement of the host molecules is required. Third, the supported system is easily incorporated into a conventional engineered format that allows large quantities of enantiomers to be processed per relatively low amounts of the separations material. Fourth, the separations system has demonstrated high loading capacity and low solvent usage, along with high throughput.

Recently, a number of applications of this technology have begun to emerge in the pharmaceutical and life science industries where more efficient and low-cost chiral separations technology is desired. The attractiveness of the technology in these areas lies in the ability to design synthetic organic ligands that can discriminate with high factors among nearly identical molecular species. Because the separation agent is a synthetic organic, it is highly selective, yet extremely rugged in its operation.

This chapter provides: (i) a brief review of the chemistry involved in chiral host–chiral guest recognition involving primary amines; (ii) a description of a nonchromatographic (equilibrium or bind-release based) separation process developed by IBC Advanced Technologies Inc. (IBC) of American Fork, Utah, USA for use in enantiomeric separations, and a comparison of this technique with chromatographic methods together with advantages and disadvantages of the methods; (iii) a summary of the economic aspects of nonchromatographic separations together with details of the operating aspects of the nonchromatographic separations system; (iv) the use of the nonchromatographic system for the separation of valine enantiomers, and (v) areas of potential industrial interest for nonchromatographic separations.

8.2 Chemistry

The similar stereochemistry of enantiomers makes their separation on a large scale a challenging problem. We and others reasoned that separations of enantiomer guests containing primary amines could be facilitated by designing host receptors whose interaction with one of the enantiomers over the other was favored from a steric standpoint [6]. This approach was successful, and has been used to show significant differentiation by a given host between enantiomers in single phase homogeneous solvents [6].

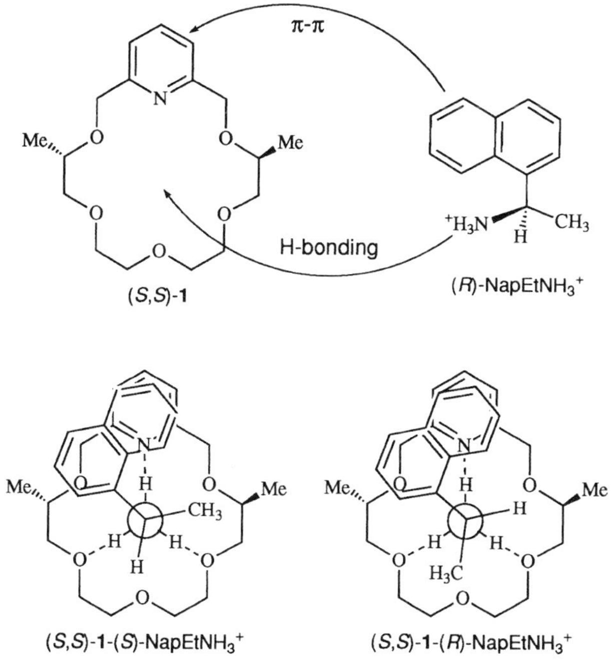

Fig. 8-1. Schematic representations of the interaction of the (R)NapEtNH$_3^+$ enantiomer guest with a chiral pyridine-18-crown-6 host (S,S)-1 and possible conformations of the (S,S)R$_2$P18C$_6$ – (R)NapEt complexes.

As shown in Fig. 8-1, primary amine-macrocycle formation occurs by a three-point interaction involving hydrogen bonding of the N–H groups of the amine and the oxygen and nitrogen atoms of the macrocycle. We used this concept to develop a number of pyridine-substituted cyclic polyethers, and studied their interaction with a number of primary organic amines. In the case illustrated in Fig. 8-1, log K (CH$_3$OH) values for the interaction of chiral ligand (S,S)-1 with (S)-NapEt and (R)-NapEt were 2.06 and 2.47, respectively [10]. Crystallographic data [11] show that the higher stability of the (S,S)-1-(R)-NapEt complex is steric in nature, as is illustrated schematically in Fig. 8-1. This selectivity is reversed if the (R,R)-1 ligand is used.

The structures of both complexes shown in Fig. 8-1 indicate that in each the primary molecular interaction is between the –NH$_3^+$ and the alternate macrocyclic ring atoms. Additional stabilization results in each case from π-π interaction of the naphthyl group from the guest and the pyridine group of the host [6, 10]. Increased differentiation between the enantiomers is accomplished by providing bulky groups at appropriate positions on the chiral host. The steric effect of these groups (CH$_3$ in the case of (S,S-1)) is sufficient to cause an appreciable difference in log K values (2.06 compared to 2.47). The effect of various hosts, host substituents, and guests on enantiomeric selectivity has been found to be appreciable in many cases [6]. Values of Δ log K ranging from 0.1 to 0.7 have been observed and reviewed previously. Since a Δ log K value of 0.6 corresponds to an α value of 4, it is apparent that these systems have potential use in separations of enantiomers containing appropriate functional groups. More recent work by IBC has shown ligand and solvent cases with even greater selectivities, as discussed later in this chapter.

8.3   Nonchromatographic Separation Process Description

Use of the concepts of molecular recognition principles as described above allows one to develop ligands that differentiate in their binding between two enantiomers. If the difference in the magnitude of binding becomes great enough (Δ log K of 0.6 or greater), one can achieve a nonchromatographic, or bind-release separation. This mode of separation confers a number of significant advantages in a separations process, which are discussed later in this chapter. The ligand chemistry allows a chiral host molecule to recognize and bind preferentially with one guest from a pair of enantiomers. The desired enantiomer (eutomer) is then released from the solid phase separations matrix and recovered in pure form. The unwanted enantiomer (distomer) can, in some cases, be isomerized in good yield to form a racemic mixture for recycling, and converted to the eutomer. In order to be useful in this method the chiral host must favor the binding of one enantiomer over the other by at least 0.6 log K units, corresponding to an α-value of 4.0 or greater.

The procedure used to accomplish enantiomer separation is shown in Fig. 8-2. It is assumed that the α value is large enough to permit essentially complete separation

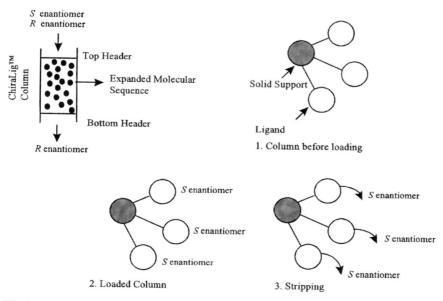

Fig. 8-2. Schematic representation of the separation of enantiomers *R* and *S* using a supported chiral macrocyclic ligand host.

in one pass through the column, though in practice this may not be the case, as will be discussed. In Fig. 8-2, the covalent attachment of the chiral host ligand to the support through a linker arm is shown schematically.

An α-value of 4.0 indicates a four-fold preference for one enantiomer over the other. The number of separations stages required to achieve 98.5 % purity for an α-value of four is three. Larger α values lead to either greater purity and/or fewer stages required for the separations. Technologies that can achieve separation stages of three or less can offer significant process economic and engineering benefits. Actual minimum enantiomeric purities required will vary from case to case.

The enantiomeric purity that can be obtained as a function of α for one, two, and three stages is given in Table 8-1. It is apparent that the higher the α value, the fewer the number of separations stages required to reach 99 % enantiomeric purity. For an α value of 5, the use of three stages allows one to obtain > 99 % purity. The required purity of the end-product defines the minimum performance requirement of the resin.

The relatively large preference for one enantiomer over another (α values $\geq$ 4) differentiates the nonchromatographic bind-release separations process from a chromatographic separations process. We will present results later in this chapter that demonstrate the use of the nonchromatographic bind-release process for the preparative scale separation of a particular enantiomer. Important preparative scale factors including solvent consumption, productivity and throughput, capital equipment, and ease of use can be positively impacted by the ability through the use of molecular recognition principles to chemically discriminate between enantiomers. A nonchromatographic system with an α value of four for a three-stage chiral separation is

Table 8-1. Enantiomeric purity obtained as a function of α values ≥ 4 and separation stages for nonchromatographic systems.

α	Number of stages	Purity obtained (%)
4	1	80
4	2	94.1
4	3	98.5
6	1	85.7
6	2	97.3
6	3	99.5
8	1	88.9
8	2	98.5
8	3	99.8
10	1	90.9
10	2	99.0
20	1	95.2
20	2	99.8

illustrated in Fig. 8-3. The two different ChiraLig™ resin particles (enantiomers of each other) are assumed in this case to have an alpha value of four for each of the *R* and *S* enantiomers in a pair.

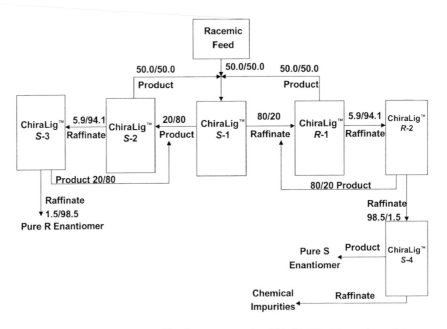

Fig. 8-3. Enantio- and chemopurification process using ChiraLig™ with α values of 4.

The separations process is described as follows. A racemic mixture is fed into the ChiraLig™ system. This feed stream is first passed through the ChiraLig™-S-1 column, selective for the *S* enantiomer. The eluate (strip solution) from this column is rich in *S* (80 % *S*, 20 % *R*). The relatively concentrated eluate is either changed in pH or solvent diluted to become similar to the pH and/or solvent of the original feed and then passed through ChiraLig™ *R*-1 and *R*-2 columns for further purification and enantiomer removal. The resulting stream from the third stage contains 98.46 % *S* enantiomer. A subsequent concentration and purification step is performed (if needed) at this stage by using a ChiraLig™-*S* column for extra purity or a different nonselective concentration column. The resulting eluted product is both optically (98.46 or 99.7 % depending on whether an extra concentration stage is included and whether ChiraLig™ is used) and chemically pure. The overall separation is efficient, minimizes the use of solvent, and is capable of producing a high-purity chiral product.

The raffinate from the ChiraLig™-*S*-1 column is rich in *R*, but still contains some (20 %) *S* enantiomer. This serves as the feed into the ChiraLig™-*S*-2 column followed by the ChiraLig™-*S*-3 column for final raffinate containing 98.46 % *R*. The eluents from the *S*-2 and *S*-3 columns are solvent or pH adjusted for recycle back into the system, so that an overall loss of *S* isomer of only ≈ 1.5 % occurs. The separations process uses the molecular recognition matrix to achieve 98.5 % purity of the *S* isomer in three stages. It also allows for potential isomerization of the 98.5 % *R*-isomer for recycle and ultimate conversion to the *S*-isomer. Increased α values above 4, as assumed of the 98.5 % *R*-isomer in this illustration, lead to higher purities and/or reduced separation stage requirements as described earlier.

8.4 Operating Aspects of Nonchromatographic Separation Systems

For an industrial-scale separation, operating and capital costs for a separations technology are critical parameters, and can make the difference between acceptance or ultimate rejection of a technology. These considerations are especially important in the pharmaceutical industry, where production costs must be minimized over the lifetime of the patent. When the drug comes off patent, production costs are even more important because the drug company must compete with other producers, and the drug price point drops. A gain in production efficiency from a superior separations process represents an additional source of competitive advantage to the drug company.

When investigating the suitability of a particular resin-bound separations process, the following factors are often important: (i) resin consumption; (ii) solvent usage; (iii) productivity–chemical, optical and volume yields; (iv) total number of separations steps; and (v) capital costs. For any particular process, these factors differ in their relative importance. However, when evaluating a new separations method it is useful to examine each of these factors. The nonchromatographic separation method compares favorably to current industry practice on these factors, including: (i) reduced number of process steps; (ii) high chemical, optical and volume yields; (iii) high-feed throughput; (iv) more open-ended solvent choice; (v) minimized solvent usage; and (vi) low resin consumption.

8.4.1 Reduced Number of Process Steps

Because the ChiraLig™ resin displays both high chemo- and enantioselectivity, the separations method allows for simultaneous chiral resolution and chemical separation, thus reducing the number of steps necessary to achieve high purity goals. The higher α-values of the ChiraLig™ resin make it possible to reach desired purities in fewer separations stages.

8.4.2 High Chemical, Optical and Volume Yields

High yields are possible due to the large capacity of the ChiraLig™ for the single enantiomer on each load cycle. A significant percentage of the available binding sites are used in each cycle to bind the eutomer.

8.4.3 High-Feed Throughput

The high selectivity of the system results in high-yield throughputs, and close to 100 % time usage of the system for feed introduction.

8.4.4 Open-Ended Solvent Choice

The ChiraLig™ resins are rugged and resistant to solvent erosion. The organic ligands are covalently bound to silica, polystyrene, or polyacrylate supports. This covalent linkage provides for long life and multiple recycle. Solvent choice is often open-ended, and the user is able to choose the best solvent in order to maximize other considerations, including achieving maximum solubility. There is the requirement, however, of either pH or solvent change to perform the stripping or elution step for each stage.

8.4.5 Minimized Solvent Usage

Solvent usage is drastically reduced due to several factors. The feed can be flowed through the ChiraLig™ columns nearly continuously and high feed concentrations can be used as needed. Rapid loading occurs due to reasonable binding kinetics. The subsequent release (elution) step is accomplished in just a few bed volumes, with rapid release of the pure enantiomer from the column. The resulting product concentrate is often of high tenor for easy collection and post-processing to the pure enantiomer. Often the main limitation to the product concentrate tenor is the solubility of the enantiomer in the elution solvent.

8.4.6 Low Resin Consumption

Resin consumption is low because of the highly efficient use of the capacity of the resin for the enantiomer during each cycle, as well as the material stability of the resin. The above benefits of the ChiraLig™ technology result in improved economics for the large-scale separation.

The use of nonchromatographic technology overcomes some of the existing issues one might have with the use of chromatography, including solvent consumption, and low overall yields. Based on the work performed in IBC's laboratories, and our commercial experience in molecular recognition technology (MRT), the scale-up of separations process systems based on ChiraLig™ is straightforward. Feasibility has been shown in the demonstration of the use of the principles described in this chapter to achieve efficient separations. Implementation of production systems involve the application of MRT scale-up principles already applied by IBC in the chemical process industry where synthetic organic ligands attached to silica or polymeric supports are in use. Design principles used for these large-scale systems can be applied to chiral applications, including column design, pre-filtration, flow metering and pumping, cycle control, construction materials, column aspect ratios, fill volumes and number, and cycle timing.

In the next section, a few illustrative examples of the use of ChiraLig™ for the analytical and three-stage preparative chiral separations involving amines and amino acids are presented and discussed.

8.5 Experimental Examples of Separations

8.5.1 Analytical Separation of Amine Enantiomers

The analytical capability of these matrices has been demonstrated for chiral amines [12, 13]. The procedure is illustrated in Fig. 8-4 for the separation of $NapEtNH_4^+$ ClO_4^-. Concentrated methanol/dichloromethane solutions of the racemic mixture were placed on a column containing the chiral macrocycle host. The enantiomers of the ammonium salts were resolved chromatographically with mixtures of methanol and dichloromethane as the mobile phase. The amounts of R and S salts in each fraction were determined by polarimetry. Because the chiral supported macrocycle interacts more strongly with S salts, the R salt passes through the column first and the S salt last, as seen in Fig. 8-4.

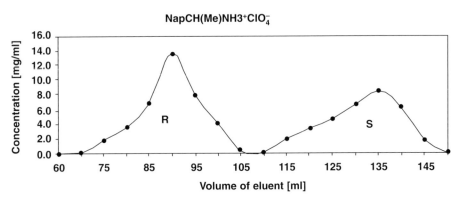

Fig. 8-4. Separation of R and S NaPEtNH$_3^+$ ClO$_4^-$ by a supported chiral macrocyclic ligand.

8.5.2 Automated Test Demonstration

In our laboratories at IBC, we have developed, synthesized and tested several chiral stationary phases to verify the enantiomeric discrimination capability. A few of these phases have been made in sufficient quantity to allow for multiple stage system testing and obtaining of the associated parameters for engineering design of process-scale systems. In this section, we discuss the results and implications for commercial-scale system operation. In order to assess the potential of MRT for meeting the industry separations needs, laboratory demonstrations are being performed at IBC. An automated, three-stage separations system was built for demonstration of actual full recovery and purification of both enantiomers from a 50/50 feed stream. One, two or three stages can be used for the separation so that percent purity values in the high 90s can be obtained with α values $\geq$ 4. The system operates in a manner similar to the process description in Section 8.3 and Fig. 8-3. The system includes up to three lead/trail columns per stage to allow for three columns in series to be operative or elution of one column in a stage while the other two columns are in load operation. Stage recycles, feedstock vessels, and other accessories are present to allow for full engineering demonstration and testing the economics, chemicals and equipment advantages of a nonchromatographic system. The compound chosen for initial evaluation was valine, $(CH_3)_2CHCH(NH_2)COOH$. This amino acid was chosen as representative of a relatively difficult separations case for racemic amino acids. In addition to α-amino acids, the technology developed can be, or is expected to be, applied to β-substituted-β-amino acids, chiral alcohols, and diols.

Enantiomer separation factors (α values) for valine and phenylalanine as well as their esters of 5–10 for phenylalanine and 4–10 for valine have been shown at the 0.1–1 g ChiraLig^{TM} scale. These α values vary as a function of solvent and other loading matrix factors (pH, salts, etc.). However, all of these cases show α values high enough to obtain reasonable enantiometric purity in less than or equal to three stages. The system with α value of $\approx$ 6 for the valine methyl ester enantiomers has the ability to load the valine onto the resin in H_2O containing LiClO$_4$ and also to elute in H_2O without the ClO_4^- present. The process flow sheet for the separation of valine enantiomers is shown in Fig. 8-5. This is an example where no organic solvents are required in either the load or elution to perform the separation.

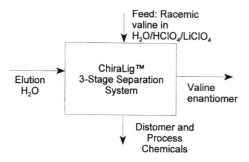

Fig. 8-5. Process flow sheet for the separation of valine enantiomers.

An understanding of how the individual stages operate can be gained by examining the specific load, wash, and elution curve data for the valine ester system. Data for the middle stage of the separation are given in Table 8-2 for a single column loading curve, in Table 8-3 for the first cycle polishing after running through three columns, in Table 8-4 for the second cycle column polishing, in Table 8-5 for the washing of a loaded column, and in Table 8-6 for the elution of a loaded column. The data in Table 8-2 for the single (lead) column loading curve show how at the beginning of the load on a fresh column both enantiomers are initially removed to below detection levels (see first three aliquots in particular). As the D-enantiomer begins preferentially to fill a significant portion of the bound ligand capacity sites, the L-enantiomer first begins to break through the column. Then, as the D-enantiomer begins to load fully and break through the column, it pushes off some of the previously bound L-enantiomer. This causes the effluent of the column to be higher in the L-enantiomer than the feed as the full column comes to equilibrium and binds the D-enantiomer in at least six-fold excess.

The results in Table 8-3 show the three-column polishing of valine ester for the first cycle of use (start-up). Three columns are sufficient to polish the D-enantiomer to below detection levels. Hence, for actual operation two columns in series may well be sufficient. However, the data in Table 8-3 also show how the preferentially rejected L-enantiomer initially breaks through the three columns in series at fairly high enantiomeric purity (up to $\geq 97\%$) even out of just the first stage during start-up or the first cycle. The data in Table 8-4 show the polishing results for two columns in series for the second cycle as the system begins to come to long-term equilibrium. Some D-enantiomer now begins to break-through the system. However, the ratio of L- to D-enantiomer breaking through the system of the total effluent volume is between 6 and 7. This is the equilibrium or α-based level due to the loading being $\approx 6:1$ in favor of the D-enantiomer.

The data in Table 8-5 show the washing of a fully loaded column (lead column) to remove the unbound feed solution remaining in the void space or volume of the

Table 8-2. Singe column (3.9 g ChiraLig®) loading curve of 24 mM D-methyl ester valine versus 25 mM L-methyl ester valine in 3 M LiClO$_4$ and 0.1 M HClO$_4$ at a flowrate of 0.4 ml min^{-1}.

Sample description	Concentration (mM)[a]	
	D-Valine-ester	L-Valine ester
Feed	25	25
0–5 ml feed effluent	b	b
5–9 ml feed effluent	b	b
9–14.25 ml feed effluent	b	0.9
14.25–19.5 ml feed effluent	b	2.8
19.5–24.6 ml feed effluent	b	6.4
24.6–33.5 ml feed effluent	2	15
33.5–53.5 ml feed effluent	10	30.5
53.5–61 ml feed effluent	17.5	32.8
61–68.5 ml feed effluent	20	34.5

[a] Analyzed by HPLC.
[b] Below detection limit of 0.2 mM.

Table 8-3. First cycle, three-column (each 3.9 g ChiraLig™) polishing curve for loading 25 mM D-methyl ester valine versus 25 mM L-methyl ester valine in 3 M LiClO$_4$ and 0.1 M HClO$_4$ at a flowrate of 0.4 ml min^{-1}.

Sample description	Concentration (mM)[a]	
	D-Valine-ester	L-Valine ester
Feed	25	25
0–70 ml feed effluent	b	2.9
70–76 ml feed effluent	b	3.0
76–81 ml feed effluent	b	9.6
81–85 ml feed effluent	b	11.7
85–89 ml feed effluent	b	14.1

[a] Analyzed by HPLC.
[b] Below detection limit of 0.2 mM.

Table 8-4. Second cycle, two-column (each 3.9 g ChiraLig™) polishing curve for loading 25 mM D-methyl ester valine versus 25 mM L-methyl ester valine in 3 M LiClO$_4$ and 0.1 M HClO$_4$ at a flowrate of 0.4 ml min^{-1}.

Sample description	Concentration (mM)[a]	
	D-Valine-ester	L-Valine ester
Feed	25	25
0–50 ml feed effluent	2.7	28.3
50–55 ml feed effluent	11.8	36.4
55–60 ml feed effluent	13.5	34.1
Σ mmol combined effluent	0.26	1.77

[a] Analyzed by HPLC.

Table 8-5. Single column (3.9 g ChiraLig™) washing curve after loading 25 mM D-methyl ester valine versus 25 mM L-methyl ester valine using a 3 M LiClO₄ and 0.1 M HClO₄ wash at a flowrate of 0.4 ml min⁻¹.

Sample description	Concentration (mM)[a]	
	D-Valine-ester	L-Valine ester
0–5 ml feed effluent	18.5	31.8
5–9 ml feed effluent	14.8	23.0
9–14 ml feed effluent	9.8	15.0
14–19 ml feed effluent	7.9	6.4

[a] Analyzed by HPLC.

Table 8-6. Single column elution of a 3.9 g ChiraLig™ column using an H₂O elution at 0.2 ml min⁻¹ following column loading (25 mM D-methyl ester valine versus 25 mM L-methyl ester valine) and washing.

Sample description	Concentration (mM)[a]	
	D-Valine-ester	L-Valine ester
0–5 ml elution effluent	22.0	8.8
5–7 ml elution effluent	28.0	5.2
7–10 ml elution effluent	29.6	1.7
10–15.5 ml elution effluent	24.7	2.8
15.5–19.5 ml elution effluent	21.5	2.1
19.5–23.5 ml elution effluent	13.8	1.0
23.5–27.5 ml elution effluent	9.2	0.8
Σ mmol combined eluent	0.57	0.09

[a] Analyzed by HPLC.

lead column. This is important since the nonselective feed would partially contaminate the bound valine ester if not washed out prior to the elution. The wash is the same as the feed matrix, so that minimal bound valine ester is lost. For purposes of collecting the wash data, the wash in Table 8-5 was collected for analysis. However, in the rest of the operation the wash volume is sent on to the trail column(s) so that this volume is not lost to the separation.

Finally, the data in Table 8-6 show the elution of the lead column. The eluent is H₂O. The driving force for the elution in this case is the lack of ClO_4^- present to act as an anion in the binding of the ammonium perchlorate salt pair. The D-enantiomer versus L-enantiomer ratio in the elution is slightly greater than 6:1, as expected by the inherent selectivity of the ligand. For this separation system, LiClO₄ is then added back to the eluent and the eluent is sent on as load to the next purification stage.

The other stages operate similarly. The α value of ≥ 6 allows for enantiomeric purity of ≥ 97 % for two stages and ≥ 99.5 % for three stages. Hence, these results demonstrate the applicability of the systems and technology to performing an enantiomeric separation in a two-to-three-stage nonchromatographic bind/release rather than chromatographic system. Full capacity of the ChiraLig™ resins are used in each cyclic use, and the corresponding expected high enantiomer throughout can thus be obtained.

8.6   Areas of Potential Industrial and Analytical Interest for Nonchromatographic Chiral Separations

Due to the benefits achieved, there are a significant number of applications areas for the chiral separations technology. Applications in the pharmaceutical industry the in three main areas: (i) analysis; (ii) drug development; and (iii) commercial production. In this industry, the need for chiral separations is to aid in each phase of the development process from discovery through preclinical and clinical development, and product launch. In the drug discovery process, the ultimate goal is the identification of a single compound for development. During this stage, extensive screening of available compounds is performed, along with animal testing. Small quantities of optically pure drug are needed at this stage for preliminary and advanced testing. Rapid turnaround time is required in order to screen candidate compounds quickly. The chiral separations phase should be readily synthesized and be relatively versatile in its ability to separate compounds from a class of drugs. For analysis of enantiomers, matrix versatility and rapid throughput are essential. The nonchromatographic matrices were shown to be very versatile and capable of separating a variety of chiral analytes in a wide range of solvents.

During the preclinical and clinical development stages, the requirements for optically pure drug quantity increases dramatically, from several grams to 10–100 kg. The drug is needed for animal studies (pharmacokinetics, metabolism, tissue distribution, and safety), and for human clinical studies in Phases I, II and III. Time is critical during this stage, and adequate supply of the separations resin must be available for production of the drug. During this phase, the purifications technology should exhibit high productivity, as kilogram quantities need to be produced in a short time period.

During product launch and steady-state production, the purifications goal is to treat large amounts of racemate (> 25 tons per year) with total process costs well under the targeted kg drug product price. Moreover, the purifications process must be robust and easy to operate.

The use of nonchromatographic ChiraLig™ separations technology provides flexibility for rapid separation and throughput of the small quantities of drug (10–100 kg) needed for the early phases of drug development. This technology also provides for highly efficient, low-cost large-scale racemate separations (> 100 kg) needed for the later stages of the drug commercialization process.

The availability of a new class of efficient separations technology for chiral drugs can have an impact on other business issues now at the forefront of the pharmaceutical industry. For example, the strategy of better drug life cycle management can be pursued by allowing a drug co. to develop and produce a single isomer of a racemic drug which would allow for application for additional patents that might extend the life of the drug by up to 20 years. The use of racemic switches is being adopted increasingly by drug firms as a management strategy as a means for innovators to counter graphic competition, and the combination of this strategy with the use of MRT adds an additional layer of intellectual property protection.

Amino acid separations represent another specific application of the technology. Amino acids are important synthesis precursors – in particular for pharmaceuticals – such as, for example, D-phenylglycine or D-parahydroxyphenylglycine in the preparation of semisynthetic penicillins. They are also used for other chiral fine chemicals and for incorporation into modified biologically active peptides. Since the unnatural amino acids cannot be obtained by fermentation or from natural sources, they must be prepared by conventional synthesis followed by racemate resolution, by asymmetric synthesis, or by biotransformation of chiral or prochiral precursors. Thus, amino acids represent an important class of compounds that can benefit from more efficient separations technology.

Specialized types of amino acids for synthesis applications represent a growing field in the biotechnology industry. Applications include peptide hormones and growth factors, immunological antigens, enzyme substrates, receptors and ligands, chemical drugs, bioactive peptides for research, combinatorial chemistry, drug discovery, pesticides, and artificial sweeteners. This market area represents a unique, high-value niche for companies, and the efficient production of speciality amino acids is a growing need.

Pharmaceuticals and intermediates represent another important class of compounds. General classes of drugs that may well lend themselves to the IBC technology include the chiral non-steroidal anti-inflammatory profen drugs, norephedryns, and intermediates for a number of important drug classes including β-blockers and racemic switch candidates.

8.7 Summary

Efficient optical purification of the classes of compounds discussed here presents a pivotal challenge to industry as we enter the 21st century. In order to meet this challenge, IBC has developed a series of novel complexing agents with high α-values for purification of enantiomers. These complexing agents are covalently attached to solid supports and the resultant ChiraLig™ materials used in non-chromatographic separations systems. We have demonstrated this technology for the separation of amines and amino acids, and discussed the potential for further open-ended application to other chiral entities. The ligands used are based on molecular recognition design principles, and can be prepared synthetically on an industrial scale. The resultant separations systems exhibit high selectivity factors, low solvent usage, and high throughput. The experimental results demonstrate the ability of the technology to perform a nonchromatographic bind-release separation in two or three stages. Full-scale, preparative enantiomeric separations systems based on this technology are expected to use engineering design principles already in practice at process industry installations currently using MRT. The technology shows great promise as an alternative to present methods of enantiomeric purification to meet the ever-present industry need for more efficient and effective separations.

References

[1] Chiral Separations: Applications and Technology, Ahuja, S., (ed.), Am. Chem. Soc.: Washington DC, **1997**.
[2] Chirality in Industry, Collins, A. N.; Sheldrake, G. N., Crosby, J., (eds.), John Wiley & Sons: New York, NY, **1996**.
[3] Stinson, S. C.; "Chiral Drug Interactions", *C & E News,* October 11, 1999; pp. 101–120.
[4] Balsehke, G.; Kraft, H. P., Markgraf, H., *Chem. Ber.* **1980**, *113,* 2318.
[5] Izatt, R. M., Zhu, C. Y., Huszthy, P., Bradshaw, J. S., *Crown Ethers: Toward Future Applications;* Cooper, S. R., (ed.); VCH: New York, NY, **1993**.
[6] Zhang, X. X., Bradshaw, J. S., Izatt, R. M., *Chem. Rev.* **1997**, *97,* 3313.
[7] Bradshaw, J. S., Bruening, R. L., Krakowiak, K. E., Tarbet, B. J., Bruening, M. L., Izatt, R. M., (the late) Christensen, J. J., *J. Chem. Soc., Chem. Commun.* **1988**, 812.
[8] Izatt, R. M., Bruening, R. L., Bruening, M. L., Tarbet, B. J., Krakowiak, K. E., Bradshaw, J. S., Christensen, J. J., *Anal. Chem.* **1988**, *60,* 1825.
[9] Izatt, R. M., Bradshaw, J. S., Bruening, R. L., Izatt, N. E., Krakowiak, K. E., *Metal Separation Technologies Beyond 2000: Integrating Novel Chemistry with Processing;* Liddell, K. C.; Chaiko, D. J. (eds), TMS: Warrendale, PA, **1999**, pp. 357–370.
[10] Davidson, R. B., Bradshaw, J. S., Jones, B. A., Dalley, N. K., Christensen, J. J., Izatt, R. M., Morin, F. G., Grant, D. M., *J. Org. Chem.* **1984**, *49,* 353.
[11] Davidson, R. B., Dalley, N. K., Izatt, R. M., Bradshaw, J. S., Campana, C. F. *Isr. J. Chem.* **1985**, *25,* 33.
[12] Huszthy, P., Bradshaw, J. S., Bordunov, A. V., Izatt, R. M., *ACH-Models Chem.* **1994**, *131,* 445.
[13] Kontös, Z., Huszthy, P., Bradshaw, J. S., Izatt, R. M., *Tetrahedron: Asymmetry,* **1999**, *10,* 2087.

9 Modeling and Simulation in SMB for Chiral Purification

Alírio E. Rodrigues and Luís S. Pais

9.1 Introduction

Simulated moving bed (SMB) is a powerful technique for preparative-scale chromatography which allows the continuous injection and separation of binary mixtures. The concept has been known since 1961 [1], the technology having been developed by UOP in the areas of petroleum refining and petrochemicals, and known as the "Sorbex" process [2, 3]. Other successful SMB processes in the carbohydrate industry are the production of high-fructose corn syrup ("Sarex" process) and the recovery of sucrose from molasses. Other companies developed alternative processes for the fructose-glucose separation [4]. Reviews on adsorptive processes are given in Rodrigues and Tondeur [5] and Rodrigues et al. [6].

The SMB concept was developed in order to overcome the limitations of conventional batch chromatography, mainly the discontinuous character of the process and its high cost due to the large eluent and adsorbent requirements. The interest for SMB operation increases for low-selectivity separations. At high selectivity, the column can be highly loaded and there is little difference between the performances of SMB and other discontinuous techniques. Generally, batch processes are more economic at very small scale if we take into account the lower cost of equipment. At larger scales, however, the savings in both solvent and chiral stationary phases make SMB technology the correct choice. In view of these properties, SMB technology is particularly appropriate for chiral separations. The resolution of enantiomers is usually a binary separation problem characterised by low selectivities and high costs of eluent and chiral stationary phases.

The SMB technology has found new applications in the areas of biotechnology, pharmaceuticals and fine chemistry [7–9]. SMB systems are available for a full range of production rates, from 10 to 1000 g per day, up to industrial scale of 5 to 50 tons per year. SMB chromatography is a useful tool for the pharmaceutical industry where, for preliminary biological tests, only a few grams of the chiral drug are needed. Furthermore, SMB can provide the two pure enantiomers, which are required for comparative biological testing [10]. On the other hand, pharmaceutical companies work with short drug development times. SMB technology, combined with proper chromatographic chiral stationary phases, can be a rapid system, easy to set up, and at the same time enhancing a high throughput of drug material [11]. However, the use of SMB technology in the pharmaceutical industry is not limited to laboratory tests. Its use at production scale is an alternative to, until now, leading techniques such as enantioselective synthesis or diastereoisomeric crystallization. In the past, large-scale chromatographic separations were limited mainly due to the high cost of the adsorbent, the high dilution of products, and the large amounts of mobile phase needed. With the introduction of the SMB technology, large-scale separations can now be carried out under cost-effective conditions.

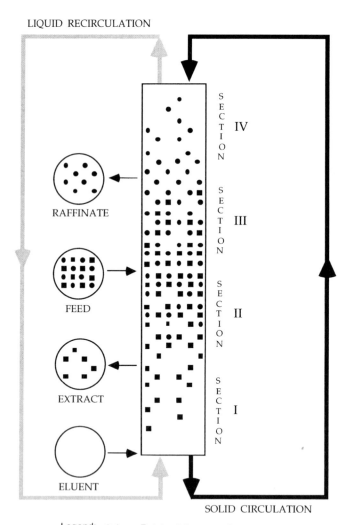

Fig. 9-1. Scheme of a true moving bed (TMB).

9.2 The SMB Concept

The principle of SMB operation can be understood by analyzing the equivalent true moving bed (TMB) process. In TMB, liquid and solid flow in opposite directions (Fig. 9-1) and are continuously recycled: the liquid flowing out of section IV is recycled to section I, while the solid coming out of section I is recycled to section IV. The feed is continuously injected in the middle of the system and two product lines can be collected: the extract, which is rich in the more retained species (B) and so is carried preferentially with the solid phase, and the raffinate, rich in the less-retained species (A) that move upwards with the liquid phase. Pure eluent is continuously injected at the inlet of section I, with the liquid recycled from the end of section IV. Because of the addition and withdrawal of the four streams (eluent, extract, feed, and raffinate), the TMB unit is divided into four sections and the liquid flow rates differ from section to section. This enables the four sections of the unit to perform different functions. In sections II and III the two components must move in opposite directions. The less retained component A must be desorbed and carried with the liquid phase, while the more retained species B must be adsorbed and carried with the solid phase. Section II is the zone of desorption of the species A, while section III is the zone of adsorption of the component B. In section IV, both components must be adsorbed in order to regenerate the eluent that will be recycled to the first zone. Section I is the zone of solid regeneration where both components must be desorbed in order to obtain a clean solid phase at the beginning of this zone. The operation of a TMB introduces problems concerning the movement of the solid phase. A uniform flow of both solid and liquid is difficult to obtain, and mechanical erosion of the adsorbent phase will also occur. In view of these difficulties, the SMB technique was developed in order to retain the process advantages of continuous and countercurrent flow without introducing the problems associated with the actual movement of the solid phase (Fig. 9-2).

In the SMB system the solid phase is fixed and the positions of the inlet and outlet streams move periodically. This shift, carried out in the same direction of the liquid phase, simulates the movement of the solid phase in the opposite direction. In the Sorbex SMB technology a rotary valve is used to change periodically the position of the eluent, extract, feed, and raffinate lines along the adsorbent bed. However, there are alternative techniques to perform the port switching. Scaling down of the Sorbex flowsheet becomes less economical than using a set of individual on-off valves connecting the inlet and outlet streams to each node between columns [12]. Recently, this technology has been applied in the pharmaceutical industry for the separation of chiral drugs [13]. Generally, for pharmaceutical applications small-sized SMB units are preferable, and a special emphasis is addressed to the versatility of this type of unit to perform different chiral separations without major design modifications.

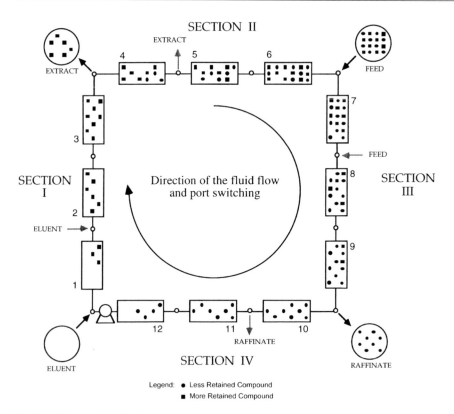

Fig. 9-2. Scheme of a Simulated Moving Bed (SMB).

9.3   Modeling of SMB Processes

The modeling of SMB can be addressed by simulating the system directly, taking into account its intermittent behavior, or by representing its operation in terms of a true countercurrent system. The first model represents the real SMB and considers the periodic switch of the injection and collection points. The second is developed assuming the equivalence with the TMB, where solid and fluid flow in opposite directions.

Models can be classified according to the description of the fluid flow as continuous-flow models (plug or axial dispersed plug flow) or as mixing cell models. Some authors considered mass transfer resistances by including a rate law, usually the linear driving force (LDF) approximation [14]; others used the equilibrium theory. Several authors carried out the modeling of the SMB considering linear equilibrium isotherms, namely for the glucose-fructose separation [15]. Nevertheless, SMB units usually operate at high feed concentrations in the region of nonlinear competitive adsorption. A constant selectivity factor model, e.g., multicomponent Langmuir isotherm, usually is used to describe this behavior. However, for many chiral separations the selectivity factor decreases with the increase of the concentration of chiral species and a concentration-dependent selectivity factor model is needed, such as the linear + Langmuir or the bi-Langmuir adsorption isotherm.

The question concerning the degree of equivalence between SMB and TMB strategies of modeling should be addressed. Three cases are analyzed for the SMB system: SMB4, constituted by four columns, one in each section; SMB8, with eight columns, two per section; and SMB12, with three columns per section. The mathematical models developed are based on the following assumptions: axial dispersed plug flow for the fluid; plug flow for the countercurrent solid flow in the TMB approach; the adsorbent particles are considered as homogeneous and mass transfer between fluid and solid is described by the LDF model.

9.3.1 The SMB Model

In the SMB operation, the countercurrent motion of fluid and solid is simulated with a discrete jump of injection and collection points in the same direction of the fluid phase. The SMB system is then a set of identical fixed-bed columns, connected in series. The transient SMB model equations are summarized below, with initial and boundary conditions, and the necessary mass balances at the nodes between each column.

- Mass balance in a volume element of the bed k:

$$\frac{\partial c_{ik}}{\partial t} = D_{L_k} \frac{\partial^2 c_{ik}}{\partial z^2} - v_k^* \frac{\partial c_{ik}}{\partial z} - \frac{(1-\varepsilon)}{\varepsilon} k(q_{ik}^* - q_{ik}) \tag{1}$$

where the subscripts i ($i = A, B$) refers to the species in the mixture, and k is the column number, c_{ik} and q_{ik} are the fluid and average adsorbed phase concentrations of species i in column k of the SMB unit, respectively, z is the axial coordinate, t is the time variable, ε is the bed porosity, v_k^* is the interstitial fluid velocity in the kth SMB column, D_{L_k} is the axial dispersion coefficient, and k is the intraparticle mass transfer coefficient.

- Mass balance in the particle:

$$\frac{\partial q_{ik}}{\partial t} = k(q_{ik}^* - q_{ik}) \tag{2}$$

where q_{ik}^* is the adsorbed phase concentration in equilibrium with c_{ik}.

- Initial conditions:

$$t = 0: \qquad c_{ik} = q_{ik} = 0 \tag{3}$$

- Boundary conditions for column k:

$$z = 0: \qquad c_{ik} - \frac{D_{Lk}}{v_k^*} \frac{dc_{ik}}{dz} = c_{ik,0} \tag{4}$$

where $c_{ik,0}$ is the inlet concentration of species i in column k.

$z = L_k$:

For a column inside a section and for extract and raffinate nodes

$$c_{ik} = c_{ik+1,0} \tag{5a}$$

For the eluent node $\qquad c_{ik} = \frac{v_I^*}{v_{IV}^*} c_{ik+1,0} \tag{5b}$

For the feed node $\qquad c_{ik} = \frac{v_{III}^*}{v_{II}^*} c_{ik+1,0} - \frac{v_F}{v_{II}^*} c_i^F \tag{5c}$

- Global balances:

Eluent node	$v_I^* = v_{IV}^* + v_E$	(6a)
Extract node	$v_{II}^* = v_I^* - v_X$	(6b)
Feed node	$v_{III}^* = v_{II}^* + v_F$	(6c)
Raffinate node	$v_{IV}^* = v_{III}^* - v_R$	(6d)

- Multicomponent adsorption equilibrium isotherm:

$$q^*_{Ak} = f_A (c_{Ak}, c_{Bk}) \text{ and } q^*_{Bk} = f_B (c_{Ak}, c_{Bk}) \tag{7}$$

Introducing the dimensionless variables $x = z/L_k$ and $\theta = t/t^*$, where t^* is the switch time interval, and L_k is the length of one SMB column, the model equations become:

$$\frac{\partial c_{ik}}{\partial \theta} = \gamma_k^* \left\{ \frac{1}{Pe_k} \frac{\partial^2 c_{ik}}{\partial x^2} - \frac{\partial c_{ik}}{\partial x} \right\} - \frac{(1-\varepsilon)}{\varepsilon} \alpha_k (q_{ik}^* - q_{ik}) \tag{8}$$

$$\frac{\partial q_{ik}}{\partial \theta} = \alpha_k (q_{ik}^* - q_{ik}) \tag{9}$$

- The initial and boundary conditions are the same presented before and, for $x = 0$ ($z = 0$), Equation (9.4) becomes:

$$c_{ik} - \frac{1}{Pe_k} \frac{dc_{ik}}{dx} = c_{ik,0} \tag{10}$$

The model parameters, in addition to the adsorption equilibrium parameters, are:

the ratio between solid and fluid volumes, $\qquad\dfrac{1-\varepsilon}{\varepsilon}$ (11)

the ratio between fluid and solid interstitial velocities, $\quad \gamma_k^* = \dfrac{v_k^*}{u_s} = \dfrac{v_k^*}{L_k/t^*}$ (12)

the Peclet number, $\qquad Pe_k = \dfrac{v_k^* L_k}{D_{Lk}}$ (13)

the number of mass transfer units, $\qquad \alpha_k = \dfrac{k\,L_k}{u_s} = k\,t^*$ (14)

where u_s is the interstitial solid velocity in the equivalent TMB model. Due to the switch of inlet and outlet lines, each column plays different functions during a whole cycle, depending on its location (section). The boundary conditions for each column change after the end of each switch time interval. This time-dependent boundary conditions leads to a cyclic steady state, instead of a real steady state as in the TMB model. This means that, after cyclic steady state is reached, the internal concentration profiles vary during a given cycle, but they are identical at the same time for two successive cycles.

9.3.2 The TMB Model

In the TMB model, the adsorbent is assumed to move in plug flow in the opposite direction of the fluid, while the inlet and outlet lines remain fixed. As a consequence, each column plays the same function, depending on its location. An equivalence between the TMB and the SMB models can be made by keeping constant the liquid velocity relative to the solid velocity, i.e., the liquid velocity in the TMB is:

$$v_j = v_j^* - u_s \qquad (15)$$

where v_j is the interstitial liquid velocity in the TMB. Also, the solid interstitial velocity in the TMB model u_s must be evaluated from the value of the switch time interval t^* of the SMB model, as

$$u_s = L_c/t^* \qquad (16)$$

where L_c is the length of one SMB column. The equations for the transient TMB model are:

- Mass balance in a volume element of the bed j:

$$\frac{\partial c_{ij}}{\partial t} = D_{Lj}\frac{\partial c_{ij}}{\partial z^2} - v_j\frac{\partial c_{ij}}{\partial z} - \frac{(1-\varepsilon)}{\varepsilon}k(q_{ij}^* - q_{ij}) \qquad (17)$$

- Mass balance in the particle:

$$\frac{\partial q_{ij}}{\partial t} = u_s\frac{\partial q_{ij}}{\partial z} + k(q_{ij}^* - q_{ij}) \qquad (18)$$

Initial conditions:

$$t = 0: \quad c_{ij} = q_{ij} = 0 \tag{19}$$

- Boundary conditions for section j:

$$z = 0: \quad c_{ij} - \frac{D_{L_j}}{v_j} \frac{dc_{ij}}{dz} = c_{ij,0} \tag{20}$$

where $c_{ij,0}$ is the inlet concentration of species i in section j.

$z = L_j$:

For the eluent node

$$c_{iIV} = \frac{v_I}{v_{IV}} c_{iI,0} \tag{21a}$$

For the extract node

$$c_{iI} = c_{iII,0} \tag{21b}$$

For the feed node

$$c_{iII} = \frac{v_{III}}{v_{II}} c_{iIII,0} - \frac{v_F}{v_{II}} c_i^F \tag{21c}$$

For the raffinate node

$$c_{iIII} = c_{iIV,0} \tag{21d}$$

And $\quad q_{iIV} = q_{iI,0}, \, q_{iI} = q_{iII,0}, \, q_{iII} = q_{iIII,0}, \, q_{iIII} = q_{iIV,0} \tag{22}$

- Global balances:

Eluent node

$$v_I = v_{IV} + v_E \tag{23a}$$

Extract node

$$v_{II} = v_I - v_X \tag{23b}$$

Feed node

$$v_{III} = v_{II} + v_F \tag{23c}$$

Raffinate node

$$v_{IV} = v_{III} - v_R \tag{23d}$$

- Multicomponent adsorption equilibrium isotherm:

$$q^*_{Aj} = f_A\,(c_{Aj},\, c_{Bj}) \text{ and } q^*_{Bj} = f_B\,(c_{Aj},\, c_{Bj}) \tag{24}$$

Introducing the dimensionless variables $x = z/L_j$ and $\theta = t/\tau_s$, with $\tau_s = L_j/u_s = N_s t^*$, where τ_s is the solid space time in a section of a TMB unit, L_j is the length of a TMB section, and N_s is the number of columns per section in a SMB unit, the model equations become:

$$\frac{\partial c_{ij}}{\partial \theta} = \gamma_j \left\{ \frac{1}{Pe_j} \frac{\partial^2 c_{ij}}{\partial x^2} - \frac{\partial c_{ij}}{\partial x} \right\} - \frac{(1-\varepsilon)}{\varepsilon} \alpha_j (q^*_{ij} - q_{ij}) \tag{25}$$

$$\frac{\partial q_{ij}}{\partial \theta} = \frac{\partial q_{ij}}{\partial x} + \alpha_j (q^*_{ij} - q_{ij}) \tag{26}$$

The initial and boundary conditions are the same presented before and, for $x = 0$, Equation (20) becomes:

$$c_{ij} - \frac{1}{Pe_j} \frac{dc_{ij}}{dx} = c_{ij,0} \tag{27}$$

The model parameters are similar to the ones presented for the SMB model, except that Pe_j and α_j are expressed in terms of the length of the TMB sections and the ratio between fluid and solid interstitial velocities is $\gamma_j = v_j/u_s$. The SMB and TMB models, defined by a set of partial differential equations, were numerically solved by using the *PDECOL* software [16] based on the method of orthogonal collocation in finite elements (OCFE). *PDECOL* implements the method of lines and uses a finite element collocation procedure for the discretization of the spatial variable which reduces the PDE system to an initial-value ODE system on the time variable. The time integration is accomplished with the ODE solver *STIFIB*, which is a modified version of the *GEARIB* ODE package developed by Hindmarsh [17]. The counter-current motion of fluid and solid in the SMB operation is achieved with a discrete jump of the injection (feed and eluent) and collection (extract and raffinate) points. Due to this switch of the inlet and outlet points, the boundary conditions for each column vary with time, changing at the end of each switch time interval. Hence, the SMB model must take into account time-dependent boundary conditions.

9.4 Simulation Results

The chromatographic resolution of bi-naphthol enantiomers was considered for simulation purposes [18]. The chiral stationary phase is 3,5-dinitrobenzoyl phenylglycine bonded to silica gel and a mixture of 72:28 (v/v) heptane/isopropanol was used as eluent. The adsorption equilibrium isotherms, measured at 25 °C, are of bi-Langmuir type and were proposed by the Separex group:

$$q_A^* = \frac{2.69\,c_A}{1+0.0336\,c_A+0.0466\,c_B} + \frac{0.10\,c_A}{1+c_A+3\,c_B} \tag{28a}$$

$$q_B^* = \frac{3.73\,c_B}{1+0.0336\,c_A+0.0466\,c_B} + \frac{0.30\,c_B}{1+c_A+3\,c_B} \tag{28b}$$

9.4.1 Equivalence Between TMB and SMB Modeling Strategies

The operating conditions and model parameters used in simulation for the TMB approach are presented in Table 9-1. The feed concentration of each enantiomer is 2.9 g L^{-1} and columns were 2.6 cm wide and 10.5 cm long. The section length was kept at 21 cm. The equivalence between TMB and SMB flow rates leads to $Q_I =$ 49.4 mL min^{-1} and $\gamma_I = 6.65$; therefore, $Q^*_I = 56.83$ mL min^{-1} and $\gamma^*_I = 7.65$. Table 9-2 presents the equivalencies in terms of model parameters that have to be made for the SMB systems with different subdivisions of the bed. Three cases are analyzed for the SMB: SMB4, SMB8 and SMB12. The length of each fixed-bed column in these cases was chosen by keeping constant the total length of each section. The value for switch time interval was then evaluated keeping constant the ratio L_c/t^*, the simulated solid velocity. Also, the number of mass transfer units per section is the same for TMB and SMB cases and is evaluated for $k = 0.1$ s^{-1}. Summarizing, all the SMB cases present the same operating conditions and model parameters at a section scale (equivalent to the TMB case), except for the degree of subdivision of the bed.

Table 9-1. Operating conditions and model parameters for the TMB approach.

TMB operation conditions:		Model parameters:	
Solid flow rate:	11.15 mL min^{-1}	Solid/fluid volumes, $(1 - \varepsilon)/\varepsilon = 1.5$	
Recycling flow rate:	27.95 mL min^{-1}	Ratio between fluid and solid velocities:	
Eluent flow rate:	21.45 mL min^{-1}	$\gamma_I = 6.65$; $\gamma_{II} = 4.23$	
Extract flow rate:	17.98 mL min^{-1}	$\gamma_{III} = 4.72$; $\gamma_{IV} = 3.76$	
Feed flow rate:	3.64 mL min^{-1}	Number of mass transfer units,	
Raffinate flow rate:	7.11 mL min^{-1}	$\alpha = 36.0$ ($k = 0.1$ s^{-1})	
		Peclet number, $Pe = 2000$	

Table 9-2. Equivalence between TMB and SMB with different subdivision of the bed.

Case	N_s	L_c	L_j	t^*	$u_s = L_c/t^*$	α	$N_s\alpha$	Pe
TMB	–	–	21	–	3.5	36	36	2000
SMB4	1	21	21	6	3.5	36	36	2000
SMB8	2	10.5	21	3	3.5	18	36	1000
SMB12	3	7	21	2	3.5	12	36	667

The cyclic steady state behavior, characteristic of a SMB operation, is shown in Fig. 9-3 for the case of 2-2-2-2 configuration in terms of the concentration of the two enantiomers in extract. Figure 9-4 shows the evolution of the internal concentration profiles for the more retained component after cyclic steady state is reached, during a switch time interval. Figure 9-5 shows the influence of the degree of subdivision of the bed in the transient concentration of extract, and makes the comparison with the TMB approach. The behavior of the SMB is predicted in three ways: the transient evolution of concentration profiles; the average concentration evaluated at each switch time interval, and the instantaneous concentration evaluated at half-time between two successive switchings. These figures show the transient evolution during the first five cycles. Although the switch time interval depends on the degree of subdivision of the bed, the duration of a full cycle will be 24 min for all SMB cases. It is clear that differences between SMB and TMB predictions are attenuated with the increase of the number of subdivisions. Figure 9-6 compares the steady state

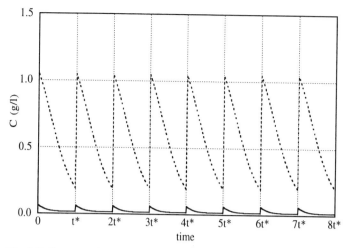

Fig. 9-3. Concentration versus time in the extract for SMB8 at cyclic steady state. --- more retained component; — less retained component.

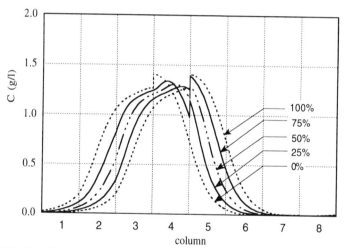

Fig. 9-4. Cyclic steady state internal concentration profiles of the more retained component during a switch time interval (start, 25 %, 50 %, 75 %, and at the end of a switch time interval) for SMB8.

internal concentration profiles, evaluated at half-time between switchings, for TMB and SMB cases. The major difference appears for the SMB4 case, while small deviations occur between SMB8 and SMB12 behaviors.

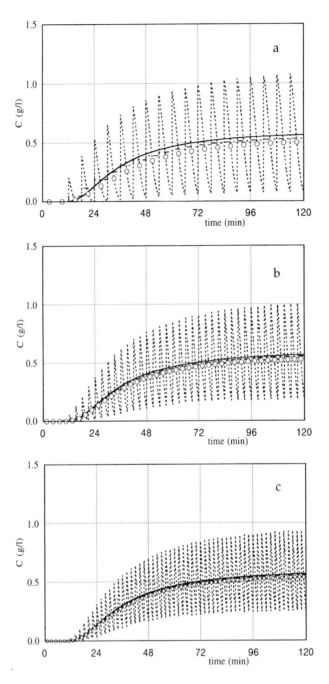

Fig. 9-5. Transient evolution (first 5 cycles) of the concentration of the more retained component in the extract for (a) SMB4, (b) SMB8, and (c) SMB12. Solid line, TMB; dotted line, SMB; stepped dotted line, SMB approach with average concentration over a switch time interval; o, SMB instantaneous concentration evaluated between switchings.

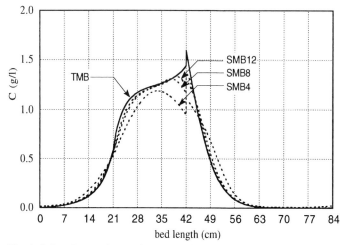

Fig. 9-6. Steady state internal concentration profiles in the TMB approach and SMB cases at half-time between switchings for the more retained component.

Nevertheless, near the feed point, there is a difference between TMB and all SMB cases due to the fact that the internal flow rates in the TMB are smaller than in the SMB, leading to a small dilution of the feed stream. As a consequence, near the feed inlet, TMB concentrations will be higher than in the SMB operation. The raffinate and extract purities in SMB units with four (95.2 % and 89.5 %), eight (98.7 % and 95.9 %) and 12 columns (99.1 % and 96.8 %) are increasing towards the one obtained in the equivalent TMB unit (99.3 % and 97.7 %). The optimum degree of subdivision of the SMB unit will depend of the difficulty of the separation and the product purity requirements. Typically, systems for the pharmaceutical industry have six to 16 columns.

9.4.2 Separation Regions

The design problem of a TMB consists on setting the flow rates in each section to obtain the desired separation. Some constraints have to be met to recover the less-adsorbed component A in the raffinate and the more retained component B in the extract. These constraints are expressed in terms of the net fluxes of components in each section (see Fig. 9-1). In section I, both species must move upwards, in sections II and III the light species must move upwards, while the net flux of the more retained component must be downwards, and in section IV the net flux of both species have to be downwards, i.e.,

$$\frac{Q_I c_{BI}}{Q_S q_{BI}} > 1 \; ; \quad \frac{Q_{II} c_{AII}}{Q_S q_{AII}} > 1 \; and \quad \frac{Q_{II} c_{BII}}{Q_S q_{BII}} < 1 \; ; \tag{29}$$

$$\frac{Q_{III} c_{AIII}}{Q_S q_{AIII}} > 1 \; and \quad \frac{Q_{III} c_{BIII}}{Q_S q_{BIII}} < 1 \; ; \quad \frac{Q_{IV} c_{AIV}}{Q_S q_{AIV}} < 1$$

where Q_I, Q_{II}, Q_{III}, Q_{IV} are the volumetric liquid flow rates in the various sections of the TMB, Q_S is the solid flow rate, c_{Aj}, c_{Bj} are the concentrations of species A and B in the liquid phase and q_{Aj}, q_{Bj} are the adsorbed concentrations of components A and B, in section j. The same constraints can be expressed in terms of fluid and solid interstitial velocities. Defining the dimensionless parameter:

$$\Gamma_{ij} = \frac{\varepsilon}{1-\varepsilon} \gamma_j \frac{c_{ij}}{q_{ij}} \tag{30}$$

the constraints become $\Gamma_{BI} > 1$; $\Gamma_{AII} > 1$ and $\Gamma_{BII} < 1$; $\Gamma_{AIII} > 1$ and $\Gamma_{BIII} < 1$; $\Gamma_{AIV} < 1$.

For the case of a binary system with linear adsorption isotherms, very simple formulas can be derived to evaluate the better TMB flow rates [19, 20]. For the linear case, the net fluxes constraints are reduced to only four inequalities, which are assumed to be satisfied by the same margin β ($\beta > 1$) and so:

$$\frac{Q_I}{Q_S K_B} = \beta \; ; \quad \frac{Q_{II}}{Q_S K_A} = \beta \; ; \quad \frac{Q_{III}}{Q_S K_B} = \frac{1}{\beta} \; ; \quad \frac{Q_{IV}}{Q_S K_A} = \frac{1}{\beta} \tag{31}$$

where K_A and K_B are the coefficients of the linear isotherms for the less and more retained species, respectively. The flow rates for TMB operation are then: $Q_E = (\alpha \beta^2 - 1) Q_{RF}$, $Q_X = (\alpha - 1) \beta^2 Q_{RF}$, $Q_F = (\alpha - \beta^2) Q_{RF}$ and $Q_R = (\alpha - 1) Q_{RF}$ where Q_E, Q_X, Q_F, and Q_R are the eluent, extract, feed, and raffinate volumetric flow rates, respectively. The volumetric flow rate in the section IV is the recycling flow rate, $Q_{RF} = K_A Q_S/\beta$ and $\alpha = K_B/K_A$ is the selectivity factor of the binary linear system. The total inlet or outlet volumetric flow rate is given by $Q_E + Q_F = Q_X + Q_R = (\alpha - 1)(1 + \beta^2) Q_{RF}$. The specification of β and the solid flow rate (or, alternatively, one of the liquid flow rates) defines all the flow rates throughout the TMB system. The β parameter has a higher limit, since the feed flow rate must be higher than zero, $1 < \beta < \sqrt{\alpha}$. The case of $\beta = 1$ corresponds to the situation where dilution of species is minimal, and the extract and raffinate product concentrations approach the feed concentrations. In fact, for $\beta = 1$, we obtain $Q_E = Q_X = Q_F = Q_R = (\alpha - 1) Q_{RF} = (K_B - K_A) Q_S$.

In the case of complete separation, the concentrations of the component A in the raffinate and of the component B in the extract are, respectively, $C_A^R = C_A^F Q_F/Q_R = C_A^F (\alpha - \beta^2)/(\alpha - 1)$ and $C_B^X = C_B^F Q_F/Q_X = C_B^F (\alpha - \beta^2)/(\alpha - 1)\beta^2$. Following the equivalence of internal flow rates, it results that the inlet and outlet flow rates are the same for the two operating modes, and

$$Q_{RF}^* = Q_{RF} + \frac{\varepsilon}{(1-\varepsilon)} Q_S = \left[\frac{K_A}{\beta} + \frac{\varepsilon}{(1-\varepsilon)} \right] Q_S = \left[1 + \frac{(1-\varepsilon)}{\varepsilon} \frac{K_A}{\beta} \right] \frac{\varepsilon V_c}{t^*} \tag{32}$$

where Q_{RF}^* is the recycling flow rate in the SMB operation.

For nonlinear systems, however, the evaluation of the flow rates is not straight-forward. Morbidelli and co-workers developed a complete design of the binary separation by SMB chromatography in the frame of Equilibrium Theory for various adsorption equilibrium isotherms: the constant selectivity stoichiometric model [21, 22], the constant selectivity Langmuir adsorption isotherm [23], the variable selectivity modified Langmuir isotherm [24], and the bi-Langmuir isotherm [25]. The region for complete separation was defined in terms of the flow rate ratios in the four sections of the equivalent TMB unit:

$$m_j = \frac{Q_j^* t^* - \varepsilon V_c}{(1-\varepsilon)V_c} \tag{33}$$

which are related to the γ_j ratios used in this work by:

$$\gamma_j = \frac{1-\varepsilon}{\varepsilon} m_j \tag{34}$$

The necessary and sufficient conditions for complete separation considering linear isotherms $q^*_i = K_i C_i$, $(i = A, B)$, are $K_B < m_1 < \infty$, $K_A < m_2 < m_3 < K_B$, and $0 < m_4 < K_A$. The region for complete separaration is the area aWb in Fig. 9-7.

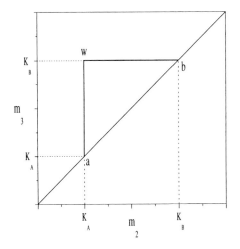

Fig. 9-7. Region for complete separation under Equilibrium Theory. Linear adsorption isotherms.

Table 9-3 presents the necessary and sufficient conditions for complete separation considering Langmuir isotherms, $q^*_i = Q b_i C_i/(1 + b_A C_A + b_B C_B)$, $(i = A, B)$. The region for complete separation for this kind of isotherms is the area aWb in Fig. 9-8. The equations presented in this table can also be used when a modified Langmuir isotherm (linear + Langmuir) is considered $q^*_i = m C_i + Q b_i C_i/(1 + b_A C_A + b_B C_B)$, $(i = A, B)$. In this case, the complete separation region must be shifted by using the

Table 9-3. Operating conditions for complete separation under Equilibrium Theory. Langmuir adsorption isotherms (see Fig. 9-8).

$$\lambda_B = m_{1,min} < m_1 < \infty; \; m_{2,min}(m_2,m_3) < m_2 < m_3 < m_{3,max}(m_2,m_3)$$

$$0 < m_4 < m_{4,max}(m_2,m_3) = \frac{1}{2}\{\lambda_A + m_3 + b_A C_A^F(m_3 - m_2)$$

$$-\sqrt{\left[\lambda_A + m_3 + b_A C_A^F(m_3 - m_2)\right]^2 - 4\lambda_A m_3}\}$$

Boundaries of the complete separation region in the (m_2, m_3) plane:

Straight line wr: $\left[\lambda_B - \omega_G(1 + b_B C_B^F)\right]m_2 + b_B C_B^F \omega_G m_3 = \omega_G(\lambda_B - \omega_G)$

Straight line wa: $\left[\lambda_B - \lambda_A(1 + b_B C_B^F)\right]m_2 + b_B C_B^F \lambda_A m_3 = \lambda_A(\lambda_B - \lambda_A)$

Curve rb: $m_3 = m_2 + \dfrac{\left(\sqrt{\lambda_B} - \sqrt{m_2}\right)^2}{b_B C_B^F}$

Straight line ab: $m_3 = m_2$

The coordinates of the intersection points are given by:

$$a\left(\lambda_A, \lambda_A\right) \quad ; \quad b\left(\lambda_B, \lambda_B\right)$$

$$r\left(\frac{\omega_G^2}{\lambda_B}, \frac{\omega_G\left[\omega_F(\lambda_B - \omega_G)(\lambda_B - \lambda_A) + \lambda_A \omega_G(\lambda_B - \omega_F)\right]}{\lambda_A \lambda_B(\lambda_B - \omega_F)}\right)$$

$$w\left(\frac{\lambda_A \omega_G}{\lambda_B}, \frac{\omega_G\left[\omega_F(\lambda_B - \lambda_A) + \lambda_A(\lambda_A - \omega_F)\right]}{\lambda_A(\lambda_B - \omega_F)}\right)$$

with $\omega_G > \omega_F > 0$, given by the roots of the quadratic equation:

$$\left(1 + b_A C_A^F + b_B C_B^F\right)\omega^2 - \left[\lambda_A(1 + b_B C_B^F) + \lambda_B(1 + b_A C_A^F)\right]\omega + \lambda_A \lambda_B = 0$$

In the above equations, C_A^F and C_B^F are the feed concentrations of species A and B, respectively, and $\lambda_i = Q b_i \quad , \quad (i = A, B)$

relation $m_j^M = m_j^L + m$ where m_j^L is the value obtained considering only the Langmuir term (by using equations in Table 9-3) and m is the linear coefficient of the linear + Langmuir isotherm. However, if mass transfer resistance is important, this region for complete separation is reduced [26–28].

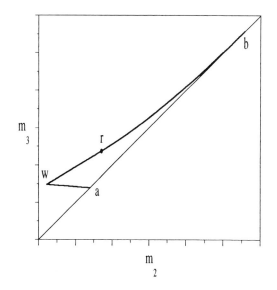

Fig. 9-8. Region for complete separation under Equilibrium Theory. Langmuir adsorption isotherms.

9.5 The Steady State TMB Model

The steady state TMB model equations are obtained from the transient TMB model equations by setting the time derivatives equal to zero in Equations (25) and (26). The steady state TMB model was solved numerically by using the COLNEW software [29]. This package solves a general class of mixed-order systems of boundary value ordinary differential equations and is a modification of the COLSYS package developed by Ascher et al. [30, 31].

9.5.1 Performance Parameters

The cyclic steady state SMB performance is characterized by four parameters: purity, recovery, solvent consumption, and adsorbent productivity. Extract (raffinate) purity is the ratio between the concentration of the more retained component (less retained) and the total concentration of the two species in the extract (raffinate). The recovery is the amount of the target species obtained in the desired product stream per total amount of the same species fed into the system. Solvent consumption is the total amount of solvent used (in eluent and feed) per unit of racemic amount treated. Productivity is the amount of racemic mixture treated per volume of adsorbent bed and per unit of time.

   The resolution of the bi-naphthol enantiomers was used for simulation purposes. A reference case relative to a 8-column configuration of the SMB, based on the values of operating variables and model parameters shown in Table 9-4 was chosen.

Table 9-4. SMB operating conditions and model parameters for the reference case.

SMB operation conditions:		Model parameters:
Feed concentration:	2.9 g/l each	Solid/fluid volumes, $(1 - \varepsilon)/\varepsilon = 1.5$
Switch time interval:	3 min	Ratio between fluid and solid velocities:
Recycling flow rate:	35.38 ml/min	$\gamma^*_I = 7.65$; $\gamma^*_{II} = 5.23$
Eluent flow rate:	21.45 ml/min	$\gamma^*_{III} = 5.72$; $\gamma^*_{IV} = 4.76$
Extract flow rate:	17.98 ml/min	Number of mass transfer units,
Feed flow rate:	3.64 ml/min	$\alpha_k = 18.0$ ($k = 0.1$ s^{-1})
Raffinate flow rate:	7.11 ml/min	Peclet number, $Pe_k = 1000$
Columns:		Column diameter: 2.6 cm
Configuration: 2 columns per section		Column length: 10.5 cm

The equivalent TMB operating conditions and model parameters for the reference case were given in Table 9-1 and Fig. 9-9 presents the corresponding steady state internal concentration profiles obtained with the simulation package. The extract and raffinate purities were 97.6 % and 99.3 %, respectively; the recoveries were 99.3 % and 97.6 % for the extract and raffinate streams. The solvent consumption was 1.19 L g^{-1} and the productivity was 68.2 g/day · L of bed.

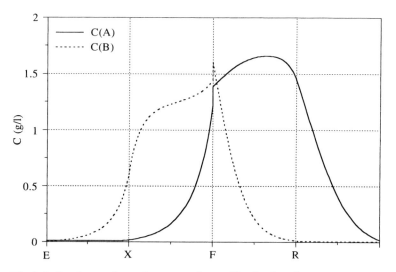

Fig. 9-9. Steady-state internal concentration profiles for the reference case.

9.5.1.1 Effect of the Switch Time Interval

The influence of the switch time interval on the purity is shown in Fig. 9-10. A change on the switch time interval will lead to a change on the equivalent solid flow rate throughout the system. In all runs the inlet and outlet flow rates, as well as the internal liquid flow rates in all the four sections of the SMB unit, are kept constant.

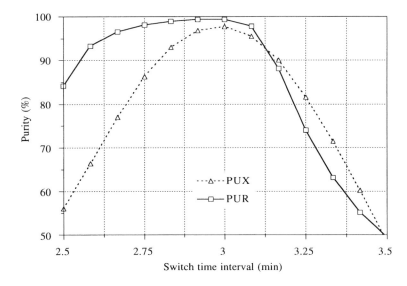

Fig. 9-10. Effect of the switch time interval on the purity.

Increasing the switch time interval is equivalent to decrease the solid flow rate and the net fluxes of components in all sections of the TMB unit will be pushed in the same direction of the liquid phase. This implies that, first, the more retained species will move upwards in section III and will contaminate the raffinate stream; and the less retained species will move upwards in section IV, will be recycled to section I, and will contaminate also the extract stream. The decrease of the switch time interval will have similar consequences. The equivalent solid flow rate will increase and the net fluxes of component in all four sections of the TMB unit will be pushed in the opposite direction of the liquid phase. This implies that, first, the less-retained species will move downwards in section II and will contaminate the extract stream; and the more retained component will also move downwards in section I, will be recycled with the solid to the section IV, and will contaminate the raffinate stream. It is possible to obtain simultaneously high purities and recoveries in a SMB, but the tuning must be carefully carried out.

9.5.1.2 Effect of the Mass Transfer Resistance on the SMB Performance

The influence of the mass transfer resistance on the purity and on the steady state internal concentration profiles are shown in Figs. 9-11 and 9-12. A higher value for the mass transfer coefficient corresponds to a situation where mass transfer resistance is less important, and a better performance of the SMB will be obtained with sharper internal concentration profiles.

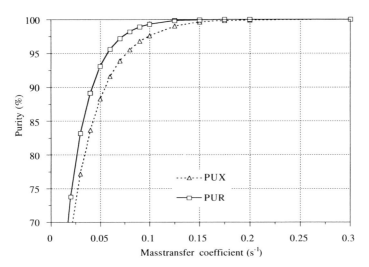

Fig. 9-11. Effect of the mass transfer resistance on purity.

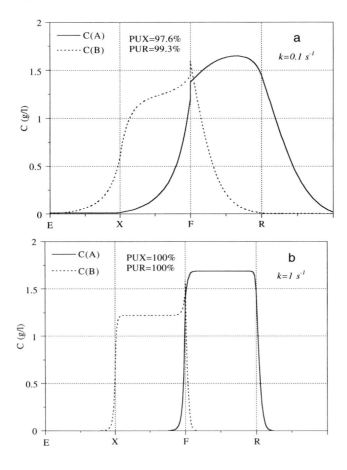

Fig. 9-12. Effect of the mass transfer resistance on the internal concentration profiles:
(a) $k = 0.1 \text{ s}^{-1}$; (b) $k = 1 \text{ s}^{-1}$.

9.5.2 Prediction of the Separation Regions

The conditions for a complete separation of a binary mixture can be defined in terms of the γ_j model parameters, which are directly related with the TMB (SMB) operating variables (fluid and solid velocities in the four sections of the TMB unit). From the constraints presented, those related to sections II and III play the crucial role on the separation performance of the TMB. It is in these central zones that the separation between the two species takes place. The role of the adjacent sections (I and IV) is to prevent cross-contamination and to allow the improvement of the continuous operation of the system by regenerating the solid and liquid phases. Taking into account these considerations, a region of complete separation in a $\gamma_{III}-\gamma_{II}$ plane can be defined. Considering that the constraints concerning sections I and IV are fulfilled, the $\gamma_{III}-\gamma_{II}$ plot is an important tool in the choice of best operating conditions.

The first case studied concerns the situation where axial dispersion and mass transfer resistances are slightly important. The value for mass transfer coefficient used in this case was $k = 0.5$ s^{-1} ($\alpha = 180$). Following the same methodology used to study the effect of the operating conditions and model parameters on the SMB performance, the $\gamma_{III}-\gamma_{II}$ plot was built, keeping constant the recycling (flow rate in section IV) and solid flow rates, and so $\gamma_{IV} = 3.76$. The total inlet or outlet flow rates were also kept constant in all simulations and equal to 25.09 mL min^{-1}. A TMB solid flow rate of 11.15 mL min^{-1} corresponds to a switch time interval of 3 min in the equivalent SMB unit; a recycling flow rate of 27.95 mL min^{-1} in the TMB corresponds to a recycling flow rate of 35.38 mL min^{-1} in the SMB. Other model parameters were solid/fluid ratio equal to 1.5 and $Pe = 2000$. The configuration was 2-2-2-2 with a section length of 21 cm. The feed concentration was 2.9 g L^{-1} of each enantiomer.

Figure 9-13 shows the $\gamma_{III}-\gamma_{II}$ plot obtained for the first case where four regions are defined: a region of complete separation, two regions where only one outlet stream is 100 % pure and a last region where neither of them is 100 % pure. The closed circles are numerical results based on the equivalence between the TMB and the SMB; the thick lines connect those results. The thin line in Fig. 9-13 has two branches. The diagonal $\gamma_{III}-\gamma_{II}$ corresponds to zero feed flow rate; therefore, γ_{III} must be higher than γ_{II}. The horizontal branch $\gamma_{III} \approx 3.76$ corresponds to zero raffinate flow rate; in this case, the extract flow rate is 25.09 mL min^{-1}.

In order to simplify the understanding of these plots, the relations between the TMB or SMB flow rates and the γ_j model parameters can be developed. The relationship between the internal liquid flow rates in the TMB unit, Q_j, and the γ_j model parameters is given by $Q_j = \dfrac{\varepsilon}{1-\varepsilon}\gamma_j Q_S$ where Q_s is the volumetric solid flow rate in the TMB system. The inlet and outlet flow rates can also be expressed in terms of the γ_j model parameters $Q_E = \dfrac{\varepsilon}{1-\varepsilon}(\gamma_I - \gamma_{IV})Q_S$, $Q_X = \dfrac{\varepsilon}{1-\varepsilon}(\gamma_I - \gamma_{II})Q_S$,

$$Q_F = \frac{\varepsilon}{1-\varepsilon}(\gamma_{III} - \gamma_{II})\,Q_S, \qquad \text{and } Q_R = \frac{\varepsilon}{1-\varepsilon}\,(\gamma_{III} - \gamma_{IV})\,Q_S.$$

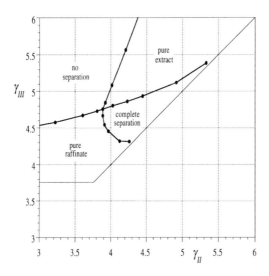

Fig. 9-13. Regions of operation of the TMB in a $\gamma_{III}\text{-}\gamma_{II}$ plot. The closed circles indicate simulation results. Mass transfer coefficient: $k = 0.5 \text{ s}^{-1}$.

Since these plots are built by keeping constant the total inlet or outlet flow rates, Q_T, we conclude that

$$Q_T = Q_E + Q_F = Q_X + Q_R = \frac{\varepsilon}{1-\varepsilon}(\gamma_I - \gamma_{II} + \gamma_{III} - \gamma_{IV})\, Q_S.$$

In addition, both the recycling and the solid flow rates are also kept constant.

Hence, γ_{IV} is also constant and equal to $\gamma_{IV} = \dfrac{1-\varepsilon}{\varepsilon}\dfrac{Q_{RF}}{Q_S}$.

Therefore, γ_I is a linear function of γ_{II} and γ_{III}: $\gamma_I = \dfrac{1-\varepsilon}{\varepsilon}\dfrac{(Q_{RF}+Q_T)}{Q_S} + \gamma_{II} - \gamma_{III}$ and,

similarly, $\gamma_I = \dfrac{(Q^*_{RF}+Q_T)t^*}{\varepsilon V_c} - 1 + \gamma_{II} - \gamma_{III}$,

where V_c is the volume of one SMB column, Q^*_{RF} is the SMB recycling flow rate, and t^* the switch time interval in the SMB operation.

The $\gamma_{III}\text{-}\gamma_{II}$ plots provide possible operating conditions that allow the separation of a binary mixture. The separation regions are built imposing that the constraints concerning sections I and IV are fulfilled. Since γ_I is a linear function of γ_{II} and γ_{III}, we must ensure that the region of complete separation of both species is not affected by the value of γ_I. In fact, in section I (between the eluent and extract nodes) the objective is to ensure that the more retained species B move upwards, in the same direction of the liquid phase, $\dfrac{\varepsilon}{1-\varepsilon}\dfrac{c_{BI}}{q_{BI}}\gamma_I > 1$. The worst situation that can occur in this section is when we are dealing with low concentrations, i.e., linear conditions. Hence, if mass transfer resistance is negligible, $\gamma_I > \dfrac{1-\varepsilon}{\varepsilon}K_B$ where K_B is the initial slope of the adsorption isotherm for the more retained species. In our case

$\gamma_I = 7.135 + \gamma_{II} - \gamma_{III}$ and, since $K_B = 4.03$, the critical value for γ_I is $\gamma_I > \gamma_I^C = 6.045$. As we concluded earlier, γ_I is constant along a straight line parallel to the diagonal $\gamma_{III} = \gamma_{II}$. Furthermore, γ_I diminishes as this straight line moves away from the diagonal $\gamma_{III} = \gamma_{II}$. Also, the vertex of the complete separation region is the furthest point from the diagonal and corresponds to the optimal conditions, because both solvent consumption and adsorbent productivity are optimized. This vertex point corresponds also to the lower value for γ_I in the complete separation region.

The vertex of the complete separation region, evaluated for $k = 0.5$ s^{-1}, is characterized by $\gamma_{II} = 3.86$ and $\gamma_{III} = 4.75$. The corresponding minimum value of γ_I is $\gamma_I = 7.135 + 3.86 - 4.75 = 6.245$, which is still higher than the critical value $\gamma_I^C = 6.045$. We conclude that, considering negligible mass transfer, within the whole separation region built under these conditions, the γ_I value does not affect the SMB performance. It should be pointed out that the presence of mass transfer resistance can influence the critical value for γ_I and the form of the separation region [32].

If mass transfer resistance is important, we may not obtain a region of 100 % purity for both species. This case is illustrated using the same operating conditions and model parameters of the previous one, except that the mass transfer coefficient is now $k = 0.1$ s^{-1} ($\alpha = 36$). Figure 9-14 emphasizes the effect of the mass transfer resistances presenting the separation regions obtained for the two values of the mass transfer coefficient, $k = 0.5$ s^{-1} ($\alpha = 180$) (open squares) and $k = 0.1$ s^{-1} ($\alpha = 36$) (closed squares), using a 99 % purity criteria. Inside the regions limited by the square points obtained numerically, both the raffinate and the extract are at least 99 % pure. It is clear from these figures that the mass transfer resistance reduces the region of separation of both species and that the region obtained for a lower mass transfer coefficient ($k = 0.1$ s^{-1}) lies inside the region obtained when mass transfer resistance is not so important ($k = 0.5$ s^{-1}).

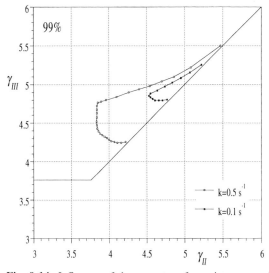

Fig. 9-14. Influence of the mass transfer resistance on the separation region: $\gamma_{III}-\gamma_{II}$ plot for a 99 % purity criteria.

The vertex of a separation region points out the better operating conditions, since it is the point where the purity criteria are fulfilled with a higher feed flow rate (and so lower eluent flow rate). Hence, in the operating conditions specified by the vertex point, both solvent consumption and adsorbent productivity are optimized. Comparing the vertex points obtained for the two values of mass transfer coefficient, we conclude that the mass transfer resistance influences the better SMB operating conditions. Moreover, this influence is emphasized when a higher purity requirement is desired [28].

9.6 Operation of the SMB Unit

The SMB pilot unit used in this work is the Licosep 12-26 (Fig. 9-15). It is a continuous chromatographic system constituted by 12 columns Superformance 300-26 (Merck) with 26 mm internal diameter and adjustable length (5–20 cm) connected in series. These have a jacket which allows operation of the SMB up to 60 °C. Each column is connected with four lines (eluent, feed, extract and raffinate lines) and 48 two-way high-pressure pneumatic valves allow the connection of the inlet-outlet lines of the columns. Consequently, there are 12 lines of each type. At a given time, only one line of each type must be opened. The eluent, extract, feed and raffinate lines are shifted after a given time period, the switch time interval, t^*. For the 12-columns SMB configuration, a whole cycle is made of 12 successive periods allowing the different lines to

Fig. 9-15. The Licosep 12-26 Simulated Moving Bed pilot unit.

return to their initial positions. A three-head membrane pump (Milroyal) is used for the recycling flow. This pump is fixed in the system, between the last and the first columns. Consequently, depending also on the injection and collection lines positions, the recycling pump can be, successively, in zone IV, III, II or I, and its flow rate must be set accordingly. Following the same notation used in the previous chapters, we define the recycling flow rate as the flow rate occurring in section IV.

9.6.1 Separation of Bi-Naphthol Enantiomers

The separation of bi-naphthol enantiomers can be performed using a Pirkle-type stationary phase, the 3,5-dinitrobenzoyl phenylglycine covalently bonded to silica gel. Eight columns (105 mm length) were packed with particle diameter of 25–40 μm. The solvent is a 72:28 (v/v) heptane : isopropanol mixture. The feed concentration is 2.9 g L^{-1} for each enantiomer. The adsorption equilibrium isotherms were determined by the Separex group and already reported in Equation (28) [33].

The operating conditions used in a SMB configuration 2-2-2-2 at 25 °C recycling flow rate 35.38 mL min^{-1}, feed 3.64 mL min^{-1}, eluent 21.45 mL min^{-1}, extract 17.98 mL min^{-1} and raffinate 7.11 mL min^{-1}. The internal concentration profiles were measured using the 6-port valve of the Licosep SMB pilot to withdraw samples from the system. The samples were collected at each half-time period, and after 40 full cycles of continuous operation. The experimental performance parameters were determined by analysis of the extract and raffinate samples collected during the whole cycle 40 (cyclic steady-state). The better purity performance was obtained for a switch time interval of $t^* = 2.87$ min, corresponding to a solid flow rate in the TMB of 11.65 mL min^{-1}. For this value, purities as high as 94.5 % in the extract and 98.9 % in the raffinate were obtained with good recoveries. A productivity of 68 g of racemic mixture processed per day and per liter of bed was achieved. The corresponding solvent consumption, was 1.2 L g^{-1} racemic mixture.

The steady state TMB model was used to simulate the SMB operation by predicting its performance and its internal concentration profiles. Model parameters are the solid/fluid ratio $(1-\varepsilon)/\varepsilon = 1.5$, $Pe = 2000$ and $\alpha = 34.44$ corresponding to $k = 0.1$ s^{-1}. The ratio between fluid and solid velocities in the TMB are: $\gamma_I = 6.31$, $\gamma_{II} = 4.00$, $\gamma_{III} = 4.47$, and $\gamma_{IV} = 3.55$. The predicted extract purity is 95.2 %; for the raffinate the purity predicted is 99.1 %.

9.6.2 Separation of Chiral Epoxide Enantiomers

The second system studied was the separation of the chiral epoxide enantiomers (1a,2,7,7a-tetrahydro-3-methoxynaphth-(2,3b)-oxirane; Sandoz Pharma) used as an intermediate in the enantioselective synthesis of optically active drugs. The SMB has been used to carry out this chiral separation [27, 34, 35]. The separation can be performed using microcrystalline cellulose triacetate as stationary phase with an average particle diameter greater than 45 μm. The eluent used was pure methanol. A packing procedure proposed by Nicoud [36] was used to fill eight SMB columns. The adsorbent bed was compacted until the column reached approximately 10 cm length. The eight columns were tested with a nonretained compound (1,3,5-tri-*tert.*-butylbenzene). Retention times are very reproducible, showing deviations smaller than 2 % and the total porosity was found to be 0.67.

A breakthrough curve with the nonretained compound was carried out to estimate the axial dispersion in the SMB column. A Peclet number of $Pe = 1000$ was found by comparing experimental and simulated results from a model which includes axial dispersion in the interparticle fluid phase, accumulation in both interparticle and intraparticle fluid phases, and assuming that the average pore concentration is equal to the bulk fluid concentration; this assumption is justified by the fact that the ratio of time constant for pore diffusion and space time in the column is of the order of 10^{-4}.

The competitive adsorption isotherms were determined experimentally for the separation of chiral epoxide enantiomers at 25 °C by the adsorption-desorption method [37]. A mass balance allows the knowledge of the concentration of each component retained in the particle, q_i^*, in equilibrium with the feed concentration, c_i^F. In fact q_i^* includes both the adsorbed phase concentration and the concentration in the fluid inside pores. This overall retained concentration q_i^* is used to be consistent with the models presented for the SMB simulations based on homogeneous particles. The bed porosity was taken as $\varepsilon = 0.4$ since the total porosity was measured as $\varepsilon_T = 0.67$ and the particle porosity of microcrystalline cellulose triacetate is $\varepsilon_p = 0.45$ [38]. This procedure provides one point of the adsorption isotherm for each component (c_i^F, q_i^*). The determination of the complete isotherm will require a set of experiments using different feed concentrations. To support the measured isotherms, a dynamic method of frontal chromatography is implemented based on the analysis of the response curves to a step change in feed concentration (adsorption) followed by the desorption of the column with pure eluent. It is well known that often the selectivity factor decreases with the increase of the concentration of chiral species and therefore the linear + Langmuir competitive isotherm was used:

$$q_A^* = 1.35C_A + \frac{7.32 \times 0.087C_A}{1 + 0.087C_A + 0.163C_B} \tag{35a}$$

$$q_B^* = 1.35C_B + \frac{7.32 \times 0.163C_B}{1 + 0.087C_A + 0.163C_B} \tag{35b}$$

A pulse of a racemic mixture (5 g L^{-1} each enantiomer) was carried out to check the adsorption model and to predict the mass transfer coefficient. The other model parameters used in simulation were $\varepsilon = 0.4$ and $Pe = 1000$. The mass transfer coefficient used to fit experimental and model predictions in the pulse experiment was $k = 0.4$ s^{-1}. Model and experimental results are compared in Figs. 9-16 and 9-17.

Considering that the separation system is fully characterized, i.e., adsorbent and mobile phases, column dimensions, SMB configuration and feed concentration, the optimization of the TMB operating conditions consists in setting the liquid flow rates in each section and also the solid flow rate. The resulting optimization problem with five variables will be certainly tedious and difficult to implement. Fortunately, the

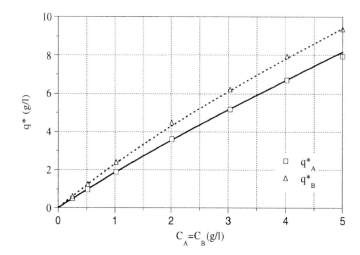

Fig. 9-16. Competitive adsorption isotherms: experimental (points) and model (lines) results.

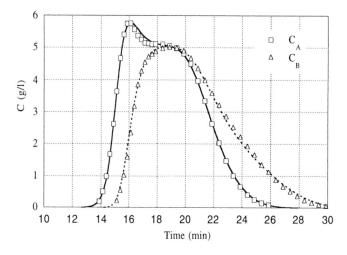

Fig. 9-17. Chromatographic response to a feed pulse. Comparison between experimental (points) and model (lines) results. (Pulse during 7.5 min; flow rate: 5 mL min^{-1}.)

optimization problem can be simplified if we take into account the functions of the different sections of a TMB system. Section I, located between the eluent and extract nodes, must provide the complete regeneration of the adsorbent phase, so the solid coming out this zone is recycled to section IV completely clean of the two components. In other words, both components A and B must move upwards, following the liquid phase. Because component B is the more retained species, we have to consider only the constraint considering this component; i.e., if the constraint is fulfilled for species B, the constraint considering the less retained component A will be always met. On the other hand, the liquid flow rate in section I is the highest flow rate in a four-section TMB system. For practical purposes, this flow rate will be limited by the system pressure-drop. Taking into account the maximum system pressure-drop accepted and the constraint considering the more retained component in section I, we can fix the liquid flow rate in this zone, as well as the switch time interval for the SMB system (related to the TMB solid flow rate). We will consider that the maximum SMB liquid flow rate allowed is $Q_I^* = 31.00$ mL min^{-1}.

The worst situation concerning the constraint in section I appears when dealing with low concentrations of the two species because it leads to bigger retention times. Since the function of section I is to regenerate completely the adsorbent phase, concentrations of both components at the beginning of this zone must be the lowest possible. Hence, the choice of the switch time interval must be made taking into account the initial slope of the proposed isotherm. The linear retention time of a component i in section j is given by the following equation:

$$t_{ij} = \frac{\varepsilon V_c}{Q_j^*}\left(1 + \frac{1-\varepsilon}{\varepsilon} K_i\right) \tag{36}$$

where K_i is the initial slope of the isotherm for component i. Considering that the SMB system under study is constituted by 9.9 cm (length) × 2.6 cm (diameter) columns ($V_c = 52.56$ mL), $\varepsilon = 0.4$, $K_B = 1.35 + 7.32 \times 0.163 = 2.543$, and using the maximum flow rate allowed in this SMB system, $Q_I^* = 31.00$ mL min^{-1}, the retention time of the more retained component in section I is $t_{BI} = 3.27$ min. Hence, the switch time interval for the SMB operation must be greater than the retention time of the more retained component in section I, if we want to fulfill the constraint previously presented for this zone. The value chosen for the switch time interval was $t^* = 3.3$ min, which corresponds to a TMB solid flow rate of $Q_S = 9.56$ mL of solid per minute.

The function of section IV, located between the raffinate and eluent nodes, is to regenerate the liquid phase, so that it can be recycled to section I as pure eluent. In other words, both components A and B must move downwards, following the solid phase. Because component A is the less-retained species, we have to consider only the constraint considering this component; i.e., if the constraint is fulfilled for species A, the constraint considering the more retained component B will be always met.

The evaluation of the retention times in section IV and the choice of the liquid flow rate for this zone (the recycling flow rate) are not straightforward as it was for section I. The worst situation, concerning the constraint in section IV, appears when dealing with nonlinear behavior because it leads to lower retention times, and we must prevent the less-retained component reaching the end of this zone before the jump of the inlet-outlet lines in the SMB operation. Since the switch time interval was already chosen and, in a situation of an effective binary separation, the concentration of the more retained component along the section IV is near zero, we suggest the choice of the liquid flow rate in section IV by using the following equation:

$$Q_{IV}^* = \frac{\varepsilon V_c}{t^*}\left(1 + \frac{1-\varepsilon}{\varepsilon} \frac{\Delta q_A^{*F}}{\Delta C_A^F}\right) \tag{37}$$

where $\Delta q_A^{*F}/\Delta C_A^F$ is the slope of the chord linking points (C_A^F, q_A^{*F}) and $(0,0)$ with $C_B = 0$. Considering that the feed concentration used is 5 g/L^{-1} of each species and $\Delta q_A^{*F}/\Delta C_A^F = 1.35 + 7.32 \times 0.087/(1 + 0.087 \times 5) = 1.794$, Equation (35) gives $Q_{IV}^* = 23.51$ mL min^{-1}.

Following the procedure presented earlier, the γ values for sections I and IV were fixed, in such a way that constraints concerning these zones were fulfilled: $\gamma_I = 3.8657$ and $\gamma_{IV} = 2.6901$. Since the liquid flow rates in sections I and IV are constants in this study, the eluent flow rate is also constant and equal to $Q_I^* - Q_{IV}^* = 7.49$ mL min^{-1}.

The original optimization problem with five variables was, by choosing the liquid flow rate in section I by pressure-drop limitations and following Equations (35) and (36) to evaluate the switch time interval and the recycling flow rate, reduced to a two-variable optimization problem: the choice of liquid flow rates in the two central sections. Table 9-5 summarizes the SMB operating conditions (and equivalent TMB conditions) used in the design of the $\gamma_{III}-\gamma_{II}$ plot.

Table 9-5. Operating conditions and model parameters for the $\gamma_{III} - \gamma_{II}$ plot.

SMB:		Equivalent TMB:
Column diameter:	$D_c = 2.6$ cm	
Column length:	$L_c = 9.9$ cm	Section length: $L_j = 2\,L_c = 19.8$ cm
Configuration:	2-2-2-2	
Bed porosity:	$\varepsilon = 0.4$	
Peclet number:	$Pe = 1000$	Peclet number: $Pe_j = 2\,Pe = 2000$
Feed concentration:	5.0 g L^{-1} each	
Switch time interval:		Solid flow rate:
$t^* = 3.3$ min		$Q_s = (1 - \varepsilon)\,V_c/t^* = 9.56$ mL min^{-1}
Flow rate in section I:		
$Q_I^* = 31$ mL min^{-1}		$Q_I = Q_I^* - Q_s\varepsilon/(1 - \varepsilon) = 24.63$ mL min^{-1}
		$\gamma_I = 3.8657$
Flow rate in section IV:		
$Q_{IV}^* = 23.51$ mL min^{-1}		$Q_{IV} = Q_{IV}^* - Q_s\varepsilon/(1 - \varepsilon) = 17.14$ mL min^{-1}
		$\gamma_{IV} = 2.6901$

Figure 9-18 presents the separation region obtained for the chiral epoxide system. Three regions are displayed: the region of complete separation obtained by the Equilibrium Theory and the regions of almost complete separation (99.5 % pure extract and raffinate) for the cases where the mass transfer coefficient is $k = 1$ and $k = 0.4$ s^{-1}. The regions for these two last cases were obtained numerically by using the steady state TMB model. The region of complete separation considering mass transfer resistance negligible was evaluated following the equations presented by Morbidelli and co-workers [24].

The case with $k = 0.4$ s^{-1} (open squares) is close to the situation where mass transfer resistance is negligible. These differences are due to mass transfer resistances as we can easily conclude by comparing the separation regions obtained for the cases with $k = 0.4$ and $k = 1$ s^{-1}. If mass transfer resistance is important, the region of complete separation can be significantly reduced from the one obtained by the Equilibrium Theory. For example, for a mass transfer coefficient of $k = 0.1$ s^{-1}, there is no separation region where extract and raffinate are 99.5 % pure.

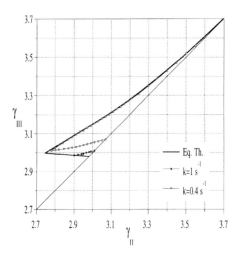

Fig. 9-18. Separation regions in a γ_{III}–γ_{II} plot: Equilibrium Theory (100 %, line), mass transfer coefficient $k = 1$ s^{-1} (99.5 %, closed squares), $k = 0.4$ s^{-1} (99.5 %, open squares).

Following the model simulations, we carried out a set of experimental runs with switch time interval $t^* = 3.30$ min, and internal liquid flow rate $Q_I^* = 33.74$ mL min^{-1}. The recycling flow rate was chosen according to our methodology as $Q_{IV}^* = 21.38$ mL min^{-1}. Since the internal liquid flow rates in these sections are fixed, the eluent flow rate is also fixed and equal to $Q_E = Q_I^* - Q_{IV}^* = 12.36$ mL min^{-1}. A feed flow rate of $Q_F = 1.00$ mL min^{-1} was used. The equivalent TMB operating conditions are $Q_I = 27.37$ mL min^{-1} ($\gamma_I = 4.2957$) and $Q_{IV} = 15.01$ mL min^{-1} ($\gamma_I = 2.3558$). To complete the choice of the SMB operating conditions we must choose the extract and raffinate flow rates. The steady state TMB package was used to evaluate the SMB performance as a function of the extract flow rate. A complete separation of both enantiomers (purities at least 99.9 %) can be obtained in a range of extract flow rates between $Q_X = 8.36$ and $Q_X = 8.95$ mL min^{-1}.

A run was carried out with an extract flow rate of $Q_X = 8.64$ mL min^{-1} (and a raffinate flow rate of $Q_R = 4.72$ mL min^{-1}). Raffinate purity close to 100 % ($PUR = 99.6$ %) was obtained, but the extract purity was lower ($PUX = 97.5$ %). The internal concentration profiles were evaluated at cyclic steady state (after 20 full cycles of continuous operation). Also evaluated were the average concentrations of both species in both extract and raffinate during a full cycle.

The steady state TMB package was used to compare the theoretical and experimental internal concentration profiles in Fig. 9-19. Figure 9-20 shows the transient evolution on the concentration of both species in the raffinate. Average concentrations over a full cycle were evaluated experimentally for cycles 3, 6, 9, 12, 15, and 18. Also shown are the corresponding SMB model predictions. The agreement between them is good and the cyclic steady-state, in terms of raffinate concentrations, is obtained after 10 full cycles.

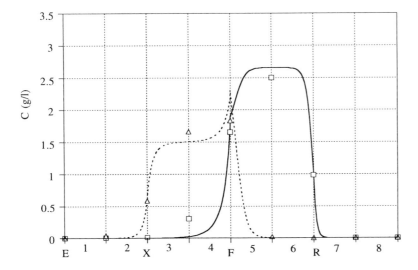

Fig. 9-19. Internal concentration profiles: experimental (points) and simulated (lines) results. Squares and full lines for component A; triangles and dashed lines for B species.

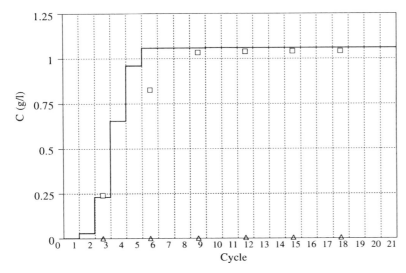

Fig. 9-20. Transient evolution of the concentration of both species in the raffinate. Points are experimental results; lines are SMB model predictions (average concentrations over each cycle).

9.7 Conclusions

SMB technology is now a mature technology adopted by pharmaceutical industry. The existence of an organized body of knowledge [39, 40] was helpful in optimizing SMB systems and making them acceptable by the industry. The future will require dynamic simulation for systems with small number of columns, e.g., configurations of the type 1-2-2-1 as encountered in some cases and also in view of process control to improve process performance.

References

[1] Broughton D. B., Gerhold C. G. (1961) Continuous Sorption Process Employing Fixed Bed of Sorbent and Moving Inlets and Outlets, U.S. Patent No. 2, 985, 589.

[2] Broughton D. B. (1968) Molex: Case History of a Process, *Chem. Eng. Prog.* 64: 60–65

[3] Broughton D. B., Neuzil R. W., Pharis J. M., Brearley C. S. (1970) The Parex Process for Recovering Paraxylene, *Chem. Eng. Prog.* 66: 70–75.

[4] Heikkilä H. (1983) Separating Sugars and Amino Acids with Chromatography, *Chem. Eng.*, January 24, 50–52.

[5] Rodrigues A. E., Tondeur D. (1981) *Percolation Processes: Theory and Applications*, NATO ASI Series, Vol. 33, Sijthoff & Noordhoff, The Netherlands.

[6] Rodrigues A. E., LeVan M. D., Tondeur D. (1989) *Adsorption: Science and Technology*, NATO ASI Series, Vol. 158, Kluwer Academic Publishers, The Netherlands.

[7] Gattuso M. J., McCulloch B., House D. W., Baumann W. M., Gottschall K. (1996) Simulated Moving Bed Technology – The Preparation of Single Enantiomer Drugs, *Pharm. Tech. Europe* 8: 20–25.

[8] Cavoy E., Deltent M .F., Lehoucq S., Miggiano D. (1997) Laboratory-Developed Simulated Moving Bed for Chiral Drug Separations. Design of the System and Separation of Tramadol Enantiomers, *J. Chromatogr. A* 769: 49–57.

[9] Francotte E., Richert P. (1997) Applications of Simulated Moving-Bed Chromatography to the Separation of the Enantiomers of Chiral Drugs, *J. Chromatogr. A* 769: 101–107.

[10] Francotte E. (1996) Chromatography as a Separation Tool for the Preparative Resolution of Racemic Compounds, in *Chiral Separations. Applications and Technology*, Ahuja S. (ed.), American Chemical Society, p. 271–308.

[11] Guest D. W. (1997) Evaluation of Simulated Moving Bed Chromatography for Pharmaceutical Process Development, *J. Chromatogr. A* 760: 159–162.

[12] Nicoud R. M. (1992) The Simulated Moving Bed: A Powerful Chromatographic Process, *LC-GC Intl.* 5: 43–47.

[13] Negawa M., Shoji F. (1992) Optical Resolution by Simulated Moving-Bed Adsorption Technology, *J. Chromatogr.* 590: 113–117.

[14] Glueckauf E. (1955) Theory of Chromatography Part 10 – Formulæ for Diffusion into Spheres and Their Application to Chromatography, *Trans. Faraday Soc.* 51: 1540–1551

[15] Ching C. B., Ruthven D. M. (1985) An Experimental Study of a Simulated Counter-Current Adsorption System – I. Isothermal Steady State Operation, *Chem. Eng. Sci.* 40: 877–885.

[16] Madsen N. K., Sincovec R. F. (1979) PDECOL: General Collocation Software for Partial Differential Equations, *ACM Trans. Math.* 5: 326–351.

[17] Hindmarsh A. C. (1976) Preliminary Documentation of GEARIB. Solution of Implicit Systems of Ordinary Differential Equations with Banded Jacobians, *Rep. UCID – 30130*, Lawrence Livermore Laboratory, Livermore.

[18] Pais L. S., Loureiro J. M., Rodrigues A. E. (1998a) Modeling Strategies for Enantiomers Separation by SMB Chromatography, *AIChEJ* 44: 561–569.

[19] Ruthven D. M., Ching C. B. (1989) Counter-Current and Simulated Counter-Current Adsorption Separation Processes, *Chem. Eng. Sci.* 44: 1011–1038.

[20] Zhong G., Guiochon G. (1996) Analytical Solution for the Linear Ideal Model of Simulated Moving Bed Chromatography, *Chem. Eng. Sci.* 51: 4307–4319.

[21] Storti G., Mazzotti M., Morbidelli M., Carrà S. (1993) Robust Design of Binary Countercurrent Adsorption Separation Processes, *AIChEJ.* 39: 471–492.

[22] Storti G., Baciocchi R., Mazzotti M., Morbidelli M. (1995) Design of Optimal Operating Conditions of Simulated Moving Bed Adsorptive Units, *Ind. Eng. Chem. Res.* 34: 288–301.

[23] Mazzotti M., Storti G., Morbidelli M. (1996) Robust Design of Countercurrent Adsorption Separation: 3. Nonstoichiometric Systems, *AIChE J* 42: 2784–2796.

[24] Mazzotti M., Storti G., Morbidelli M. (1997) Optimal Operation of Simulated Moving Bed Units for Nonlinear Chromatographic Separations, *J. Chromatogr. A* 769: 3–24.

[25] Gentilini A., Migliorini C., Mazzotti M., Morbidelli M. (1998) Optimal Operation of Simulated Moving-Bed Units for Non-Linear Chromatographic Separations. II. Bi-Langmuir Isotherm, *J. Chromatogr. A* 805: 37–44.

[26] Pais L. S., Loureiro J. M., Rodrigues A. E. (1997b) Modeling, Simulation and Operation of a Simulated Moving Bed for Continuous Chromatographic Separation of 1,1'-bi-2-naphthol Enantiomers, *J. Chromatogr. A* 769: 25–35.

[27] Pais L. S., Loureiro J. M., Rodrigues A. E. (1998b) Separation of Enantiomers of a Chiral Epoxide by Simulated Moving Bed Chromatography, *J. Chromatogr. A* 827: 215–233.

[28] Pais L. S., Rodrigues A. E. (1998) Separation of Enantiomers by SMB Chromatography: Strategies of Modeling and Process Performance, *Fundamentals of Adsorption 6: Proceedings of the Sixth International Conference of Fundamentals of Adsorption*, F Meunier (ed.), Elsevier, Paris, p. 371–376.

[29] Bader G., Ascher U. (1987) A New Basis Implementation for a Mixed Order Boundary Value ODE Solver, *SIAM J. Sci. Stat. Comput.* 8: 483–500.

[30] Ascher U., Christiansen J., Russell R. D. (1979) A Collocation Solver for Mixed Order Systems of Boundary Value Problems, *Math Comput* 33: 659–679.

[31] Ascher U., Christiansen J., Russell R. D. (1981) Collocation Software for Boundary-Value ODEs. *ACM Trans. Math Software* 7: 209–222.

[32] Azevêdo D. C. S., Rodrigues A. E. (1999) Design of a Simulated Moving Bed Separator in the Presence of Mass Transfer Resistances, *AIChE J* 45: 956–966.

[33] Pais L. S., Loureiro J. M., Rodrigues A. E. (1997a) Separation of 1,1'-bi-2-naphthol Enantiomers by Continuous Chromatography in Simulated Moving Bed, *Chem. Eng. Sci.* 52: 245– 257.

[34] Nicoud R. M., Fuchs G., Adam P., Bailly M., Küsters E., Antia F., Reuille R., Schmid E. (1993) Preparative Scale Enantioseparation of a Chiral Epoxide: Comparison of Liquid Chromatography and Simulated Moving Bed Adsorption Technology, *Chirality* 5: 267–271.

[35] Küsters E., Gerber G., Antia F. D. (1995) Enantioseparation of a Chiral Epoxide by SMB Chromatography using Chiralcel-OD, *Chromatographia* 40: 387–393.

[36] Nicoud R. M. (1993) A Packing Procedure Suitable for High Flow Rate and High Stability Columns Using Cellulose Triacetate, *LC-GC Int.* 6: 636–637.

[37] Nicoud R. M., Seidel-Morgenstern A. (1993) Adsorption Isotherms: Experimental Determination and Application to Preparative Chromatography *Simulated Moving Bed: Basics and Applications*, R. M Nicoud (ed.), Institut National Polytechnique de Lorraine, Nancy, France, p. 4–34.

[38] Lim B. G., Ching C. B., Tan R. (1995) Determination of Competitive Adsorption Isotherms of Enantiomers on a Dual-site Adsorbent, *Sep. Techno.* 5: 213–228.

[39] Charton F., Nicoud R. M. (1995) Complete Design of a Simulated Moving Bed, *J. Chromatogr. A* 702: 97–112.

[40] Pröll T., Küsters E. (1998) Optimization Strategy for Simulated Moving Bed Systems, *J. Chromatogr. A* 800: 135–150.

10 The Use of SMB for the Manufacture of Enantiopure Drug Substances: From Principle to cGMP Compliance

S.R. Perrin and R.M. Nicoud

10.1 Introduction

10.1.1 FDA as the Driving Force: Enantiopure Drugs and Compliance

In May 1992, the U.S. Food and Drug Administration (FDA) issued a policy statement for the development of stereoisomeric drugs [1]. The statement focused on the study and pharmaceutical development of single enantiomers (enantiopure) and racemic drugs. The policy stated: "Now that technological advances (large scale chiral separation procedures or asymmetric syntheses) permit production of many single enantiomers on a commercial scale, it is appropriate to consider what FDA's policy with respect to stereoisomeric mixture should be". Furthermore, the manufacturer should develop quantitative assays for stereoisomeric drugs in the early stages of drug development to assess the efficacy of the drug. The message from the FDA was clear and urged the pharmaceutical industry to evaluate enantiopure drugs alongside racemic drugs as new candidates for the future [2, 3]. Consequently, worldwide scientific and regulatory agencies (European Union, Canada, and Japan) soon released guidelines for the development and manufacture of enantiopure drugs [4–7].

It was apparent that the FDA recognized the ability of the pharmaceutical industry to develop chiral assays. With the advent of chiral stationary phases (CSPs) in the early 1980s [8, 9], the tools required to resolve enantiomers were entrenched, thus enabling the researcher the ability to quantify, characterize, and identify stereoisomers. Given these tools, the researcher can assess the pharmacology or toxicology and pharmacokinetic properties of enantiopure drugs for potential interconversion, absorption, distribution, and excretion of the individual enantiomers.

Despite widespread adherence to the 1992 guidelines, the FDA soon toughened its stance in January 1997 by first proposing and then removing the drug Seldane (terfenadine), manufactured by Hoechst Marion Roussel [10–12]. The rule change by the FDA regarding extended market exclusivity and new chemical entities were based on pharmacological data, and resulted in a new interpretation of fixed combination dosages. The example of Seldane was due to the rapid metabolism of terfenadine to a pharmacologically active carboxylic acid metabolite (fexofenadine) and the inhibition of terfenadine metabolism by drugs such as ketoconazole [13, 14]. The result was later associated with significant toxicity [15].

In November 1997, the Department of Health and Human Services along with the International Conference on Harmonisation (ICH) released a draft guidance for the selection of test procedures, which included chiral drugs. For the development of an enantiopure drug substance, acceptable criteria shall include, if possible, an enantioselective assay. This assay should be part of the specification for the identification of an enantiopure drug substance and related enantioenriched impurities [16].

These and other FDA policy decisions launched the pharmaceutical industry and academia into a new era of developing stereoselective processes for the manufacture of enantiopure active pharmaceutical ingredients (APIs).

10.1.1.1 Market Exclusivity: Newly Approved Drug Substances

The FDA had originally argued that the individual enantiomers for a previously approved racemic drug should not be considered as a new chemical entity. In January 1997, following the proposal to remove Seldane, the FDA asked for public comment on their proposal to grant 5-year marketing exclusivity to developers of single-enantiomers of previously approved racemic drugs [17]. This would be a change from FDA's policy of treating chiral switches as a mere reformulation of existing approved drugs, and therefore, entitled to only 3-year marketing exclusivity.

New drug applications (NDA) as defined in the Code of Federal Regulations (CFR), was revised as of April 1, 1998, by the FDA for approval to market a new drug or an antibiotic drug [18]. The FDA's final ruling defines new drug product exclusivity for a drug product containing a new chemical entity that was approved after September 24, 1984. The same active moiety cannot be resubmitted as a new chemical entity for a period of 5 years from the date of approval of the first NDA [19].

These policy decisions by the FDA were the driving force for chiral switches and the commercial development of chromatographic processes such as simulated moving bed (SMB) technology. Due to technological advances such as SMB and the commercial availability of CSPs in bulk quantities for process-scale purification of enantiopure drugs, the production of many single enantiomers now exists on a commercial scale.

10.1.1.2 Fixed-Combination Dosage: Enantiopure Drug Substances

The FDA's perspective with respect to the manufacturing of enantiopure drugs has raised several issues. One issue is the acceptable control of a stereoselective synthesis or purification of enantiopure drugs where discrepancy in pharmacological and toxicological assessment during clinical evaluation is due to enantioenrichment of one enantiomer in the presence of another. Other issues are when the properties of enantiopure drugs have revealed instances in which both enantiomers exhibited similar pharmacological activities, while others represented a combined pharmacological effect. The pharmacologically different properties, which may occur due to in vivo interconversion during drug absorption, gave rise to the FDA policy [20].

The FDA's policy regarding fixed-combination dosage where the racemic drug is used with a combination of different stereoisomeric forms defines that two or more drugs may be combined into a single dosage form only when each component contributes. The claimed effects of the combined drug and the dosage of each should be deemed as a safe combination and effective. An interpretation of the FDA's policy forewarns pharmaceutical companies that develop or manufactures enantiopure drugs. Unless the "component or components" represent a fixed dose combination, parallel studies for the racemate and the resulting enantiopure drug should be implemented to ensure compliance.

Although in many cases an enantiopure drug can be safer than the racemate, the advantages are clear. The final formulation of the drug product could be reduced in-half, potential side effects could be minimized, and the resulting pharmokinetic and pharmacodynamic studies could clearly determine the efficacy of the active pharmaceutical ingredient (API) [21].

Due to FDA policies, this was a pivotal point for the pharmaceutical industry and established the onslaught of mergers for the development of enantiopure drugs.

10.1.1.3 Pharmaceutical Industry: Mergers

Pharmaceutical manufacturers began to develop technologies either to resolve or selectively synthesize enantiopure drugs. The justification was that the active enantiopure drug would prove to be more efficacious, and this would allow drug companies to extend expiring originator patents.

Merger and acquisition activity is heating up in the fine chemicals and pharmaceutical intermediates arena as many companies seek to acquire cGMP facilities. They are targeting this high-margin business, a result of increasing outsourcing of the production of intermediates by large pharmaceutical companies. The emergence of "virtual" pharmaceutical companies is also driving demand for fine chemicals. These small companies have no other option but to outsource production, since they have no manufacturing facilities of their own.

With the combined lure of high-margin business and attractive growth prospects, chemical companies of all types and sizes are actively seeking acquisitions. Major firms have been very aggressively acquiring or setting up alliances.

As markets for enantiopure drugs continue to develop, the pharmaceutical industry, fine chemical companies, and academic chemists are prospecting for new enantioselective technologies to produce them.

10.2 Chromatographic processes

10.2.1 SMB: Comparisons to Batch Chromatography

The purification of "value-added" pharmaceuticals in the past required multiple chromatographic steps for batch purification processes. The design and optimization of these processes were often cumbersome and the operations were fundamentally complex. Individual batch processes requires optimization between chromatographic efficiency and enantioselectivity, which results in major economic ramifications. An additional problem was the extremely short time for development of the purification process. Commercial constraints demand that the time interval between non-optimized laboratory bench purification and the first process-scale production for clinical trials are kept to a minimum. Therefore, rapid process design and optimization methods based on computer aided simulation of an SMB process will assist at this stage.

Chromatography is a separation technique widely adopted not only at the analytical scale, but also at the preparative and large industrial scale. In this framework, SMB technology is an established technique for continuous chromatographic manufacture of enantiopure drugs and other pharmaceuticals. Although SMB units are relatively complicated, engineering issues at the scale of pharmaceutical production of enantiopure drugs are well understood.

There are many advantages of the SMB technology compared to batch preparative chromatography. The process is continuous and the solvent requirement is minimal compared to batch. In SMB, the whole stationary phase is used for the separation while in batch chromatography only a small part of the column is involved in the separation. This allows optimization of productivity with respect to the stationary phase.

At the current time, there is considerable interest in the preparative applications of liquid chromatography. In order to enhance the chromatographic process, attention is now focused on the choice of the operating mode [22]. SMB offers an alternative to classical processes (batch elution chromatography) in order to minimize solvent consumption and to maximize productivity where expensive stationary phases are used.

SMB technology was introduced in the late 1950s [23] and has mainly been applied to large-scale productions in the petrochemical and sugar industries [24]. SMB is recognized as a very efficient technology, but was ignored by the pharmaceutical and fine chemical industries mainly for patent reasons and complexity of concept.

In 1993, shortly after the FDA announced their first policy statement on enantiopure drugs, separations of pharmaceutical compounds were performed using SMB technology [25, 26]. Other applications now include fine chemistry, cosmetics, and perfume industry [27].

10.2.2 Illustrations of SMB Processes

The concept of using continuous chromatography for the separation of stereoisomers or "optical isomers" is very old and was probably proposed for the first time by Martin and Kuhn in 1941 [28]. The suggested implementation was different from today's SMB technology, though the basic concept is the same. The chromatographic media is moved continuously in a conveyor belt, the feed is injected continuously at a fixed point, and the pure enantiomers are recovered at fixed points. In the idea of Martin and Kuhn, benefits were taken from the possibility of modulating the adsorption of the products at different temperatures.

Hotier and Balannec published one of the first real proposals to use SMB as a production tool for the pharmaceutical industry, and thus to scale down a process already used on a production scale [29]. The first commercially available plant (Licosep) SMB system was offered by Separex in 1991 and was exhibited for the first time in June 1991 during the Achema Exhibition. The system consisted of 24 stainless steel columns with adjustable lengths between a few centimeters up to almost 1 meter, HPLC pumps, and mulitpositional valves. To improve the robustness of the system, a rotary valve replaced two-way valves and the pumps were modified.

Early scientific publications detailed experimental results for the enantioseparation of phenyl-ethyl alcohol [30] and on threonine [31]. Sandoz released the first announcement for the development and use of SMB on an industrial scale [32], this separation has been described by Nicoud et al., 1993 [25].

Following these announcements, the first wave of publications addressing the use of SMB for the manufacture of pharmaceutical products of interest was published. The separation of a chiral hetrazepine [26], WEB 2170 6-(2-chlorophenyl)-8-9-dihydro-1-methyl-8-[(morpholinyl)-carbonyl]-4H,7H-cyclopenta[4,5]-thieno[3,2-f][1,2,4]triazolo[4,3-a][1,4]diazepine. WEB 2170 is a chiral hetrazepine from Boehringer-Ingelheim. The enantioseparation of WEB 2170 was performed using cellulose triacetate (CTA) from Merck (Darmstadt) as the CSP and with pure methanol as eluent.

A precursor in the synthesis of a promising calcium sensitizing agent from E. Merck [33], a chiral thiadiazin-2-one EMD 53986, 3,6-Dihydro-5-[1,2,3,4-tetrahydro-6-quinolyl]-6-methyl-2H-1,3,4-thiadiazin-2-one [26]. The study was performed using Celluspher®, a CSP prepared from cellulose tri(p-methylbenzoate) according to a patent from Ciba-Geigy [34]. The spherical particles had a mean particle diameter of 20–43 µm and the mobile phase was pure methanol.

Other examples of enantioseparations include the separation of the antihelmintic drug, prazinquatel [35], which used a 4-column SMB system composed of columns of 12.5 mm i.d. packed with CTA and with methanol as the eluent. Ikeda and Murata separated the enantiomers of β-blockers [36].

A second wave of publications describes SMB in simplistic terms, while integrating economical calculations [37–40]. Moreover, some users (researchers in pharmaceutical companies) disclosed the design and applications of SMB in their own laboratories [41–45].

The first modeling software which allowed for the optimization of nonlinear separations by SMB was presented in the early 1990s [46]. Today, numerous publications from academia allows one to have a better understanding of the SMB system [47–51]. Industry now has the practical tools for modeling SMB for quick and efficient process optimization [41, 52].

The Novasep team in 1994, successfully resolved 2 kg of racemic binaphthol per day on a Pirkle-Type 3,5-DNBPG CSP (Merck KGaA, Germany) using a Licosep 8-200 SMB system (Summary report on the BRITE-EURAM project BRE2-CT92-0337).

10.3 SMB as a Development Tool

10.3.1 Basic Principles and Technical Aspects

In opposition to the "usual" (elution) chromatography, SMB is a continuous process, and is thus much more adapted to large-scale production. Moreover, SMB is based on a countercurrent contact between the liquid and the adsorbent, which leads to lower eluent consumption.

The easiest way to understand the SMB concept is to consider a true moving bed (TMB) as described in Figure 10.1, in which a countercurrent contact is promoted between the solid and liquid phases. The solid phase moves down the column due to gravity and exits the system in Zone I. The liquid (eluent) stream follows exactly the opposite direction. It is recycled from Zone IV to Zone I. The feed, containing components A and B are injected at the middle of the column, and the fresh eluent is replenished in Zone I.

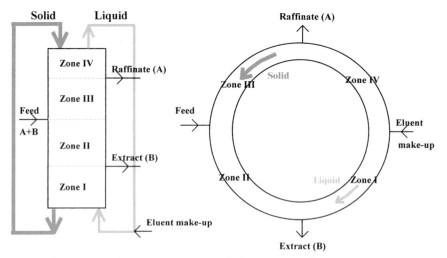

Figure 10.1. True moving bed (TMB): two equivalent representations.

The affinity of A and B for the solid phase are different, B is more retained than A. Thus, it is possible to choose flowrates in order to make A move upward and B to move downward, leading to a spatial separation. This system requires inlet lines for the feed and eluent and outlet lines for the raffinate A and extract B.

The classical moving bed is made of four different zones in which different constraints must be fulfilled [53].

- Zone 1: between the eluent make-up and the extract points where the more retained product (B) must be completely desorbed.
- Zone II: between the extract and the feed points where the less-retained product (A) must be completely desorbed.
- Zone III: between the feed and the raffinate points where the more retained product (B) must be completely adsorbed.
- Zone IV: between the raffinate and the eluent make-up points where the less-retained product (A) must be completely adsorbed.

All the internal flowrates are related to the inlet/outlet flowrates by simple mass balances:

$$Q_{II} = Q_I - Q_{Ext} \qquad Q_{III} = Q_{II} + Q_{Feed} \tag{1}$$

$$Q_{IV} = Q_{III} - Q_{Raff} \qquad Q_I = Q_{IV} + Q_{El}$$

In addition, the inlet/outlet flowrates are related by:

$$Q_{Ext} + Q_{Raff} = Q_{Feed} + Q_{El} \tag{2}$$

In fact, it is extremely difficult to operate a TMB because it involves circulation of a solid adsorbent. Thus, the concept must be implemented in a different way where the benefit of a true countercurrent operation can be achieved by using several fixed-bed columns in series with an appropriate shift of the injection and collection points between the columns. This is the SMB implementation as presented in Fig. 10.2.

In this mode, the solid is no longer moving. The shifting of the inlet and outlet lines only simulates solid flow, and the solid flowrate downward is directly linked to the shift period. Proper selection of flowrates is required to stabilize the different fronts of species A and B in the proper zones. The adequate choice of the flowrates requires a minimum knowledge of the physico-chemical properties of the system. The influence of adsorption isotherms and plate numbers is simulated by the software.

A SMB, whatever the number of zones, consist of 4 to 24 columns with 3 to 5 pumps and a number of valves, which allow the columns to be connected to different lines. Further discussions are restricted to the classical 4-zone SMB. There are different ways to connect columns in order to build a SMB. An important option is linked to the presence or absence of a recycling pump in order to build a SMB. The examples are shown in Fig. 10.3.

As shown in Fig. 10.3a, the most classical way is to use a recycling pump that is located between two columns (for instance 12 and 1). The recycling pump, which is

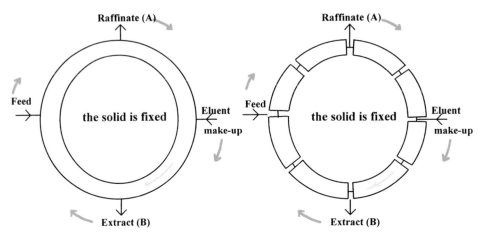

Figure 10.2. Principle of the simulated moving bed (SMB), left: inlet and outlet lines are shifted discontinuously; right: inlet and outlet lines are shifted simultaneously.

fixed with respect to the columns moves with respect to the zones and is alternatively located in Zone IV, III, II and I. As the flowrates in the different zones are different, the pump flowrate varies from cycle to cycle. It should be pointed out that this constraint is perfectly mastered and that this design is relatively simple and is used with small differences on all the large-scale units.

A more serious limit to this implementation is due to the volume of the recycling pump and associated equipment such as flowmeters and pressure sensors. As the pump moves with respect to the zones, its volume leads to a dead volume dissymmetry, which can lead to a decrease extract and raffinate purities. This decrease can be significant for SMB with short columns and/or compounds with low retention. However, it can be easily overcome by using a shorter column or asynchronous shift of the inlets/outlets [54, 55]. This last solution is extremely efficient and does not induce extra costs because it is a purely software solution.

In a different implementation (Fig. 10.3b), the recycling pump is fixed with respect to the zones. It is always located between zones IV and I where no solutes are present. In order to implement this idea, additional valves are needed, which makes the system more complex than the previous one. Its main interest is found when physical modulation is used, as in the supercritical fluid SMB, for which it can be shown that a great interest could be taken from a higher pressure in zone I [56]. The only way to obtain this result is to maintain the recycling pump immediately before zone I.

As illustrated in Fig. 10.3c, a final solution is to use the eluent pump instead of the recycling pump. This implementation may enable the setup to be simplified, but more valves are required than option a), and a drawback is that one outlet must be recycled to the eluent tank.

Whichever, the retained design (type a, b, or c), there are always different options to control the outlet flowrates. A pump, analog valve, and flowmeter or pressure

a)

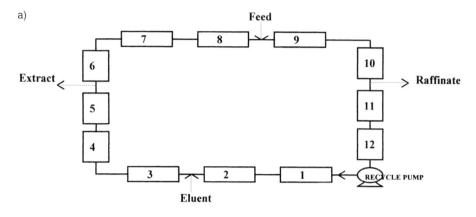

b)

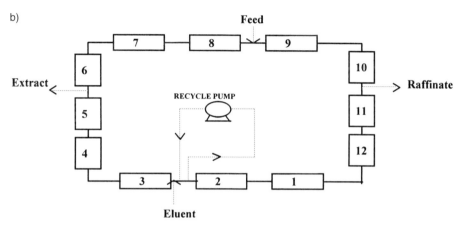

c)

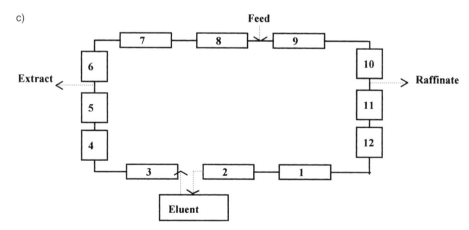

Figure 10.3 Recycling options in SMB systems. a) Recycling pump fixed with respect to the columns. b) Recycling pump fixed with respect to the zones (between Zone IV and Zone I). c) No recycling pump.

control can all be used to control outlet flow rates. Our experience led us to the design of a system using a recycling pump, which is fixed with respect to the columns as shown in Fig. 10.3a (reliability, minimum number of valves). To use pumps instead of valves to deliver the flowrates, and to counterbalance the effect of the recycling pump by the use of an asynchronic shift of the inlet/outlet lines.

10.3.2 Operating Conditions

The steps when designing a SMB which would allow one to process a given amount of feed per unit time have been described in detail [46, 57]. The procedure described was based on modeling of nonlinear chromatography. The procedure is rigorous, versatile and mainly requires the determination of competitive adsorption isotherms. If the adequate tools and methods are used, the procedure is not tedious and requires less work than is sometimes claimed. The procedure is briefly described below.

10.3.2.1 Step A: Acquisition of Relevant Physico-Chemical Parameters

In order to determine rigorously the process parameters, a few relevant parameters are to be experimentally determined on a laboratory scale column.

Equilibrium adsorption isotherms: It would be impossible to design a distillation unit without knowledge of the boiling points and liquid-vapor equilibrium. Similarly, the calculation of chromatographic processes requires the knowledge of the partition of the solutes between the liquid and the solid phases at a given temperature at equilibrium (adsorption isotherm). In the case of a multicomponent mixture, there is usually a competition between the various compounds for the accession to the adsorption sites. Consequently, the concentration of a given species on the stationary phase $\overline{C}_i$: does not only depend on C_i but on all liquid phase concentrations. Each adsorption isotherm is a relation of the following type:

$$\overline{C}_i = \bar{f}_i(C_1, C_2, ...) \tag{3}$$

The adsorption isotherm becomes linear at low concentrations[1]:

$$\overline{C}_i = \overline{K}_i \cdot C_i \tag{4}$$

The knowledge of these adsorption isotherms allows quantification of the respective affinity for the stationary phase with respect to the different solutes. Many different isotherm equations have been described in the literature, and experimental methods allowing their determination are reviewed by [58]. As a first approximation, modified competitive Langmuir isotherms can often he used:

---

[1] Note that the initial slope of the adsorption isotherm can be easily obtained from the knowledge of the retention time associated to a small injection performed on a column, as this retention time is given by: $t_R = t_0 \cdot \left(1 + \frac{1-\varepsilon}{\varepsilon} \cdot \overline{K}\right)$ where $t_0 = \varepsilon \cdot V/Q$ is the "zero-retention time" based on the external bed porosity ε (commonly, ε is about 0.36–0.4).

$$\overline{C}_A = \lambda \cdot C_A + \frac{\tilde{K}_A \cdot \overline{N} \cdot C_A}{1 + \tilde{K}_A \cdot C_A + \tilde{K}_B \cdot C_B}$$

$$\overline{C}_A = \lambda \cdot C_B + \frac{\tilde{K}_B \cdot \overline{N} \cdot C_B}{1 + \tilde{K}_A \cdot C_A + \tilde{K}_B \cdot C_B}$$
(5)

Using this model, the initial slopes of the adsorption isotherms are given by:

$$\overline{K}_A = \lambda + \tilde{K}_A \cdot \overline{N} \quad \text{and} \quad \overline{K}_B = \lambda + \tilde{K}_B \cdot \overline{N}$$

If adsorption data are not available and/or if a quick evaluation is required, the parameters of the isotherms can be set to:

- λ is set to about 0.5 (internal porosity)
- $\tilde{K}_A$ and $\tilde{K}_B$ derived from the knowledge analytical retention times[1]
- $\overline{N}$, the saturation capacity, taken in the range:
  - ✓ 100–300 g L^{-1} in the case of silica
  - ✓ 50–200 g L^{-1} in the case of C_{18} or related stationary phases.
  - ✓ 25–50 g L^{-1} in the case of bulk polymeric chiral phases.
  - ✓ 10–20 g L^{-1} in the case of silica-based chiral stationary phases (cellulosic, amylosic)
  - ✓ 1–5 g L^{-1} for other silica-based chiral stationary phases

Column efficiency (number of theoretical plates): As in batch chromatography, one needs to determine the efficiency of the column in order to evaluate the dispersion of the fronts due to hydrodynamics dispersion or kinetics limitations. The relationship of N proportional to L can be expressed in terms of the equation for height equivalent to a theoretical plate (HETP) as:

$$H = \frac{L}{N}$$
(6)

where L is the column length and N is the number of theoretical plates.

HETP can be related to the experimental parameters through the Van Deemter [59] or Knox [60] equations. It is possible to describe the dependence of H on u since H is a function of the interstitial mobile phase velocity u. In the case of preparative chromatography, where relatively high velocities are used, these equations can very often be simplified into a linear relation [61, 62].

$$H = a + b \cdot u$$
(7)

Parameters a and b are related to the diffusion coefficient of solutes in the mobile phase, bed porosity, and mass transfer coefficients. They can be determined from the knowledge of two chromatograms obtained at different velocities. If H is unknown, b can be estimated as 3 to 5 times of the mean particle size, where a is highly dependent on the packing and solutes. Then, the parameters can be derived from a single analytical chromatogram.

Estimation of the pressure-drop: The system is designed to work within a given pressure limit; thus, one needs a relation giving the pressure-drop in the column (per unit length). Darcy's law gives the relation of $\Delta P/L$ versus the mobile phase velocity u. However, the Kozeny-Carman equation is best adapted for laminar flows as described:

$$\frac{\Delta P}{L} = h_k \cdot \frac{36}{d_p^2} \cdot \left(\frac{1-\varepsilon}{\varepsilon}\right)^2 \cdot \mu \cdot u = \frac{\varphi}{d_p^2} \cdot u = \Phi \cdot u \tag{8}$$

where h_k is the Kozeny coefficient (close to 4.5), μ is the eluent viscosity, and u is the linear velocity.

10.3.2.2   Step B: Calculation of TMB

For given feed composition, eluent, and stationary phase, the flowrates of a TMB to allow processing a given flowrate of feed are calculated based on the knowledge of the adsorption isotherms.

Optimum flowrates, resulting in high productivity and low eluent consumption, are estimated first for an "ideal system", which means that kinetic and hydrodynamic dispersive effects are assumed to be negligible [46]. This procedure has recently been improved [57].

In order to present the results in a normalized form, it is convenient to define the reduced flowrates as:

$$m_i = \frac{Q_i}{\overline{Q}}, i = I \text{ to } IV \tag{9}$$

where Q_i are the flowrates in zone i and $\overline{Q}$ is the solid flow rate.

Linear case: This case is met when the adsorption isotherm is considered linear, which means operation under diluted conditions. Taking into account the saturation capacities of the CSP, this behavior is usually met for concentrations around or below 1 g L^{-1} for separation of enantiomers.

A more general criterion for linearity can be derived noting that the denominator of the Langmuirian adsorption isotherms must approach 1, and consequently:

$$\tilde{K}_A \cdot C_A^F + \tilde{K}_B \cdot C_B^F \leq 0.1 \tag{10}$$

Assuming a linear behavior, the conditions that have to be fulfilled by the different flowrates can be shown to be:

$$m_{IV} \leq \overline{K}_A \quad m_{III} \leq \overline{K}_B$$
$$m_{II} \geq \overline{K}_A \quad m_I \geq \overline{K}_B \tag{11}$$

It is convenient to define a parameter $\gamma \geq 1$ and to replace Equation (11) by:

$$m_{IV} = \overline{K}_A \cdot \frac{1}{\gamma} \quad m_{III} = \overline{K}_B \cdot \frac{1}{\gamma}$$

$$m_{II} = \overline{K}_A \cdot \gamma \quad m_I = \overline{K}_B \cdot \gamma \tag{12}$$

γ can be considered as a safety factor, if it is equal to 1, the system works at its optimum productivity, but it will be very sensitive to any deviation regarding the flowrates. If it is greater than 1, the system is less efficient in term of specific productivity or eluent consumption, but it is less sensitive to possible perturbations.

Taking into account Equation (12) and the definition of the normalized flow rates, one can derive:

$$\overline{Q} = \frac{Q_F}{\overline{K}_B \cdot \dfrac{1}{\gamma} - \overline{K}_A \cdot \gamma} \quad Q_{Ext} = \overline{Q} \cdot \left(\overline{K}_B - \overline{K}_A \right) \cdot \gamma \tag{13}$$

$$Q_{Raf} = \overline{Q} \cdot \left(\overline{K}_B - \overline{K}_A \right) / \gamma \quad Q_I = \overline{Q} \cdot \overline{K}_B \cdot \gamma$$

As the feed flowrate is known, all the TMB flowrates are calculated from Equation (13), for a given γ value. From Equation (13), it follows that γ must fall in this range:

$$1 \leq \gamma < \sqrt{\frac{\overline{K}_B}{\overline{K}_A}} \tag{14}$$

There are no general rules allowing selection of the correct γ value. The correct selection is a result of technico-economical optimization; however, for a first guess γ can be set to $\gamma = 1.02$.

At this point, all the flowrates are known and only the total number of plates required has to be estimated. This estimation is determined by numerical simulation. Experience shows that SMB equivalent to 500 plates solves almost all problems.

Nonlinear case: The calculation of the flowrates is much more complex, and it is beyond the scope of this chapter to present it in detail. However, as a useful tool, Morbidelli and coworkers [48–50, 63], applied the solutions to the equations of the equilibrium theory (when all the dispersion phenomena are neglected) to a four-zone TMB. The solutions are explicit for m_I and m_{IV}

$$m_I > (m_I)_{min} = \overline{K}_B \tag{15}$$

$$m_{IV} < (m_{IV})_{max} = \frac{1}{2} \left\{ \begin{array}{c} \overline{K}_A + m_{III} - 2\lambda + \check{K}_A C_A^F (m_{III} - m_{II}) \\ -\sqrt{\left[\overline{K}_A + m_{III} - 2\lambda + \check{K}_A C_A^F (m_{III} - m_{II}) \right]^2 - 4\overline{N}\check{K}_A (m_{III} - \lambda)} \end{array} \right\} + \lambda \tag{16}$$

As illustrated in Figure 10.4, the conditions on m_{II} and m_{III} are independent of those on m_I and m_{IV} and can be visualized in the (m_{II}, m_{III}) plane.

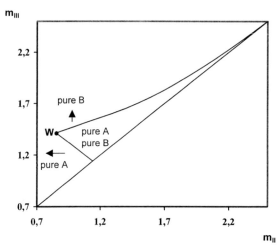

Figure 10.4. Complete triangular separation region.

The optimum of the complete triangular separation region (maximizing productivity and minimizing eluent consumption) is the point w corresponding to:[2]

$$(m_{II})_{opt} = \lambda + \frac{\tilde{K}_A}{\tilde{K}_B}\omega_G \tag{17}$$

$$(m_{III})_{opt} = \lambda + \frac{\omega_G\left[\omega_F(\overline{K}_B - \overline{K}_A) + \overline{N}\tilde{K}_A(\overline{N}\tilde{K}_A - \omega_F)\right]}{\overline{N}\tilde{K}_A(\overline{N}\tilde{K}_B - \omega_F)} \tag{18}$$

where $\omega_G > \omega_F > 0$ are the square roots of equation (19):

$$\left(1 + \tilde{K}_A C_A^F + \tilde{K}_B C_B^F\right)\omega^2 - \left[\overline{N}\tilde{K}_A(1 + \tilde{K}_B C_B^F) + \overline{N}\tilde{K}_B(1 + \tilde{K}_A C_A^F)\right]\omega + \overline{N}^2\tilde{K}_A\tilde{K}_B = 0 \tag{19}$$

The above conditions are not robust because they are at the limit of the complete separation zone. The equilibrium theory neglects the dispersion phenomena and therefore the purity obtained under these flowrate conditions would be less than 100 % on a TMB system. Complex simulation software, which takes into account the dispersion phenomena, gives a more robust system with higher purities [57].

---

[2] It has been shown recently [57] that this assertion is in fact wrong for the productivity. However, the solution is quite close to the optimum.

The total number of plates (N) required can easily be obtained based on the knowledge of the selectivity $\alpha_{B/A} = K_B/K_A$: about 100 plates are required for large selectivities (greater than 2) and abut 600 plates are required for small selectivities (about 1.1–1.2). Typical values of 200–400 plates allow high purity with significant productivity for intermediate selectivity.

10.3.2.3 Step C: Calculation of SMB

Using SMB flowrates simply derived from those of the TMB, the SMB behavior is simulated according to the number of columns and equivalent number of plates per column. The SMB raffinate, extract, and eluent flowrates are identical to those of the TMB feed. The SMB recycling flowrate Q_R (zone I) is given by:

$$Q_R = Q_I + \frac{\varepsilon}{1-\varepsilon} \cdot \overline{Q} \tag{20}$$

Moreover, the shift period is derived from:

$$\Delta T = \frac{(1-\varepsilon) \cdot V_{col}}{\overline{Q}} \tag{21}$$

where V_{col} is the volume of a single column.

All of the flow rates as well as the shift period are known. The remaining parameters to be determined are:

- **Number of columns:** the number of columns and equivalent number of plates per column are determined in order to obtain the expected purities and yield. For a first estimation, choose total number of 8 columns, that is to say 2 columns per zone, and use the number plates proposed in the TMB calculation.
- **Column length and diameter:** these are set according to the minimum number of plates required and the maximum acceptable pressure-drop, and calculated by taking into account the constraints given by the Van Deemter and Darcy's laws.

10.4 Example of process design

10.4.1 Manufacture of Enantiopure Drug Substances

In order to illustrate an example of process design for the manufacture of enantiopure drug substances on an industrial SMB system, consider manufacturing 10 ton/year of an enantiopure drug. The racemic drug by definition is a 50:50 mixture of each enantiomer (products A and B). The goal is to process enantiopure drug substances in order to obtain 99 % purity for both the extract and the raffinate.

Two different feed concentrations or streams are possible to illustrate this example:

- Feed stream a: contains a total concentration of 1 g L^{-1}
- Feed stream b: contains a total concentration of 10 g L^{-1}

The issue is to design an industrial SMB system capable of processing these two feed streams, and to compare them in order to select the most adapted solution. The working pressure should not exceed 10 bar.

10.4.1.1 Gathering Physico-Chemical Parameters

Figure 10.5 shows two analytical injections of a racemic mixture on an analytical column ($V = 4.15$ mL (i.e. 250×4.6 mm); $Q = 1$ mL min^{-1} and 2 mL min^{-1}).

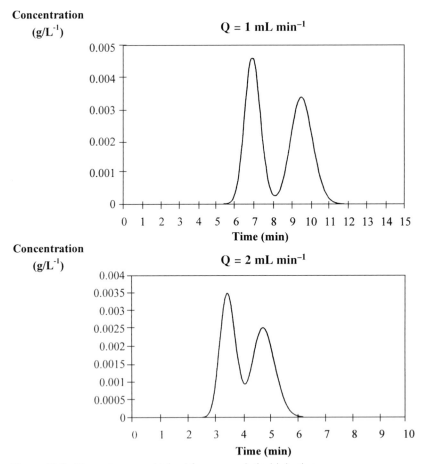

Figure 10.5. Chromatograms obtained from an analytical injection.

At a flow rate of 1 ml min^{-1}, the zero retention time on the analytical column is:

$$t_0 = \frac{\varepsilon \cdot V}{Q} = \frac{0.4 \cdot 4.15}{1} = 1.66 \quad \text{min}$$

The retention times of the two products being $t_R(A) = 6.9$ min and $t_R(B) = 9.5$ min (Fig. 10.5), one obtains:

$$\overline{K}_A = \frac{t_R(A) - t_0}{t_0} \cdot \frac{\varepsilon}{1-\varepsilon} = \frac{6.9 - 1.66}{1.66} \cdot \frac{0.4}{1-0.4} = 2.1$$

$$\overline{K}_B = \frac{t_R(B) - t_0}{t_0} \cdot \frac{\varepsilon}{1-\varepsilon} = \frac{9.5 - 1.66}{1.66} \cdot \frac{0.4}{1-0.4} = 3.15$$

Note that knowledge of the initial slopes of the adsorption isotherms gives some constraint to be fullfilled between parameters λ, N, and $\overline{K}$. In order to fit the adsorption isotherms, frontal analysis has performed with the pure components at 1, 25, 50, 75 and 100 g L^{-1} on the analytical column at 1 ml min^{-1}.

Results (breakthrough times) are given in Table 10.1.

Table 10.1 Retention times associated with breakthrough curves (A and B injected separately).

Concentration (g L^{-1})	Retention time front A (min)	Retention time front B (min)
1	6.9	9.5
25	6.2	7.5
50	5.7	6.5
75	5.3	5.9
100	5.1	5.5

The retention times obtained at a concentration of 1 g L^{-1} are identical to the analytical retention times. Therefore, system behavior is linear at concentrations below 1 g L^{-1}. When the concentration increases, the retention decreases, which are consistent with a Langmuir-type behavior.

It has been shown that retention times obtained by breakthrough curves for single component solutions is given by [58]:

$$t_R = t_0 \left[1 + \frac{1-\varepsilon}{\varepsilon} \cdot \frac{\overline{C}(feed)}{C(feed)} \right] \tag{22}$$

$\overline{C}$: concentration on the solid phase in equilibrium with the feed concentration.

Knowing the experimental retention times, the previous equation allows the calculation of "experimental" concentration on the solid phase. Parameters of adsorption isotherms, can then be determined by fitting experimental and calculated concentrations.

Correlation between experimental and calculated concentrations is obtained using a modified Langmuir adsorption isotherm with the following parameters:

$$\lambda = 0.738 \quad \bar{N} = 120$$

$$\tilde{K}_A = 0.0113 \quad \tilde{K}_B = 0.02$$

In Table 10.2, this correlation is shown, comparing solid phase concentration calculated from the retention times of the fronts, and using the adsorption isotherm equation.

Table 10.2 Correlation: experimental and calculated concentrations in solid phase

Concentration (g L^{-1})	t_R (A)	t_R (B)	$\bar{C}_A$ (exp.)	$\bar{C}_B$ (exp.)	$\bar{C}_A$ (Theoret.)	$\bar{C}_B$ (Theoret.)
1	6.9	9.5	2.10	3.15	2.08	3.10
25	6.2	7.5	45.5	58.6	45.0	58.6
50	5.7	6.5	81.1	97.1	80.4	97.2
75	5.3	5.9	109.6	127.7	110.6	127.7
100	5.1	5.5	138.1	154.2	137.7	154.2

The Van Deemter curve (HETP vs fluid velocity) is estimated from the two analytical chromatograms. At a flow rate of 1 ml min^{-1}, the liquid velocity is 0.001 m s^{-1} the number of theoretical plates associated to the second peak is about 250. The efficiency at 2 mL min^{-1} drops to 150.

The Van Deemter curve is represented by:

$$HETP = \frac{L}{N} = 0.0003 + 0.7 \cdot u \tag{23}$$

$$L \text{ in } \boldsymbol{m} \quad u \text{ in } \boldsymbol{m/s}$$

For accurate determination of pressure drop, the flowrate is measured at 20 ml min^{-1}. A pressure drop of 5.5 bar was measured, allowing Darcy's law to be expressed as:

$$\frac{\Delta P}{L} = 1097 \cdot u \tag{24}$$

$$L \text{ in } m\ u \text{ in } m/s\ \Delta P \text{ in } bar$$

10.4.1.2 SMB: Linear Conditions

Calculation of TMB flowrates: To calculate TMB flowrates, linear behavior of the adsorption isotherms for a feed concentration of 1 g L^{-1} is assessed. To check this point, we will use the criterion given in Equation (10).

At a feed concentration of 1 g L^{-1} which means that $C_A^F = C_B^F = 0.5$, the criterion $\tilde{K}_A \cdot C_A^F + \tilde{K}_B \cdot C_B^F \leq 0.1$ equals approximately 0.02, and the system behaves linearly.

Equation 13 is used to calculate the SMB flow rates. To ensure a throughput of 10 tons/year of the racemic mixture we need a feed flow rate of 1250 L h⁻¹. The γ value must fulfill $1 \leq \gamma \leq 1.22$ with our recommended value of 1.02.

The reduced flow rates given by Equation (12) are thus:

$$m_I = 3.213 \qquad m_{II} = 2.142 \qquad m_{III} = 3.088 \qquad m_{IV} = 2.059$$

leading to the following internal flow rates for the TMB:

$$\overline{Q} = 1321 \text{ l h}^{-1}$$
$$Q_I = 4244 \text{ l h}^{-1} \qquad Q_{II} = 2829 \text{ l h}^{-1} \qquad Q_{III} = 4079 \text{ l h}^{-1} \qquad Q_{IV} = 2719 \text{ l h}^{-1}$$

The inlet-outlet flow rates are then given by:

$$Q_{ext} = 1415 \text{ l h}^{-1} \qquad Q_{Feed} = 1250 \text{ l h}^{-1} \qquad Q_{Raf} = 1360 \text{ l h}^{-1}$$

Calculation of the SMB flowrates: The flow rate in each zone of a TMB is related to the flow rate of a SMB by Equation (20):

$$Q_I^{SMB} = Q_I^{TMB} + \frac{\varepsilon}{1-\varepsilon} \cdot \overline{Q}.$$

Consequently, one obtains for the SMB:

$$Q_I = 5125 \text{ l h}^{-1} \qquad Q_{II} = 3710 \text{ l h}^{-1} \qquad Q_{III} = 4960 \text{ l h}^{-1} \qquad Q_{IV} = 3600 \text{ l h}^{-1}$$

the mean value of the flow rates in the different zone is thus approximately $Q_{mean} = 4350 \text{ l h}^{-1}$.

Calculation of the SMB system: The SMB system requires between 200 and 400 theoretical plates for an enantioselectivity of 1.5. The working system pressure is 10 bar. With the equations for the HETP and the pressure-drop, we have two equations for two unknowns.

$$HETP = \frac{L}{N} = \frac{L}{300} = 0.0003 + 0.7 \cdot u$$

$$\frac{\Delta P}{L} = \frac{10}{L} = 1097 \cdot u$$

To solve the total column length for the SMB system, one eliminates the fluid velocity:

$$\frac{L}{300} = 0.0003 + 0.7 \cdot \frac{10}{L \cdot 1097}$$

$$L^2 - 0.09 \cdot L - 1.91 = 0$$

The total column length is:
L = 1.43 m (8 columns of 18 cm length)

The velocity is then:
$u = 0.0064$ m s^{-1}

The column diameter is then determined using the mean flow rate in the SMB.

$$\varnothing = \sqrt{\frac{4 \cdot Q_{mean}}{\pi \cdot u}} = 0.49 \; m$$

and Equation (21) gives the shift period: $\Delta T = 0.92$ min.

Using simulation software (essentially solving the mass balance equation in the transitory regime), one can show that the internal profiles of the products in the system are given by the profiles presented in Figure 10.6. Samples were taken between columns at half periods.

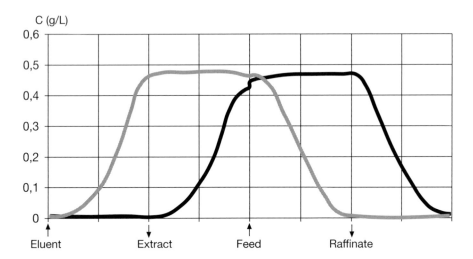

Figure 10.6. Internal profile (simulated) obtained for the separation under linear conditions.

This simulation shows that the purity of the raffinate collected over a complete cycle is about 97.0 % and the extract is about 98.5 %. These results are close to the target and adequate purity can be reached with minor changes in flowrates. Further details will be given in the next section.

10.4.1.3 SMB: Nonlinear Conditions

Calculation of the SMB flowrates: The criterion $\tilde{K}_A \cdot C_A^F + \tilde{K}_B \cdot C_B^F \le 0.1$ (Equation 10) equaling now approximately 0.22, the system is operating under nonlinear adsorption isotherm conditions.

Conditions of nonlinear chromatography prevail, and the set of Equations (15–19) leads to the following optimum reduced flowrates:

$$\omega_G = 2.2116 \qquad \omega_F = 1.2785$$
$$m_I = 3.14 \qquad m_{II} = 1.988 \qquad m_{III} = 2.831 \qquad m_{IV} = 2.022$$

and the region of complete separation is given in Figure 10.7.

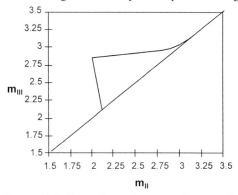

Figure 10.7. Zone of complete separation according to the equilibrium theory.

Knowing that the feed flow rate should equal $Q_{feed} = 125\ \mathrm{l\ h^{-1}}$, one can estimate the required solid flow rate: $\bar{Q} = Q_{Feed}/(m_{III}-m_{II}) = 149\ \mathrm{l\ h^{-1}}$ leading to all the internal flow rates:

$$Q_I = 466\ \mathrm{l\ h^{-1}} \qquad Q_u = 295\ \mathrm{l\ h^{-1}} \qquad Q_m = 420\ \mathrm{l\ h^{-1}} \qquad Q_{IV} = 300\ \mathrm{l\ h^{-1}}$$

Calculation of SMB flowrates: The flow rate in each zone of a SMB is related to the flow rate of a TMB by Equation (20):

$$Q_I^{SMB} = Q_I^{TMB} + \frac{\varepsilon}{1-\varepsilon} \cdot \bar{Q}$$

Consequently, one obtains for the SMB:

$$Q_I = 564\ \mathrm{l\ h^{-1}} \qquad Q_{II} = 394\ \mathrm{l\ h^{-1}} \qquad Q_{III} = 519\ \mathrm{l\ h^{-1}} \qquad Q_{IV} = 399\ \mathrm{l\ h^{-1}}$$

the mean value of the flow rates in the different zone is thus approximately $Q_{mean} = 469\ \mathrm{l\ h^{-1}}$.

Calculation of the SMB system: As the enantioselectivity is 1.5, we need between 200 and 400 theoretical plates in the system with a working pressure of 10 bar. The system constraints are similar to those of the linear system, one has:

$$L = 1.43 \text{ m (8 columns of 18 cm length)} \quad u = 0.0064 \text{ m s}^{-1}$$

The column diameter is then determined using the mean flow rate in the SMB.

$$\varnothing = \sqrt{\frac{4 \cdot Q_{mean}}{\pi \cdot u}} = 0.16 \, m$$

and Equation (21) gives the shift period: $\Delta T = 0.89$ min.

Using simulation software (essentially solving the mass balance equation in the transitory regime), one can show that the internal profiles of the products in the system are given by the profiles presented in Figure 10.8. Samples are taken between the column at half period.

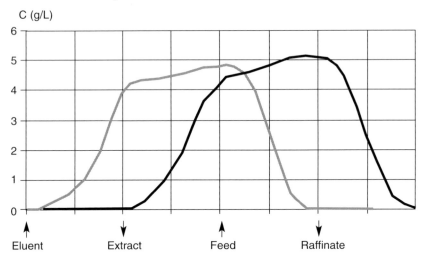

Figure 10.8. Internal profile (simulated) obtained for the separation under nonlinear condition.

According to the simulation of the process, the purity of the raffinate product was 98.0 % and the extract was slightly below 97.3 %. These results are quite close to the target purity of 99 %. To verify these constraints, the operating conditions were adjusted.

Two possible options allow the increase in purity of the recovered streams:

- Increase the number of plates equivalent to the system (adjusting column length, or fluid velocity).
- Increase the shift periods between the different zones (change internal flow rates).

Choosing the second option, we would increase the flowrates in zones I and decrease the flowrate in zone IV. Then adjust the flowrates in zones II and III (the feed flowrate is slightly decreased to improve the margin in these two zones). With the assistance of the simulation program, we can determine to what degree the flowrates should be decreased or increased to obtain the target purity.

The final flowrates are determined by replacing the initial set of flowrates:

$$Q_I = 567 \ 1 \ h^{-1} \qquad Q_{II} = 395 \ 1 \ h^{-1} \qquad Q_{III} = 521 \ 1 \ h^{-1} \qquad Q_{IV} = 400 \ 1 \ h^{-1}$$

by:

$$Q_I = 587 \ 1 \ h^{-1} \qquad Q_{II} = 409 \ 1 \ h^{-1} \qquad Q_{III} = 519 \ 1 \ h^{-1} \qquad Q_{IV} = 395 \ 1 \ h^{-1}$$

which leads to an extract purity of 99.4 % and to a raffinate purity of 99.4 %. The new flow rates are thus very close from the initial one.

10.5 SMB as a Production Tool

10.5.1 cGMP Compliance

Current Good Manufacturing Practices (cGMP) and compliance are the requirements found in the legislation, regulation, and administrative provisions for drug manufacturing processes. It defines the facilities, equipment, manufacturing processes, and packaging of a drug substance or product. It assures that such drug substance or product meets the requirements as to safety, efficacy, and purity. Under cGMP, the FDA provides guidelines for the manufacturing, processing, and packaging of all drug substances, which include enantiopure drugs as API. The API or finished dosage form of the drug is intended to provide pharmacological activity for the diagnosis, cure, treatment, or prevention of a disease. APIs include substances manufactured by processes such as: chemical synthesis, fermentation, and biotechnology methods, purification process such as chromatography, and isolation and recovery from natural products.

The FDA recognizes that at certain early production stages, applying stringent controls may not be feasible or necessary. The stringency of controls, such as the extent of written instructions, in-process controls, sampling, testing, monitoring and documentation, in the manufacture of API's increase as the process proceeds from early intermediate stages to final synthesis and purification.

10.5.1.1 Manufacturing and Process Controls

In March of 1998, the FDA announced a draft guidance document for Industry for the manufacturing, process, or holding of APIs [64]. We shall apply our interpretation of cGMP compliance for the manufacture of enantiopure drug substances to the SMB process.

10.5.1.2 Solvent Recovery

The draft document address the issue of solvent recovered from a process and the use of these solvents in the same process or reused for different processes. It requires that recovery procedures be validated to ensure cross-contamination between recovered solvents and monitoring of the solvent composition at suitable intervals during the process.

Figure 10.9 illustrates an industrial SMB system. The system is a closed-loop continuous SMB system composed of a chromatographic and solvent recycling unit. The total SMB system consists of 8 HPLC columns, which range from 200 to 1000 mm in diameter. An eluent tank for preparation of a binary mobile phase and monitoring of composition and back-up solvent tank for addition of a single solvent. A feed preparation tank is used for the preparation of a solution of the racemic drug substance before pumping to a holding feed tank. To complete the closed-loop system, two falling film evaporators are for the concentration of extract and raffinate and recycling of the eluent.

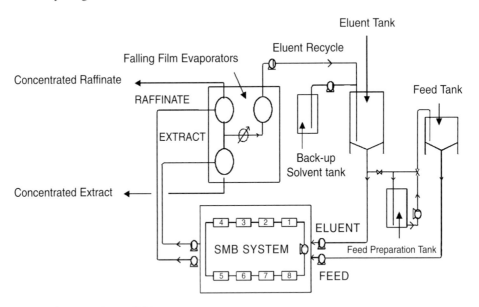

Figure 10.9 Closed loop SMB system.

With the SMB system, solvents are recovered, continuously monitored, and analyzed by an automated software system. The chromatographic SMB process ensures that the solvents meet appropriate standards and cross-contamination is eliminated.

Another important parameter is the eluent composition. Binary mixtures (and obviously pure solvents) should be preferred to complex mixtures, since new systems perform an on-line analysis of the composition of binary eluents. These eluent systems allow the automatic eluent recycling, with a reduced number of controls.

10.5.1.3 In-Process Testing

Another issue during manufacturing is in-process controls and sampling of APIs. The guidance document defines sampling methods for in-process materials, intermediates, and APIs. These sampling protocols are based on validated data and controlled sampling practices. The sampling technique requires controls to prevent cross-contamination with other APIs or intermediates with procedures to ensure integrity of the in-process samples after collection [65].

Due to the nature of the SMB process, in-process samples of the unwanted enantiomer and the enantiopure drug substance can be sampled at controlled times during the continuous process to assess the enantiomeric and chemical purity. One can monitor the process without system shutdown by diverting either the extract or the raffinate streams. Further monitoring of the receiving tanks can also be accomplished.

10.5.1.4 Calculation of Yields and Definition of Batch

The guidance document requires calculation of actual yields and percentages of expected yields. The yield should be recorded at the conclusion of each phase of manufacturing of an API. The expected yield and ranges are established during process validation or from a pilot-scale production run [66].

For a continuous SMB process, the specific identified amount or batch produced is defined by unit of time in such a way that ensures a homogeneous material and quality within specified limits. In the case of a continuous SMB production run a batch is defined by the amount produced in a fixed time interval. A time limitation during manufacturing using SMB is established by the same fixed time interval as the batch. The duration of the production phase is thus established, which does not affect the quality of the drug substance [66].

10.5.2 Process Validation

Process validation is a requirement of the cGMP Regulations for Finished Pharmaceuticals [66]. The Global Harmonization Task Force Study Group #3 issued the most recent guidelines on process validation [67]. Therefore, our task is to address some of the issues concerning the industrial SMB system during process validation of enantiopure drug substances and the issues they pose in the pharmaceutical industry.

The design of the system must take into account possible variation of critical control parameters that could affect performance. The maximum performance of the process should be defined by a reasonable safety margin. In order to comply with cGMP guidelines, established validation protocols, and parameters should allow the process to achieve reproducible purity and yield under stressed conditions. This implies that the industrial SMB system must be stressed to simulate worst-case conditions for process validation.

Process validation requires documented evidence that a process will provide a high degree of assurance that it consistently produce an enantiopure drug substance meeting its predetermined specifications and qualities characteristics. To validate an SMB process, a range of critical process parameters are established based on research or pilot-scale batches that encompasses values that are capable of producing enantiopure drug substances with acceptable quality attributes. To establish worst-case conditions a validation protocol is a written plan demonstrating how validation will be conducted. The protocol for manufacturing an enantiopure drug substance should identify the processing equipment. In our case, this represents the industrial SMB system as a single unit, from the chromatography to evaporation. Critical process parameters such as the influence of the feed concentration change in internal flowrates, column efficiency, and evaluation of different batches of CSPs should be obtained and acceptable test results established. Data must be collected for extended operating ranges or target ranges during routine production. Establishment of monitoring points for in-process sampling and test data required for evaluation of product characteristics.

In addition, the protocol should specify a sufficient number of process runs to prove consistency of the process, and provide an accurate measure of variability among successive runs. The number of batches should depend on the extent of validation and complexity of the process or importance of any process changes. Furthermore, the protocol should address the quality of materials used in the process from starting materials to new and recovered solvents, and evidence of the performance and reliability of equipment and systems.

Process validation should be extended to those steps determined to be critical to the quality and purity of the enantiopure drug. Establishing impurity profiles is an important aspect of process validation. One should consider chemical purity, enantiomeric excess by quantitative assays for impurity profiles, physical characteristics such as particle size, polymorphic forms, moisture and solvent content, and homogeneity. In principle, the SMB process validation should provide conclusive evidence that the levels of contaminants (chemical impurities, enantioenrichment of unwanted enantiomer) is reduced as processing proceeds during the purification process.

In order to illustrate the critical process parameters of SMB process validation, we will consider the separation of the racemic drug as described in Process design. The study represents the effect of the influence of feed concentration, number of plates and retention factor on the second eluting enantiomer. The simulation of the process for different values of feed concentration is performed and the variations of the extract and raffinate purities are shown in Fig. 10.10.

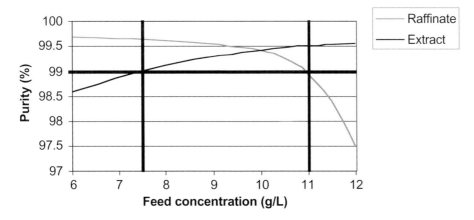

Figure 10.10 Influence of the feed concentration on extract and raffinate purity.

A target purity of 99 % was established for both extract and raffinate. According to the simulation results, one can predict that a variation of the feed concentration range between 7.5 and 11 g L^{-1} will meet the required purity. The system was designed for a feed concentration equal to 10 g L^{-1}. The influence of change in feed concentration on the purity of both extract and raffinate illustrates the robustness of SMB, and that the process tolerates fluctuations when critical parameters are stressed during process validation.

A second simulation study was performed to measure the effect on both extract and raffinate purities of a loss of chromatographic efficiency (Fig. 10.11).

The graph in Fig. 10.11 shows that the SMB can tolerate a loss of 13 % chromatographic efficiency and still reach a purity of greater than 98 %. The industrial SMB system was designed to operate with 300 theoretical plates without any modification of the operating flowrates.

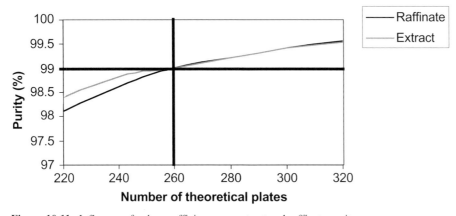

Figure 10.11. Influence of column efficiency on extract and raffinate purity.

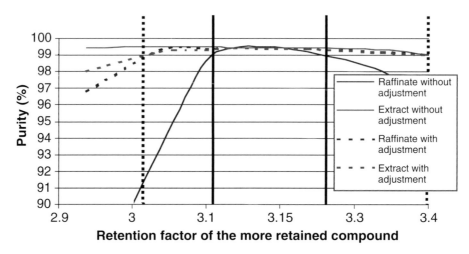

Figure 10.12. Influence of retention factor of the more retained compound on extract and raffinate purity. Solid line: without adjustment of the operating flow rates; dotted lines: with adjustment of the operating flow rates.

Finally, simulation studies were performed to evaluate the influence of change in eluent consumption or variations due to different lots of CSPs on the retention of the second peak. The variation in retention of the second peak is another critical parameter on resulting purity of the extract and raffinate.

The effect on purity and the influence of retention factor by adjusting operating flowrates is illustrated in Fig. 10.12.

The graph in Fig. 10.12 shows that the purity decreases very quickly below acceptable levels as retention factor of the more retained enantiomer decreases. However, with minor adjustment of the SMB internal flow rates, a variation of more than 10 % of the retention factor of the more retained enantiomer still meets required purity, productivity, and eluent consumption. Control of critical parameters such as retention factors can be made without modification of the feed and eluent flowrates.

Using computer-aided numerical calculations, one can readily simulate and identify critical parameters for process validation. Thus, one can evaluate the robustness of the process during its design. To ensure performance, optimization of the process and evaluation of critical parameters can be determined before actual operating conditions.

10.6 SMB Accepted for Manufacturing

10.6.1 Practical Implications for Manufacturing

SMB is now accepted as a real production tool. For instance, the Belgium pharmaceutical company U.C.B. Pharma announced recently the use of SMB for performing multi-ton scale purification of an enantiopure drug substance. The concept of large-scale purification of enantiomers using chromatographic techniques has moved from a dream to a reality within the last few years.

A major advantage of SMB technology compared to the cost of stereoselective synthesis is that manufacture of a racemic drug substance has already been proven to be straightforward, inexpensive, and less time-consuming. Coupling non-stereoselective synthesis with an SMB process is an economical way to produce both enantiomers with high purity and recovery. The main object for any pharmaceutical manufacturer is the rapid delivery of enantiopure drugs to market.

The interests of SMB for performing large-scale separations of enantiopure drugs has been recognized (very short development time, extremely high probability of success, and attractive purification cost) [68]. Several pharmaceutical and fine chemical companies have already developed SMB processes. However, because of strong confidentiality constraints, public information is limited, and some of the major announcements are summarized below:

- **DAICEL announcement:** Daicel has announced an investment in a SMB production plant with columns of 100 mm i.d. to be installed in their facilities at Arai (Japan). This plant produces hundreds of kg per year for a drug currently manufactured by Nissan Chemicals (*Japan Chemical Week,* December 4, 1997).
- **UCB announcement:** UCB Pharma (Belgium) announced in 1997 its decision to install a SMB made of columns of 45 cm i.d. in order to perform large-scale manufacturing for a promising new class of drugs. UCB decided to replace a classical chemical process used in the pharmaceutical industry by SMB technology.
- **AEROJET:** AEROJET (California) has announced in 1999 its decision to invest in a SMB system using columns of 80 cm in diameter.

10.7 Conclusions

The evolution of FDA policies continues to be a significant driving force on the global pharmaceutical market. Several pharmaceutical firms have made new discoveries while evaluating enantiopure drugs originally discovered and marketed as racemates by others. These pharmaceutical firms have merged, or other companies have appropriated portfolios of patents based on chiral switches. Thus, the FDA continues to establish policies to control market exclusivity as it applies to cGMP guidelines defined for the development and manufacture of enantiopure drug substances.

Preparative chromatography has been used for chiral separations for years, but examples of multi-kg separations (and hence larger ones) were rare until recently. The development of SMB techniques (both hardware and simulation software) has made major breakthroughs in this field. The ability of SMB as a development tool has allowed the pharmaceutical manufacturer to obtain kilo grams quantities of enantiopure drug substances as well benefit from the economics of large-scale production.

Process validation is the procedure that allows one to establish the critical operating parameters of a manufacturing process. Hence, the constraints imposed by the FDA as part of process control and validation of an SMB process. The total industrial SMB system, as described, is a continuous closed-loop chromatographic process, from the chromatographic to recycling unit and, with the use of numerical simulation software allows the pharmaceutical manufacturer rapidly to design and develop worst-case studies.

From the position of the FDA, acceptance of SMB as a viable tool for cGMP manufacturing of enantiopure drug substances, there shall be no compromise, it must be properly engineered, and follow established guidelines.

References

[1] FDA's Policy Statement on the Development of New Stereoisomeric Drugs (Stereoisomeric Drug Policy) Fed. Regist. (1992) 57 FR22249.
[2] Agranat, I. and Cancer, H., Drug Discover Today, 4 No. 7, (1999) 313–321.
[3] Crosby, J. (1997) in Chirality in Industry II: Developments in the Commercial Manufacture and Application of Optically Active Compounds (Collins, A.N., Sheldrake, G.N. and Crosby, J., eds), pp 1–10, Wiley.
[4] Committee for Proprietary Medicinal Product (1993) Note for Guidance: Investigation of Chiral Active Substances III/3501/91.
[5] Daniels, J.M., Nestmann, E.R. and Kerr, A. Drug Inf. J., 31, (1997) 639–646.
[6] Blumenstein, J.J. (1997) in Chirality in Industry II: Developments in the Commercial Manufacture and Application of optically Active Compounds (Collins, A.N., Sheldrake, G.N. and Crosby, J., eds), pp 11–18, Wiley.
[7] Shindo, H. and Caldwell, J., (1995) Chirality, 7, 349–352.
[8] Francotte, E., J. Chromator. A., 666 (1994) 565–601.
[9] Welch, C.J., J. Chromatogr. A, 666 (1994) 3–26.
[10] International Conference on Harmonisation; Draft Guidance on Specifications: Test Procedures and Acceptance Criteria for New Drug Substances and New Drug Products: Chemical Substances; Notice, Fed Regist. Docket No. 97D-0448, 1997.
[11] Draft Guidance for Industry on Manufacturing, Processing, or Holding Active Pharmaceutical Ingredients; Availability; Notice, Fed Regist. Docket No. 98-0193, 1998.
[12] Current Good Manufacturing Practice for Finished Pharmaceuticals, Sampling and Testing of In-Process Materials and Drug Products, (1998), CFR, Title 21, Part 211, Volume 4, Section 2ll.ll0.
[13] Zamani, K., Conner, D.P, Weems, H.B., Yang and Cantilena, L.R, Chirality, 3, (1991) 467–470.
[14] Okerholm, R.A., Weiner, D.L., Hook, R.H., Walker, B.J., Biopharm. Drug Dispos., 2, (1981)185–190.

[15] Monahan, B.P., Ferguson, C.L., Killeavy, E.S., Lloyd, B.K, J., Cantilena, L.R., J. Am. Med. Assoc., 264, (1990) 2788–2790.

[16] Proposal to Withdraw Approval of Two New Drug Applications and One Abbreviated New Drug Application; Terfenadine, Hoechst Marion Rousssel, Inc and Baker Norton Pharmaceuticals, Fed Regist., Docket No. 96N-0512, 1997, 1998.   Part 216 Pharmacy Compounding, Drug Products Withdrawn or Removed From the Market for Reasons of Safety or Effectiveness, Cite: 216.24.

[17] Policy on Period of Marketing Exclusivity for Newly Approved Drug Products with Enantiomer Active Ingredients; Request for Comments, (1997) Docket No. 97N-0002.

[18] Applications for FDA Approval to Market a New Drug or an Antibiotic Drug, Content and format of an Application, CFR, Title 21, Volume 5, Part 314, Section 314.50 (1998).

[19] Applications for FDA Approval to Market a New Drug, New Drug Product Exclusivity, (1999) CFR, Title 21, Part 314, Volume 5, Section 314.108.

[20] Fixed-Combination Prescription Drug for Humans, (1999), CFR, Title 21, Part 300, Volume 5, Section 300.50.

[21] Caldwell, J., (1995) Regulation of Chiral Drugs, Pharmaceutical News, 2, 22–23. Caldwell, J. (1995) Chiral Pharmacology and the Regulation of New Drugs, Chem. and Industry, 5, 176–179.

[22] Nicoud R.M. and M. Bailly, in Proceedings of the 9th Symposium on Preparative and Industrial Chromatography "Prep 92", INPL, Nancy, France, 1992, pp. 205–220.

[23] Broughton D.B., US Patent 2985 589(1961).

[24] Balannec B. and Hotier G., "From batch to countercurrent chromatography", in Preparative and Production Scale Chromatography, G. Ganetsos and P.E. Barker (Editors), Marcel Dekker, New York, 1993.

[25] Nicoud R.M., Fuchs G., Adam P., Bailly M., Küsters E., Antia, F.D., Reuille R. and Schmid E., Chirality, 5 (1993) 267. Nicoud R.M., LC-GC Int., volume 6, Number 10, October 1993.

[26] Nicoud R.M., Bailly M., Kinkel J.N., Devant R., Hampe T., and Küsters E., in R.M. Nicoud (Editor), Simulated Moving Bed: Basics and applications, INPL, Nancy, France, 1993, pp. 65–88.

[27] Blehaut J., Charton F., Nicoud R.M., LC-GC Intl, Vol. 9, No. 4, pp. 228–238, 1996.

[28] Martin and Kuhn, Ztschr. elektrokem., bd. 47, no 3, 1941.

[29] Hotier G. and Balannec G., Xvème conférence des Arts Chimiques, 5/8 Décembre 1989.

[30] Negawa M. and Shoji F., J. Chrom., 590, 113–117 (1992).

[31] Fuchs G., Nicoud R.M. and M. Bailly, in Proceedings of the 9th Symposium on Preparative and Industrial Chromatography "Prep 92", INPL, Nancy, France, 1992, pp. 205-220.

[32] Nicoud R.M., Chem. Eng. 1992, 21.

[33] Devant R.M., Jonas R., Schultre M., Keil A. and Charton F., J. Prakt. Chem., 339, pp. 315–321, 1997.

[34] Francotte E., Wolf R.M., Chirality, 3, p. 43, 1991.

[35] Ching C.B., Lim B.G., Lee E.J.D., Ng S.C., J. of Chrom., 634, pp. 215–219, 1993.

[36] Ikeda H. and Murata K., 4th Symposium on Chiral Discrimination, September 1993.

[37] Nicoud R.M., Proceedings of the Chiral Europe 96, Strasbourg, 14–15 October 1996, Spring Innovation Ltd Publisher, Cheshire, SK7 1BA, 1996.

[38] Kinkel J.N., Proceedings of Chiral Europe 95, 28–29 Sept. 1995, London, Published by Spring Innovation Ltd, Cheshire, SK7 1BA, England, 1995.

[39] Gattuso M.J., Chimica Oggi, September 1995.

[40] Gattuso M.J., Mc Culloch, House D.W., Bauman W.M. and Gottschall K., Pharmaceutical technology Europe, June 1996.

[41] Guest D.W. – Evaluation of simulated moving bed chromatography for pharmaceutical process development., J. of Chromatography A, 760 (1997), pp 159–162.

[42] Cavoy E., Deltent M.F., Lehoucq S., Miggiano D., J. of Chrom., 769, pp. 49–57, 1997.

[43] Küsters E., Proceedings of Chiral Europe 95, Published by Spring Innovation Ltd., Stockport, England, 1995.

[44] Francotte E., Proceedings of Chiral Europe 96, Strasbourg 14–15 October 1996, Published by Spring Innovations Ltd, Stocport, England, 1996.

[45] Nagamatsu, Proceedings of Chiral Europe 96, Strasbourg 14–15 October 1996, Published by Spring Innovations Ltd, Stocport, England, 1996.

[46] Charton F., Nicoud R.M., J. of Chrom. A., 702, pp. 97–112, 1995.

[47] Pais L.S., Loureiro J.M., Rodrigues A.E., Chem. Eng. Sci., 52, p. 245, 1997.

[48] Mazzotti M., Pedeferri M.P., Morbidelli M., Proceedings of Chiral Europe 96, Strasbourg 14–15 October 1996, Published by Spring Innovations Ltd, Stocport, England, 1996.
[49] Mazzotti M., Storti G., Morbidelli M., J. Chromatogr. A., 769 (1997), 3.
[50] Aiche J., 42 (1996), 2734.
[51] Strube et al., J. of Chrom. A, 769, pp. 81–92, 1997.
[52] Nagamatsu S. and Makino S., poster presented during the International Symposium on Chiral Discrimination, Nagoya, 1997.
[53] Ruthren and Chuthren D.M., Chinsc. B., Chem. Eng. Sci., 44 (5), 1011–1038, 1989.
[54] Hotier G. and Nicoud R.M., US Patent 5,578,216.
[55] Hotier G., Cohen C., Couenne N., and Nicoud R.M., US Patent 5,578,216.
[56] Nicoud et al., US patent 5,422,07 1995.
[57] Biressi G., Ludemann-Hombourger O., Nicoud R.M., Morbidelli M., J. Chromatogr. A, submitted for publication (1999).
[58] Seidel-Morgenstern A. and Nicoud R.M., Isolation and purification, vol. 2, p. 165–200, 1996.
[59] Van Deemter J.J., F.J. Zuiderweg and A. Klinkenberg, Chem. Eng. Sci., 5 (1956) 271.
[60] Knox J.H., J. Chromatogr. Sci., 15 (1977) 352.
[61] Horvath C. and H.J. Lin, J. Chromatogr., 149 (1978) 43.
[62] Endele R., I. Halasz and K. Unger, J. Chromatogr., 99 (1974) 377.
[63] Mazzotti M., Storti G., Morbidelli M., AIChE J., 40 (1994), 1825.
[64] Current Good Manufacturing Practice for Finished Pharmaceuticals, Calculation of Yield, (1998), CFR, Title 21, Part 211, Volume 4, Section 211.103.
[65] Current Good Manufacturing Practice for Finished Pharmaceuticals, Time Limitations on Production, (1998), CFR, Title 21, Part 211, Volume 4, Section 211.111.
[66] Quality System Regulation, Production and Process Controls, (1998), CFR, Title 21, Part 820, Volume 8, Section 820.75.
[67] Medical Devices: Draft Global Harmonization Task Force Study Group 3 Process Validation Guidance, (1998), Fed. Regist., Docket No. 98D-0508.
[68] Nicoud R.M., Pharmaceutical Technology Europe (1999) 11 [3], 36–44 and 11 [4] 28–34.

FDA Documents

Current Good Manufacturing Practices for Finished Pharmaceuticals, U.S. Food and Drug Administration, 21 CFR 210 and 211

FDA Guideline for Submitting Documentation for the Stability of Human Drugs and Biologics, Center for Drugs and Biologics, February 1987

FDA Guideline for Submitting Documentation in Drug Applications for the Manufacture of Drug Substances, Center for Drugs and Biologics, February 1987

FDA Guide to Inspection of Bulk Pharmaceutical Chemicals, Office of Regional Operations and Center for Drug Evaluation and Research, September 1991, Reformatted May 1994

FDA Biotechnology Inspection Guide, Office of Regional Operations, November 1991

FDA Guideline to Inspection of Validation of Cleaning Processes, Office of Regulatory Affairs, July 1993

HHS/FDA "International Conference on Harmonisation; Stability Testing of New Drug Substances and Products; Guideline; Availability; Notice," Federal Register, September 22, 1994

HHS/FDA "International Conference on Harmonisation, Guidelines Availability: Impurities in New Drug Substances; Notice, Federal Register, January 4, 1996

An FDA Perspective on Bulk Pharmaceutical Chemicals," Edmund M. Fry, Pharmaceutical Technology, February 1984, Pages 48–53

11 Electrophoretically Driven Preparative Chiral Separations using Cyclodextrins

Apryll M. Stalcup

11.1 Introduction

The separation of enantiomers is a very important topic to the pharmaceutical industry. It is well recognized that the biological activities and bioavailabilities of enantiomers often differ [1]. To further complicate matters, the pharmacokinetic profile of the racemate is often not just the sum of the profiles of the individual enantiomers. In many cases, one enantiomer has the desired pharmacological activity, whereas the other enantiomer may be responsible for undesirable side-effects. What often gets lost however is the fact that, in some cases, one enantiomer may be inert and, in many cases, both enantiomers may have therapeutic value, though not for the same disease state. It is also possible for one enantiomer to mediate the harmful effects of the other enantiomer. For instance, in the case of indacrinone, one enantiomer is a diuretic but causes uric acid retention, whereas the other enantiomer causes uric acid elimination. Thus, administration of a mixture of enantiomers, although not necessarily racemic, may have therapeutic value.

Despite tremendous advances in stereospecific synthesis, chiral separations will continue to be important because of possible racemization along the synthetic pathway [2], during storage or in vivo (e.g., ibuprofen [3]). While analytical methods are necessary and have become almost routine, economical methods for preparative and semipreparative scale-chiral separation remains largely unexplored. Yet, preparative chiral separations may be particularly important in an R&D setting where only small amounts of material may be required to initiate screening prior to developing a potentially more costly stereospecific synthetic strategy. In addition, the pharmaceutical industry has a critical need for methods which produce pure enantiomers for reference materials.

During the past two decades, significant progress has been made in chromatographic chiral separation technology. However, the bioavailability of drug substances dictates that the compounds be water-soluble, and many are ionized at physiological pH. The pK_as of many drugs (see Table 11-1) are well outside the safe operating range for silica-based media, and almost all high-performance liquid chromatography (HPLC) chiral stationary phases currently available commercially are on silica substrates. In addition, most preparative liquid chromatography (LC)

Table 11-1. Examples of pK$_a$s for various chiral drugs.

Drug	Class	pK$_a$
Albuterol	Bronchodilator	9.3
Bupivacaine	Anesthetic	8.1
Chloroquine	Antimalarial	10.8, 8.4
Chlorpheniramine	Antihistamine	9.2

chiral chromatographic separations are performed using organic mobile phases in which many drug substances have limited solubility. Hence, there is a critical need for alternative methods for preparative chiral separations.

Chiral additives have been shown to be very effective for chiral separations by capillary electrophoresis (CE) [4, 5]. Indeed, it may be argued that there has been considerably more research activity in chiral separations by CE than by LC methods since the introduction of the former technique. Chiral additives in CE have several advantages, some of which are highlighted in Table 11-2.

Table 11-2. Advantages of chiral additives in Ce.

- Additive can be readily changed
- Variety of chiral selectors available
- Rapid sreening of chiral selectors/conditions/analytes
- Small amounts of background electrolyte required
- Small amounts of chiral additive required
- Can use "counter current" processes
- Multiple complexation possible
- No pre equilibration

While most discussion of resolution in CE focuses on the tremendous efficiencies (e.g., narrow peak widths) achievable with capillary columns, it should be noted that resolution is also a function of differences in selectivity. Unlike HPLC, where flow is unidirectional, CE using chiral additives can exploit true countercurrent migration of oppositely charged analytes and additives. For instance, in Fig. 11-1, an electropherogram [6] obtained with minimal cathodic electro-osmotic flow directed away from the detector, the cationic analytes only reach the detector through complexation with the anionic cyclodextrin. Inhibition of complexation through the addition of methanol amplifies chiral recognition because the analyte effectively experiences a "longer" column as its own intrinsic electrophoretic mobility carries it back up the column when it is in the uncomplexed state.

CE is generally more suited to analytical separations than to preparative-scale separations. However, given the success of CE methods for chiral separations, it seems reasonable to explore the utility of preparative electrophoretic methods to chiral separations. Thus, the purpose of this work is to highlight some of the developments in the application of preparative electrophoresis to chiral separations. Both batch and continuous processes will be examined.

Important in all of this preparative electrophoretic work is the recognition that CE has been used in the method development of these preparative electrophoretic

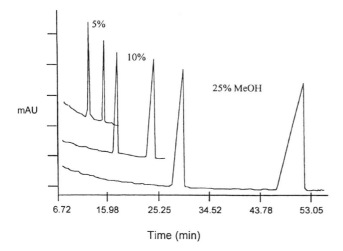

Fig. 11-1. Effect of the addition of methanol on the enantiomeric separation of terbutaline using 2 % sulfated cyclodextrin in 25 mM phosphate buffer (pH 3).

methods. An analogous relationship may be seen in the use of chiral mobile phase additives in thin-layer chromatography (TLC) for screening potential chiral selectors for immobilization in chiral stationary phases for HPLC.

In considering the applicability of preparative classical electrophoretic methods to chiral separations, it should be noted that practitioners in the art of classical electrophoresis have been particularly inventive in designing novel separation strategies. For instance, pH, ionic strength and density gradients have all been used. Isoelectric focusing and isotachophoresis are well-established separation modes in classical electrophoresis and are also being implemented in CE separations [7, 8]. These trends are also reflected in the preparative electrophoretic approaches discussed here.

11.2 Classical Electrophoretic Chiral Separations: Batch Processes

Classical gel electrophoresis has been used extensively for protein and nucleic acid purification and characterization [9, 10], but has not been used routinely for small molecule separations, other than for polypeptides. A comparison between TLC and electrophoresis reveals that while detection is usually accomplished off-line in both electrophoretic and TLC methods, the analyte remains localized in the TLC bed and the mobile phase is immediately removed subsequent to chromatographic development. In contrast, in gel electrophoresis, the gel matrix serves primarily as an anticonvective medium and is usually designed to minimize interactions with the solute, excluding molecular sieving effects. In addition, the presence of bulk liquid in the post-run gel no doubt contributes to the solute diffusivity problem, thereby reducing efficiency and complicating detection. Hence, separation of small molecules by classical gel electrophoresis is generally not done. However, solute diffusion may be reduced through complexation with bulky additives.

Righetti and co-workers [11] were one of the first to demonstrate the utility of classical isoelectric focusing for the chiral separation of small molecules in a slab gel configuration. In their system, dansylated amino acids were resolved enantiomerically through complexation with β-cyclodextrin. Preferential complexation between the cyclodextrin and the derivatized amino acid induced as much as a 0.1 pH unit difference in the pK_bs of the dansyl group.

Stalcup et al. [12] also demonstrated that chiral analytes complexed with a bulky chiral additive (e.g. sulfated cyclodextrin, MW ~2000–2500 Da, depending on degree of substitution, DS), with reasonably large binding constants (~10^3 m^{-1}) [13] could be resolved enantiomerically using classical gel electrophoresis. Initial work used a tube agarose gel containing sulfated cyclodextrin as the chiral additive. Mechanical support and cooling for the gel was provided by a condenser from an organic synthetic glassware kit (Fig. 11-2). Although 10 mg of racemate were loaded onto the gel and significant enantiomeric enrichment was obtained, recovery of the analyte required extrusion and slicing of the gel with subsequent extraction of the individual slices followed by chiral analysis of the extracts, a fairly labor-intensive process.

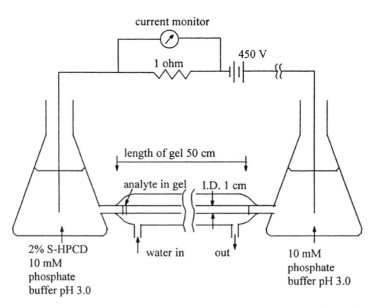

Fig. 11-2. Schematic for preparative gel electrophoresis using a condenser for mechanical support and cooling.

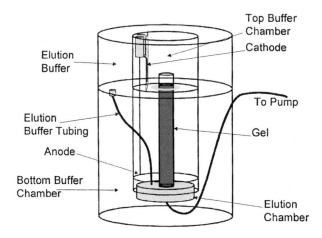

Fig. 11-3. Mini-prep continuous elution electrophoretic cell.

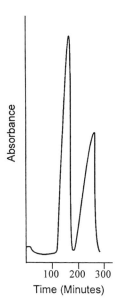

Time (Minutes)

Fig. 11-4. UV trace of piperoxan enantiomers eluting from mini-prep electrophoresis cell.

Stalcup and co-workers [14] adapted this method to a continuous elution mini-prep electrophoresis apparatus shown in Fig. 11-3. In this apparatus, the end of the electrophoretic gel is continuously washed with elution buffer. The eluent can then be monitored using an HPLC detector (Fig. 11-4) and sent to a fraction collector where the purified enantiomers, as well as the chiral additive, may be recovered. In this system, the gel configuration was approximately 100 mm × 7 mm, and was air-cooled. The number of theoretical plates obtained for 0.5 mg of piperoxan with this gel was approximately 200. A larger, water-cooled gel was able to handle 15 mg of terbutaline. However, run times were on the order of 20 h and thus represents the probable limitations of this approach.

In contrast to most classical electrophoretic separations of biologically derived samples, the racemic mixtures separated on these gels should be relatively pure, and sample degradation should not contaminate the gels. Hence, a significant advantage to the apparatus used in Fig. 11-3 is that the gel can be used several times. Indeed, Table 11-3 illustrates the results for four consecutive runs obtained on the same gel. Thus, in essence, the gel may serve as a surrogate column, suggesting that classical electrophoresis may provide a low-cost alternative to chiral chromatography.

Table 11-3. Migration times for piperoxan enantiomers for four consecutive runs on the same agarose gel using sulfated cyclodextrin as the chiral selector.

Run	T1 (min.)	T2 (min.)
1	133	167
2	136	172
3	129	168
4	137	179
Average	134	172
SD	3.6	5.4

Ultimately, however, it should be noted that these examples of classical gel electrophoretic separations are batch processes and therefore limited in sample throughput. To achieve true preparative-scale separations by electrophoresis, it becomes necessary to convert to continuous processes.

11.3 Classical Electrophoretic Chiral Separations: Continuous Processes

Preparative continuous free flow electrophoresis was first reported in 1958 [15]. As in the case of classical gel electrophoresis, most of the work done in this area has been primarily in the purification of biopolymers. Continuous free flow electrophoresis for the separation of small molecules has remained relatively unexplored [16], although this is beginning to change.

In preparative continuous free flow electrophoresis, continuous buffer and sample feed are introduced at one end of a thin, rectangular electrophoresis chamber. A schematic is presented in Fig. 11-5. The sample stream is usually introduced through a single port while buffer is introduced through several ports, essentially producing a buffer "curtain". Because the buffer streams are introduced independently, it is fairly easy to establish a variety of gradients (e.g., pH, density, ionic strength) across the buffer "curtain".

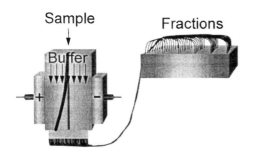

Sample **Fractions**

Fig. 11-5. Schematic of continuous free flow electrophoresis apparatus.

When an electric field is imposed perpendicular to the flow, differential interaction between the various solutes and the electric field produce a lateral displacement of individual analyte streams between the two electrodes (Fig. 11-5). Thus, the separations are accomplished in free solution. Individual fractions are collected through an array of closely spaced ports evenly placed across the other end of the chamber.

In the presence of a buffer with constant composition across the electrophoretic chamber, the angle of deflection (Θ) of the solute in the electric field is dependent upon the intrinsic electrophoretic mobility of the solute (μ_i), the linear velocity of the buffer (v) and the current through the chamber (I) and can be described as [17]:

$$\tan \Theta = \frac{\mu_i i}{q \kappa v} \tag{1}$$

where q is the cross-section of the separation chamber and κ is the specific conductance of the buffer.

For the separation of enantiomers, we are interested in $\Theta_1 - \Theta_2$. Substituting a = I/$q\kappa v$, using the expression relating the apparent mobility of an analyte to its binding constant with a chiral additive

$$\Theta_1 - \Theta_2 \approx a(\mu_1 - \mu_2) \tag{2}$$

and the concentration of the additive, and using a series expansion of tan Θ, to a first approximation, the difference in the angle of deflection for the enantiomers can be expressed as

$$\Theta_1 - \Theta_2 \approx a[\frac{[\mu_f - \mu_c][K_1 - K_2][CA]}{(1 + K_1[CA])(1 + K_2[CA])}] \tag{3}$$

where the subscripts f and c refer to the free and complexed analyte, respectively, and the numbered subscripts refer to the two enantiomers, 1 and 2. Because the mobilities of the free enantiomers are the same and assuming, to a first approximation, that the mobilities of the complexes formed by each of the enantiomers with the cyclodextrin are the same, Equation (3) predicts that, as in CE, separation depends upon differences in the mobilities of the free and complexed state and differences in the binding constants, mediated by the dimensions of the chamber as well as the specific conductance and linear velocity of the buffer.

Despite the use of density and pH gradients, cooling and performance in microgravitational environments (e.g. the space shuttle) [18], convection and heat dissipation contributed to flow stream instability which was parasitic to the desired separations and limited the utility of this approach.

Recent innovations [19] have circumvented the heat dissipation and sample stream distortion inherent in most of the previous designs. In one apparatus, developed by R&S Technologies, Inc. (Wakefield, RI, USA), Teflon capillary tubes are aligned close to each other in the electrophoretic chamber. Coolant is pumped through the Teflon capillary tubes during the electrophoretic run while the electrophoretic separation is accomplished in the interstitial volume between the Teflon tubes.

Continuous free flow electrophoresis has been used for the separation of biopolymers (e.g. ovalbumin and lysozyme) [20] as well as smaller inorganic species (e.g. $[Co^{III}(sepulchrate)]^{3+}$ and $[Co^{III}(CN)_6]^{3-}$) [21]. Sample processing rates of 15 mg h^{-1} were reported for a mixture of Amaranth (MW: 804) and Patent Blue VF (MW: 1159) [22].

Three basic approaches have been used for chiral separations by continuous free flow electrophoresis. Thormann and co-workers [23] used 2-hydroxypropyl-b-cyclodextrin as an additive for the enantiomeric enrichment of methadone in an Octopus continuous free flow electrophoresis apparatus. In this work, both zone and isotachophoretic electrophoresis was used. Processing rates were on the order of 10–20 mg h^{-1}, which represents a significant improvement in sample throughput relative to CE or the earlier gel work. The authors realized higher enantiomeric purities with interrupted buffer flow than with continuous buffer flow, and suggested the potential of multistage continuous free flow for achieving even higher purities.

Glukhovskiy and Vigh [24] also used 2-hydroxypropyl-β-cyclodextrin as an additive, but their strategy involved isoelectric focusing. These authors developed the theoretical framework and effectively demonstrated the synergism between CE and continuous free flow electrophoresis. In this work, also using an Octopus continuous free flow apparatus, they were able to establish a pH gradient between 3.5 to 3.6 across the electrophoretic chamber by using polydisperse ampholytes or Bier's serine-propionic acid binary buffers in the buffer stream. As in Righetti's earlier work, complexation with the cyclodextrin additive induced sufficient differences in the pI of various dansylated amino acid enantiomers that complete enantioresolution was obtained. Although production rates were somewhat lower (~1.3 mg h^{-1}) than achieved by Thormann and co-workers, the enantiomeric purity was significantly higher.

In a different approach, Stalcup and co-workers [25] used sulfated β-cyclodextrin for the enantioseparation of piperoxan in work directly derived from earlier CE and classical gel results. Their results were obtained using a continuous free flow apparatus developed by R&S Technologies, Inc. Processing rates on the order of 4.5 mg h^{-1} were reported.

Several issues important from a processing standpoint were addressed in this work. With the exception of the single isomer derivatized cyclodextrins developed by Vigh [26], almost all commercially available derivatized cyclodextrins are complex mixtures of homologues and isomers. Each component, in all probability, has different affinities for the two enantiomers. In the case of neutral cyclodextrins, each cyclodextrin component has the same electrophoretic mobility (e.g. migrates only with the electro-osmostic flow) and the potential complexes should also have fairly similar mobilities. Therefore, chiral additive polydispersity should not contribute significantly to sample band dispersion. However, in the case of cyclodextrins functionalized with ionizable moieties, different degrees of substitution should produce ions with significantly different electrophoretic mobilities. In addition, for analytes interacting through electrostatic attraction, substitution patterns may also significantly impact affinity. Thus, solute sample stream dispersion may be significantly aggravated by chiral additive polydispersity. Figure 11-6 shows the distribution obtained for piperoxan in the presence and the absence of sulfated cyclodextrin. As can be seen from the figure, the number of vials across which the individual piperoxan enantiomers are distributed is about the same number as the piperoxan racemate. Thus, polydispersity of the cyclodextrin does not appear to be an issue with regard to bandwidth.

Table 11-4. Distribution of piperoxan enantiomers in CFFE vials.

Day	V	Vial$_1$	Vial$_2$	W$_1$	W$_2$	i	R
1	200	12	15	8	7	186	0.4
2	200	12	15	9	8	185	0.35
22	200	12	15	8	9	186	0.35
22	200	12	15	9	7	189	0.38
22	200	12	15	9	7	189	0.38
2	160	13	16	8	8	151	0.35
2	180	12	15	10	8	169	0.33

V: voltage.
Vial: vial containing max concentration.
W: number of vials containing enantiomer.
i: current.

With respect to method robustness, Table 11-4 shows results obtained on several different days during which a variety of buffer conditions were used. As can be seen from the table, the vial corresponding to the maximum concentration of the individual enantiomers as well as the number of vials containing piperoxan is fairly constant.

As in CE, changing system variables (e.g., pH, ionic strength, additive concentration) is very easy in any of the continuous free flow electrophoresis systems reported here because all the interactions take place in free solution. Indeed, changing system variables may be easier in continuous free flow electrophoresis systems than in a CE system because there are essentially no wall effects. Of course, changing system variables in the continuous free flow electrophoresis apparatus may also

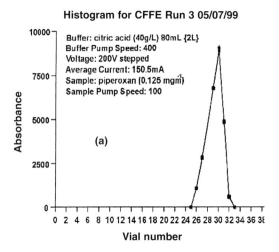

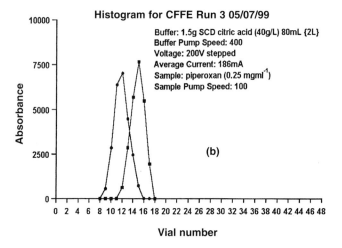

Fig. 11-6. Histograms showing the distribution of piperoxan enantiomers in the absence (a) and presence (b) of sulfated cyclodextrin in continuous free flow electrophoresis.

be easier than in a chromatographic system because there is no solid sorbent that is subject to degradation or that needs to be pre-equilibrated with the mobile phase.

11.4 Conclusions

Clearly, chiral separations, particularly preparative, present such a challenging problem that no single technology can provide complete satisfaction. Much of the activity in chiral separations by CE may be attributed to the advantages of CE relative to liquid chromatography (e.g., efficiency, rapid method development, ease of changing chiral selector, etc.). However, it must be noted that the true countercurrent processes possible in CE also allow selectivity to be manipulated to a much greater extent and with greater ease than is generally feasible in liquid chromatography using immobilized chiral selectors. In principle, any of the chiral entities enantiomerically resolved by CE should be amenable to preparative electrophoretic methods.

An important consideration for the ultimate economic viability of any preparative electrophoretic approach is the potential recovery of the chiral additive. Because the electrophoretic separation depends only upon the stability of the chiral additive itself and not the combined stability of an immobilized chiral selector, a spacer and an underlying substrate, as in the case of cyclodextrins immobilized on a silica chromatographic support, preparative electrophoretic separations have the potential to be more robust than analogous chromatographic methods. Although still in its infancy, preparative chiral electrophoresis represents an important technological advance in chiral separations, and has the potential to complement preparative chiral chromatographic methods as well as chiral CE complements chiral analytical chromatography.

Acknowledgments

The author would like to acknowledge R& S Technologies, Inc. (Wakefield, RI, USA) for the loan of the continuous free flow electrophoresis system, and Cerestar, Inc. for the donation of the sulfated cyclodextrin. The author would also like to thank Drs. Chris Welch and Prabha Painuly for helpful discussions.

References

[1] W.D. Hooer and M.S. Qing. *Clin Pharmacol. Ther.*, **48** (1990) 633.
[2] D. W. Armstrong, L. F. He, T. Yu, J. T. Lee and Y. S. Liu. *Tet. Assym.*, **10** (1999) 37.
[3] W. J. Wechter, D. G. Loughhead, R.J. Reischer, G. J. Van Giessen and D. G. Kaiser. *Biochem. Biophys. Res. Comm.*, **61** (1974) 833.
[4] T. J. Ward. *Anal. Chem.*, **66** (1994) 632A.
[5] H. Nishi and S. Terabe. *J. Chromatogr. A*, **694** (1995) 245.
[6] S. R. Gratz and A. M. Stalcup. *Anal. Chem.*, **70** (1998) 5166.
[7] X. W. Yao and F. E. Regnier. *J. Chromatogr.*, **632** (1993) 185.
[8] E. Kenndler. *Chromatographia*, **30** (1990) 713.

 [9] R. C. Allen and B. Budowle. *Gel electrophoresis of proteins and nucleic acids: selected techniques.* Walter de Gruyter, Berlin (1994).

[10] T. J. Bruno. *Chromatographic and electrophoretic methods.* Prentice Hall, NJ (1991)   pp. 141-169.

[11] P. G. Righetti, C. Ettori, P. Chafey and J. P. Wahrmann. *Electrophoresis*, **11** (1990) 1.

[12] A. M. Stalcup, K. H. Gahm, S. R. Gratz and R. M. C. Sutton. *Anal. Chem..*, **70** (1998) 144.

[13] N. M. Agyei. Master's Thesis, University of Hawaii, 1994.

[14] R. M. C. Sutton, S. R. Gratz and A. M. Stalcup. *Analyst*, **123** (1998) 1477.

[15] M. C. Roman and P. R. Brown. *Anal. Chem.*, **68** (1994) 86A.

[16] R. M. C. Sutton and A. M. Stalcup. Preparative Electrophoresis. *Encyclopedia of Separation Science*, Academic Press Ltd (1999) in press.

[17] M. C. Roman and P. R. Brown. *Anal. Chem.*, **66** (1994) A86.

[18] R. E. Allen, P. H. Rhodes and R. Snyder. *Sep. Purif. Meth.*, **6** (1977) 1.

[19] Y. Tarnopolsky. U. S. Patent 5,104,505 (1992).

[20] P. Painuly and M. C. Roman. *Appl. Theor. Electroph.*, **3** (1993) 119.

[21] M. E. Ketterer, G. E. Kozerski, R. Ritacco and P. Painuly. *Sep. Sci. Tech.*, **32** (1997) 641.

[22] Y. Tarnopolsky, M. Roman and P. R. Brown. *Sep. Sci. Tech.*, **28** (1993) 719.

[23] P. Hoffmann, H. Wagner, G. Weber, M. Lanz, J. Caslavska and W. Thormann. *Anal. Chem.*, **71** (1999) 1840.

[24] P. Glukhovskiy and G. Vigh. *Anal. Chem.*, **71** (1999) 3814.

[25] A. M. Stalcup, R. M. C. Sutton and J. Vela-Rodrigo. Unpublished results.

[26] J. B. Vincent, D. M. Kirby, T. V. Nguyen and G. Vigh. *Anal. Chem.*, **69** (1997) 4419.

12 Sub- and Supercritical Fluid Chromatography for Enantiomer Separations[*]

Karen W. Phinney

12.1 Introduction

Growing recognition of the role of chirality in biological activity and subsequent regulatory guidelines for chiral drug development have spurred tremendous growth in chiral separations technology. Applications of this technology range from pharmacokinetic studies during drug development to assessment of asymmetric synthesis strategies for the expanding list of drugs marketed in single-isomer form. Chromatographic methods have proven to be the most reliable and versatile analytical techniques for measurement of stereochemical composition, although capillary electrophoresis (CE) has become a viable alternative for certain applications [1, 2]. Chiral chromatographic methods are not limited to analytical-scale applications. Preparative scale chromatography is often the most efficient approach to resolve sufficient quantities of enantiomers for drug discovery activities [3, 4] .

Development and commercialization of chiral stationary phases (CSPs) for liquid chromatography (LC) revolutionized the field of enantiomer separations. Suddenly, the analyst could choose from a variety of immobilized chiral selectors to achieve direct enantiomer resolution and significantly reduce the need for chiral derivatizing agents [5]. However, a number of limitations of this approach also became apparent. First, prediction of CSP selectivity remains elusive, and method development often requires a trial-and-error approach. Second, successful chiral resolution often occurs within a narrow range of mobile phase compositions, and parameters that yield successful enantiomeric separation may not be suitable for resolving the enantiomers from other sample components. Finally, chromatographic efficiency of CSPs tends to be inferior to the efficiency of nonchiral stationary phases for LC, and broad chromatographic peaks can hinder quantification of desired analytes [1].

Supercritical fluid chromatography (SFC) provides a means of minimizing the limitations of CSPs developed for LC while retaining the impressive chiral selectivity that has been achieved through the evolution of CSPs during the past two decades [6, 7]. The use of supercritical fluids as eluents for chromatographic separations was

---

* Contribution of the National Institute of Standards and Technology. Not subject to copyright.

first reported more than 30 years ago, but most of the growth in SFC has occurred very recently. In fact, the dramatic increase in enantiomeric separations provided the perfect opportunity for a resurgence of SFC [8]. Numerous reports have now illustrated the advantages that can be realized by utilizing SFC as an alternative to LC for chiral separations, including increased efficiency, simplified method development, and reduced analysis time [9]. This revival has also been facilitated by the re-emergence of commercial instrumentation for SFC.

12.2 Sub- and Supercritical Fluid Chromatography

Myths and misconceptions about the characteristics of supercritical fluids have slowed their application to chromatographic separations. While these fluids do have interesting properties, they are not "super" fluids, and they are not suitable for all types of separations. An understanding of the fundamental behavior of supercritical fluids is key to identifying appropriate applications [10].

Table 12-1. Physical parameters of selected supercritical fluids.

Fluid	Critical temperature T_c (°C)	Critical pressure P_c (MPa)
CO_2	31.3	7.39
N_2O	36.5	7.34
NH_3	132.5	11.40
SF_6	45.5	3.76
C_5H_{12}	196.6	3.37
CHF_3	25.9	4.75

12.2.1 Properties of Supercritical Fluids

By definition, a supercritical fluid exists when both the temperature and pressure of the system exceed the critical values, T_c and P_c. The critical parameters of some fluids are listed in Table 12-1 [11]. Supercritical fluids have physical properties that position them between liquids and gases. Like gases, supercritical fluids are highly compressible, and properties of the fluid including density and viscosity can be manipulated by changes in pressure and temperature. Under the conditions used for most chromatographic separations, solute diffusion coefficients are often an order of magnitude higher in supercritical fluids than in traditional liquids, and viscosities are lower than those of liquids [12]. Supercritical fluids can be comprised of a single component, but binary and ternary fluid systems are also possible. At temperatures below T_c and pressures above P_c, the fluid becomes a liquid. However, many of the desirable properties of supercritical fluids, including low viscosity and high diffusivity, are retained under these subcritical conditions. At temperatures above T_c and pressures below P_c, the fluid becomes a gas. Therefore, supercritical fluids can be viewed as part of a continuum between liquids and gases [13].

12.2.2 Supercritical Fluids as Mobile Phases

The use of supercritical fluids as eluents for chromatographic separations was first reported by Klesper et al. in 1962. They reported the separation of porphyrin mixtures using supercritical chlorofluoromethanes as eluents [14]. Several years later, Giddings et al. demonstrated the separation of solutes such as carotenoids and sterols that were not amenable to separation by gas chromatography (GC). Compression of certain gases reportedly produced eluents with liquid-like solvent properties [15]. Much of the early work in SFC utilized either ammonia (NH_3) or carbon dioxide (CO_2), but compressed ammonia proved to be too hazardous for most users. Carbon dioxide has been the primary component of most eluent systems in SFC because of its modest critical parameters, moderate cost, and low toxicity. Early work greatly overestimated the elution power of pure CO_2. Giddings proposed that the solvent strength of supercritical CO_2 approximated that of isopropyl alcohol [16], a belief which led to considerable confusion and disappointment in early applications of SFC. A number of studies with solvatochromic dyes have now revealed that pure CO_2 is actually similar to pentane or hexane in solvent strength [17] and is, therefore, not a suitable eluent for most polar compounds.

The elution power of CO_2 and other fluids can be altered through the incorporation of an organic modifier. Most common organic solvents, such as methanol, acetonitrile, and dichloromethane, can be used as modifiers in SFC [18]. The addition of modifiers increases the critical parameters for the fluid. At the near-ambient temperatures ($T < T_c$) used for many SFC separations, the modified eluent may actually be in the subcritical (liquid) state. However, solute diffusion coefficients remain higher in subcritical fluids than in traditional liquids [19]. In binary eluent systems, separation of the eluent into two phases is possible. Phase diagrams for many binary systems such as methanol-CO_2 are available, and chromatographic parameters that result in phase separation are easily avoided [20]. In some instances, a very polar substance may be added to the modifier to form a ternary eluent system. These additives are generally acids or bases and are used to elute certain components or to improve chromatographic peak shape [21, 22]. Common additives include isopropylamine and trifluoroacetic acid.

The use of both sub- and supercritical fluids as eluents yields mobile phases with increased diffusivity and decreased viscosity relative to liquid eluents [23]. These properties enhance chromatographic efficiency and improve resolution. Higher efficiency in SFC shifts the optimum flowrate to higher values so that analysis time can be reduced without compromising resolution [12]. The low viscosity of the eluent also reduces the pressure-drop across the chromatographic column and facilitates the use of long columns or multiple columns connected in series [24]. Manipulation of mobile phase composition, temperature, and pressure can be used to achieve the desired selectivity in SFC [25]. Rapid column equilibration in SFC is an additional advantage over liquid chromatographic methods [26]. Replacement of liquid eluents with sub- and supercritical fluids also reduces solvent consumption.

12.2.3 Instrumentation for SFC

SFC has been performed with either open capillary columns similar to those used in GC or packed columns transferred from LC, and the instrumentation requirements differ for these two approaches [12]. This chapter will focus on the use of packed column technology because of its dominance in the area of pharmaceutical compound separations. Current commercial instrumentation for packed column SFC utilizes many of the same components as traditional LC instruments, including pumps, injection valves, and detectors. In fact, most modern packed column SFC instruments can also be used to perform LC separations, and many of the same stationary phases can be used in both LC and SFC [9].

Certain modifications to the chromatographic system are necessary to accommodate the compressibility of the eluent in SFC [27]. Carbon dioxide and other fluids comprising the bulk of the mobile phase are liquefied gases supplied in cylinders. Chilling of the pump head is necessary to ensure that the fluid remains in the liquid state. A second pump can be used to deliver a modifier if binary eluents are desired. When necessary, mobile phase additives are incorporated by adding them to the modifier. Automatic injection systems are available for SFC, but injection valves must be modified to permit introduction of the sample into a high-pressure environment. Injection volumes of 5–20 µL are common. The most noticeable difference between LC and SFC is the addition of a back-pressure regulator after the detector to control outlet pressure and prevent expansion of the eluent into a gas. Electronic back-pressure regulators allow independent control of pressure and mobile phase composition [12]. Measurement of UV absorbance remains the most common detection method for packed column SFC. The position of the back-pressure regulator necessitates the use of detection cells capable of withstanding high pressure.

12.3 Advantages of SFC for Chiral Separations

The high diffusivity and low viscosity of sub- and supercritical fluids make them particularly attractive eluents for enantiomeric separations. Mourier et al. first exploited sub- and supercritical eluents for the separation of phosphine oxides on a brush-type chiral stationary phase [28]. They compared analysis time and resolution per unit time for separations performed by LC and SFC. Although selectivity (α) was comparable in LC and SFC for the compounds studied, resolution was consistently higher in SFC, and analysis times were shorter compared to LC. They also explored the influence of the type of modifier on selectivity in SFC.

Mourier's report was quickly followed by successful enantiomeric resolutions on stationary phases bearing other types of chiral selectors, including native and derivatized cyclodextrins and derivatized polysaccharides. Many chiral compounds of pharmaceutical interest have now been resolved by packed column SFC, including antimalarials, β-blockers, and antivirals. A summary is provided in Table 12-2. Most of the applications have utilized modified CO_2 as the eluent.

Table 12-2. Selected applications of chiral SFC to pharmaceutical compounds[a].

Chiral stationary phase	Compounds resolved
Cellulose derivatives	β-Blockers, benzodiazepines, NSAIDs, barbiturates
Amylose derivatives	NSAIDs, protease inhibitors, β-blockers, benzodiazepines
Brush-type	Antimalarials, NSAIDs, β-blockers, bronchodilators
Cyclodextrins and derivatives	Phosphine oxides, NSAIDs, anticonvulsants
Macrocyclic antibiotics	Bronchodilators, β-blockers

[a] For a more comprehensive listing, see. ref. [12].

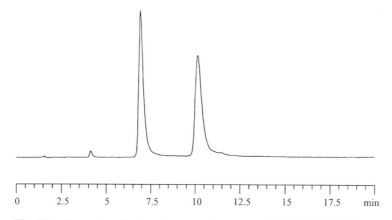

Fig. 12-1. Separation of primaquine enantiomers on a Chiralcel OD CSP. Chromatographic conditions: 20 % methanol with 0.5 % isopropylamine in carbon dioxide, 2.0 mL min⁻¹, 15 MPa, 30 °C.

Some of the initial enthusiasm surrounding chiral SFC was tempered by the fact that many of the same separations had already been achieved by LC [29]. Therefore, researchers were reluctant to add SFC to their analytical laboratories. In some instances, SFC does yield separations that can not be achieved on the same CSP in LC [30, 31]. The enantioseparation of primaquine, an antimalarial compound, on a Chiralcel OD CSP is illustrated in Fig. 12-1 [32]. This compound was not resolved on the same CSP in LC [33]. The reverse situation, where a separation obtained in LC may not be observed on the same CSP in SFC, can also occur [34]. These disparities seem to be related to differences in analyte-eluent and eluent-CSP interactions between the two techniques. The occasional failures of SFC do not diminish the unique advantages of sub- and supercritical eluents for enantiomeric separations.

12.3.1 Increased Efficiency

The efficiency of many CSPs increases dramatically when liquid eluents are replaced with sub- or supercritical fluids. During a comparison of LC and SFC performed with a Chiralcel OD CSP, Lynam and Nicolas reported that the number of theoretical plates obtained was three to five times higher in SFC than in LC [26]. The separation of metoprolol enantiomers by LC and SFC on a Chiralcel OD CSP is illustrated in Fig. 12-2. Although impressive selectivity is achieved by both techniques, resolution is higher in SFC (R_s = 12.7) than in LC (R_s = 4.8), and the higher flowrate in SFC reduces the analysis time. The increased efficiency of SFC also improves peak symmetry.

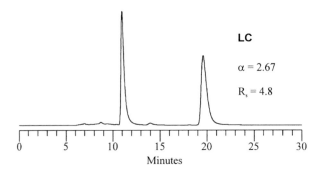

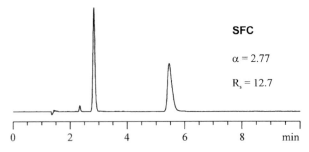

Fig. 12-2. Separation of metoprolol enantiomers by LC and SFC on a Chiralcel OD CSP. Chromatographic conditions for LC: 20 % 2-propanol in hexane, with 0.1 % diethylamine, 0.5 mL min^{-1}. Chromatographic conditions for SFC: 20 % methanol with 0.5 % isopropylamine in carbon dioxide, 2.0 mL min^{-1}, 15 MPa, 30 °C.

The difference in resolution between LC and SFC can be significant enough to turn a marginal LC separation into a viable chromatographic method in SFC. Stringham et al. reported that packed column SFC yielded satisfactory chiral analysis methods for all four compounds related to synthesis of an antiviral drug candidate [35]. Successful LC methods with the same CSP could only be developed for two of the candidate compounds. Samples for pharmaceutical analysis rarely consist only of the target analyte. Ideally, achiral and chiral analysis can be performed concurrently, and the increased resolution power of SFC reduces the likelihood that unexpected sample components will interfere with the analysis.

12.3.2 Rapid Method Development

Method development remains the most challenging aspect of chiral chromatographic analysis, and the need for rapid method development is particularly acute in the pharmaceutical industry. To complicate matters, even structurally similar compounds may not be resolved under the same chromatographic conditions, or even on the same CSP. Rapid column equilibration in SFC speeds the column screening process, and automated systems accommodating multiple CSPs and modifiers now permit unattended method optimization in SFC [36]. Because more compounds are likely to be resolved with a single set of parameters in SFC than in LC, the analyst stands a greater chance of success on the first try in SFC [37]. The increased resolution obtained in SFC may also reduce the number of columns that must be evaluated to achieve the desired separation.

12.3.3 Column Coupling

In some instances, a single CSP may not provide the desired separation. Compounds having multiple chiral centers often present analytical challenges because co-elution of enantiomers and diastereomers can occur. Drug formulations typically contain excipients, preservatives, or other active ingredients that can cause chromatographic interferences. Serial coupling of multiple chiral columns or coupling of achiral and chiral columns resolves some of these issues, and this is readily performed in SFC as a consequence of the reduced pressure-drop across the column. Mobile phase compatibility issues, often a problem when coupling columns in LC, are eliminated in SFC. Coupling of multiple CSPs of the same type was used as a brute force approach by Mathre et al. to resolve the four possible stereoisomers of a drug candidate [38]. Kot et al. reported serial coupling of three different CSPs in SFC. The column triplet provided a multipurpose chromatographic system for the enantioresolution of a wide variety of racemic compounds [39]. This approach should be used with caution, however, because opposing selectivities of the coupled CSPs may effectively cancel out a separation that would normally be achieved on a single CSP [40]. The separation of components of a cough syrup on a Chiralpak AD CSP alone, and on a coupled achiral/chiral column system is illustrated in Fig. 12-3. Peak overlap of the two active ingredients occurred when only the chiral column was used. No changes in chromatographic parameters were necessary to accommodate the coupled column system [41]. Column coupling in SFC can provide tremendous advantages over LC for challenging separations.

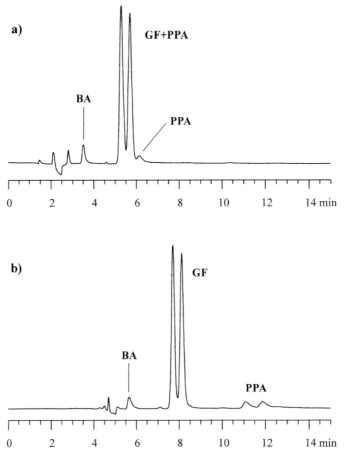

Fig. 12-3. Separation of ingredients of an expectorant syrup, on a Chiralpak AD CSP (a) and the coupled cyano/Chiralpak AD CSP (b). Chromatographic conditions: 10 % methanol with 0.5 % iso-propylamine in carbon dioxide, 2.0 mL min⁻¹, 15 MPa, 30 °C. Peaks are benzoic acid (BA), guaifenesin (GF), and phenylpropanolamine (PPA).

12.3.4 Preparative Separations

Packed column SFC has also been applied to preparative-scale separations [42]. In comparison to preparative LC, SFC offers reduced solvent consumption and easier product recovery [43]. Whatley [44] described the preparative-scale resolution of potassium channel blockers. Increased resolution in SFC improved peak symmetry and allowed higher sample throughput when compared to LC. The enhanced resolution obtained in SFC also increases the enantiomeric purity of the fractions collected. Currently, the major obstacle to widespread use of preparative SFC has been the cost and complexity of the instrumentation.

12.4   Chiral Stationary Phases in SFC

The low polarity of CO_2-based eluents makes SFC a normal phase technique. Therefore, it is not surprising that most of the successful applications of chiral SFC have utilized CSPs designed for normal phase LC. However, some exceptions have been noted. Specific applications of various CSPs are outlined in the next sections.

12.4.1   Brush-type

The brush-type (Pirkle-type) CSPs have been used predominantly under normal phase conditions in LC. The chiral selector typically incorporates π-acidic and/or π-basic functionality, and the chiral interactions between the analyte and the CSP include dipole–dipole interactions, π–π interactions, hydrogen bonding, and steric hindrance. The concept of reciprocity has been used to facilitate the rational design of chiral selectors having the desired selectivity [45].

As noted earlier, the first report of chiral packed column SFC utilized a brush-type CSP for the separation of phosphine oxides [28]. The CSP consisted of (*R*)-*N*-(3,5-dinitrobenzoyl)phenylglycine covalently bonded to aminopropyl silica. Macaudière and co-workers resolved a series of aromatic amides on the same type of CSP. They postulated that the chiral recognition mechanisms were identical in SFC and LC, and that the methods were interchangeable [46]. Their results contrast significantly with the observations of Siret et al. for a CSP based on a tyrosine derivative (ChyRoSine-A) [47]. Enantioseparation of β-blockers was achieved on the ChyRoSine-A CSP with a modified CO_2 eluent in SFC, but the same compounds were not resolved in LC. Modeling studies have suggested that CO_2 interacts with solutes having certain structural features to enhance chiral recognition in SFC [48]. A short (5 cm) column packed with ChyRoSine-A yielded separations of β-blocker enantiomers in less than 2 min [49].

A new brush-type CSP, the Whelk-O 1, was used by Blum et al. for the analytical and preparative-scale separations of racemic pharmaceutical compounds, including verapamil and ketoprofen. A comparison of LC and SFC revealed the superiority of SFC in terms of efficiency and speed of method development [50]. The Whelk-O 1 selector and its homologues have also been incorporated into polysiloxanes. The resulting polymers were coated on silica and thermally immobilized. Higher efficiencies were observed when these CSPs were used with sub- and supercritical fluids as eluents, and a greater number of compounds were resolved in SFC compared to LC. Compounds such as flurbiprofen, warfarin, and benzoin were enantioresolved with a modified CO_2 eluent [37].

12.4.2 Cyclodextrins

Cyclodextrins are cyclic oligosaccharides comprised of glucose units joined through α-1,4 linkages. In LC, chiral recognition is believed to involve the formation of inclusion complexes between the analyte and the hydrophobic cavity of the cyclodextrin. The use of aqueous-organic mobile phases facilitates complex formation. Under normal phase conditions, the apolar component of the mobile phase occupies the cyclodextrin cavity, reducing opportunities for chiral complexation [51]. Cyclodextrin-based CSPs have also been used in the polar organic mode. In this mode, the analyte is believed to interact primarily with the hydroxyl groups along the mouth of the cyclodextrin [52].

Macaudière et al. first reported the enantiomeric separation of racemic phosphine oxides and amides on native cyclodextrin-based CSPs under subcritical conditions [53]. The separations obtained were indicative of inclusion complexation. When the CO_2–methanol eluent used in SFC was replaced with hexane-ethanol in LC, reduced selectivity was observed. The authors proposed that the smaller size of the CO_2 molecule made it less likely than hexane to compete with the analyte for the cyclodextrin cavity.

Comparisons of LC and SFC have also been performed on naphthylethylcarbamoylated-β-cyclodextrin CSPs. These multimodal CSPs can be used in conjunction with normal phase, reversed phase, and polar organic eluents. Discrete sets of chiral compounds tend to be resolved in each of the three mobile phase modes in LC. As demonstrated by Williams et al., separations obtained in each of the different mobile phase modes in LC could be replicated with a simple CO_2-methanol eluent in SFC [54]. Separation of tropicamide enantiomers on a Cyclobond I SN CSP with a modified CO_2 eluent is illustrated in Fig. 12-4. An aqueous-organic mobile phase was required for enantioresolution of the same compound on the Cyclobond I SN CSP in LC. In this case, SFC offered a means of simplifying method development for the derivatized cyclodextrin CSPs. Higher resolution was also achieved in SFC.

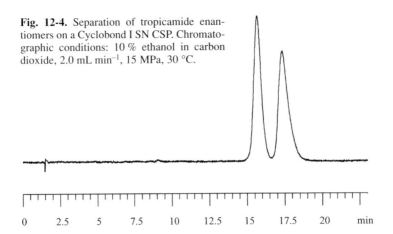

Fig. 12-4. Separation of tropicamide enantiomers on a Cyclobond I SN CSP. Chromatographic conditions: 10 % ethanol in carbon dioxide, 2.0 mL min^{-1}, 15 MPa, 30 °C.

0   2.5   5   7.5   10   12.5   15   17.5   20   min

12.4.3 Derivatized Polysaccharides

Polysaccharide-based CSPs incorporate derivatives of cellulose and amylose adsorbed on silica gel. The selectivity of these CSPs depends upon the nature of the substituents introduced during the derivatization process. The secondary structure of the modified polysaccharide is believed to play a role in selectivity, but the chiral recognition mechanisms have not been fully elucidated [55].

Several research groups have reported the separation of β-blocker enantiomers on cellulose tris(3,5-dimethylphenylcarbamate) CSPs [56, 57]. Bargmann-Leyder et al. performed a detailed comparison of LC and SFC on a cellulosic CSP (Chiralcel OD) [58]. Although resolution was generally higher in SFC than in LC for this family of compounds, differences in selectivity between the two techniques were observed, and these discrepancies seemed to be compound specific. Examination of a series of propranolol analogues provided additional insight into differences in the chiral recognition mechanisms operative in LC and SFC.

Benzodiazepine enantiomers have also been resolved on the Chiralcel OD CSP. Wang et al. utilized this CSP to determine the enantiomeric composition of camazepam and its metabolites [59]. SFC provided improved resolution of the compounds of interest in a shorter period of time than LC. Phinney et al. demonstrated the separation of a series of achiral and chiral benzodiazepines. An amino column was coupled in series with the Chiralcel OD CSP to achieve the desired separation [41].

The selectivity of another cellulose-based CSP, Chiralcel OJ, has also been examined in SFC [60]. Separations of racemic drugs such as benoxaprofen, temazepam, and mephobarbital were obtained. Acetonitrile proved to be a better modifier than methanol for some of the compounds investigated. The four optical isomers of a calcium channel blocker were resolved by Siret et al. on the Chiralcel OJ CSP [30]. In LC, two CSPs were required to perform the same separation.

Derivatized amylose is the basis for the Chiralpak AD CSP. This CSP has been utilized for the resolution of ibuprofen and flurbiprofen, as well as other members of the family of nonsteroidal inflammatory drugs (NSAIDs) [39, 61]. Ibuprofen was not resolved on the Chiralpak AD CSP in LC. Pressure-related effects on stereoselectivity were observed by Bargmann-Leyder et al. on a Chiralpak AD CSP [58]. No corresponding effect of pressure on selectivity was observed with a Chiralcel OD CSP. The authors speculated that the helical conformation of the amylose-based CSP is more flexible than that of the cellulose-based CSP.

12.4.4 Macrocyclic Antibiotics

This relatively new class of CSPs incorporates glycopeptides attached covalently to silica gel. These CSPs can be used in the normal phase, reversed phase, and polar organic modes in LC [62]. Various functional groups on the macrocyclic antibiotic molecule provide opportunities for π-π complexation, hydrogen bonding, and steric interactions between the analyte and the chiral selector. Association of the analyte with a hydrophobic cleft of the macrocycle may also play a role in chiral recognition.

The macrocyclic antibiotic-based CSPs have not been used extensively in SFC. Two macrocyclic antibiotic CSPs, Chirobiotic T and Chirobiotic V, were included in a study of various CSPs in SFC. At least partial resolution of approximately half of the 44 test compounds could be obtained on these two CSPs in SFC [63]. A high concentration of modifier was necessary to elute some of the analytes. Enantioresolution of derivatized amino acids was also demonstrated in the same study. However, a complex modifier comprised of methanol, water, and glycerol was required for separations performed on the Chirobiotic T CSP. The separation of coumachlor enantiomers on a vancomycin-based CSP (Chirobiotic V) in SFC is illustrated in Fig. 12-5 [32].

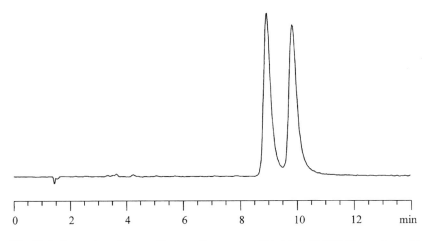

Fig. 12-5. Separation of coumachlor enantiomers on a Chirobiotic V CSP. Chromatographic conditions: 15 % methanol in carbon dioxide, 2.0 mL min^{-1}, 20 MPa, 30 °C.

12.4.5 Other CSPs

Macaudière and co-workers performed a comparison of LC and SFC on a polymer based-CSP (Chiralpak OT) [64]. The chromatographic behavior of this CSP seemed to be quite different in SFC than in LC, although satisfactory separations were achieved with both techniques. The chiral recognition mechanisms may be altered by the nature (hexane-based or CO_2-based) of the eluent.

A CSP based on the adsorption of a chiral anthrylamine on porous graphitic carbon successfully resolved the enantiomers of tropic acid derivatives and anti-inflammatory agents in SFC [65]. The carbon-based CSP produced superior results when compared to an analogous silica-based CSP. Occasional washing of the column was necessary to remove highly retained substances.

12.5 Method Development in Chiral SFC

Rapid method development remains one of the principal advantages of SFC for enantiomeric separations, particularly when the analyst is faced with a structurally diverse array of analytes. However, the applicability of SFC for a given separation should be assessed prior to attempting method development. Berger [12] has suggested that solubility of the analyte in methanol or a less polar solvent can be used to gauge the likelihood of success. Compounds requiring aqueous conditions are generally not good candidates for SFC [66].

12.5.1 Stationary Phase Selection

Column selection remains the most important factor in successful enantiomeric separations. The CSPs most likely to be effective in SFC are those that have been employed under normal phase conditions in LC. In fact, the tremendous body of knowledge that has been accumulated for LC can also guide column selection in SFC [66]. The likelihood of success with a particular CSP can generally be gauged after one or two injections [67]. If no evidence of separation is observed, another CSP should be investigated.

12.5.2 Modifiers

The nature of the modifier and the modifier concentration impact both retention and selectivity in packed column SFC. SFC offers considerable flexibility in modifier selection because nearly all commonly used organic modifiers, including methanol and acetonitrile, are miscible with CO_2. In contrast, methanol and acetonitrile are rarely used as modifiers in normal phase LC because they are immiscible with hexane [68].

In general, retention decreases as the modifier concentration increases because the modifier competes with the analytes for sites on the stationary phase. The effect on retention of changes in modifier concentration seems to be more pronounced for CSPs than for achiral stationary phases in SFC, and peak shapes are apt to degrade rapidly at low modifier concentrations [12]. Efficiency tends to decrease as the modifier concentration increases because analyte diffusion is slowed by the increased viscosity of the eluent [39].

Methanol remains the most widely used modifier because it produces highly efficient separations, but it does not always produce the highest selectivity [8]. Recent studies have provided insight into the role of the modifier in enantioselectivity in SFC [69]. Blackwell and Stringham examined a series of phenylalanine analogues on a brush-type CSP and developed a model that allowed prediction of selectivity based on the bulk solvation parameters of various modifiers [70]. Careful choice of modifiers can be used to mask or enhance particular molecular interactions and ultimately provide control of selectivity [71].

Modifier additives also play a role in method optimization and are typically added to the modifier at concentrations less than 1 % (v/v). Additives can provide increased efficiency by minimizing undesirable interactions between the analyte and the CSP, and may be necessary to elute certain types of compounds. The type of additive (acidic or basic) that will produce the best results depends upon the functionality of the analyte [72]. Certain additives are strongly retained on the stationary phase, and their effect may persist even after they are removed from the eluent [22]. The impact of both modifiers and additives can also be affected by the proximity of the operating conditions to the critical point of the eluent [73].

12.5.3 Temperature

Temperature can also be used to optimize enantioselectivity in SFC. The selectivity of most CSPs increases as temperature decreases. For this reason, most chiral separations in SFC are performed at ambient or subambient temperatures [50, 74]. Subambient temperatures are particularly useful for compounds having low conformational stability [75]. Stringham and Blackwell explored the concept of entropically driven separations [76]. As temperature increased, enantioselectivity decreased until the enantiomers co-eluted at the isoelution temperature. Further increases in temperature resulted in reversal of elution order of the enantiomers. The temperature limitations of the CSP should be considered before working at elevated temperatures.

12.5.4 Pressure

Changes in pressure typically have a greater impact on retention than on selectivity. Most studies of CSPs have indicated little effect of pressure on stereoselectivity [28, 31]. However, Bargmann-Leyder et al. reported pressure-related changes in selectivity for an amylose-based CSP, though the magnitude of the pressure effect was not the same for all the compounds studied [58]. Pressures in the range of 15–20 MPa are common for chiral SFC.

12.5.5 Flowrate

Optimum flowrates are higher in packed column SFC than in LC. Flowrates as high as 5.0 mL min^{-1} generally do not dramatically reduce efficiency in SFC [12]. Biermanns and co-workers reported the separation of β-blockers at a flowrate of 4.0 mL min^{-1}, a rate eight times higher than the flowrate recommended for LC [56]. No deterioration of column performance was observed.

12.6 Conclusions

Enantiomeric separations have proven to be one of the most successful applications of packed column SFC. Despite initial reluctance, many analysts now use SFC routinely for both analytical and preparative chiral separations. Additional studies of chiral recognition in SFC and continued improvements in instrumentation will ensure a prominent role for SFC in chiral separations methodology in the future.

Disclaimer

Certain commercial equipment, instruments, or materials are identified in this report to specify adequately the experimental procedure. Such identification does not imply recommendation or endorsement by the National Institute of Standards and Technology, nor does it imply that the materials or equipment identified are necessarily the best available for the purpose.

References

[1] T. J. Wozniak, R. J. Bopp, E. C. Jensen, *J. Pharm. Biomed. Anal.*, 9 (1991) 363–382.
[2] K. D. Altria, *J. Chromatogr. A*, 856 (1999) 443–463.
[3] L. Miller, C. Orihuela, R. Fronek, D. Honda, O. Dapremont, *J. Chromatogr. A*, 849 (1999) 309–317.
[4] G. J. Terfloth, *LC-GC*, 17 (1999) 400–405.
[5] D. W. Armstrong, *Anal. Chem.*, 59 (1987) 84A–91A.
[6] D. W. Armstrong, *LC-GC*, (1997) S20–S28.
[7] K. L. Williams, L. C. Sander, *J. Chromatogr. A*, 785 (1997) 149–158.
[8] K. W. Phinney, *Anal. Chem.*, in press.
[9] T. L. Chester, J. D. Pinkston, D. E. Raynie, *Anal. Chem.*, 70 (1998) 301R–319R.
[10] R. M. Smith, *J. Chromatogr. A*, 856 (1999) 83–115.
[11] CRC Handbook of Chemistry and Physics, CRC Press, Boca Raton, 1978.
[12] T. A. Berger, Packed Column SFC, Royal Society of Chemistry, Cambridge, 1995.
[13] P. J. Schoenmakers, in R. M. Smith (ed.), Supercritical Fluid Chromatography, Royal Society of Chemistry, London, 1988, pp. 102–136.
[14] E. Klesper, A. H. Corwin, D. A. Turner, *J. Org. Chem.*, 27 (1962) 700–701.
[15] L. McLaren, M. N. Myers, J. C. Giddings, *Science*, (1968) 197–199.
[16] J. C. Giddings, M. N. Myers, L. McLaren, R. A. Keller, *Science*, 162 (1968) 67–73.
[17] J. F. Deye, T. A. Berger, A. G. Anderson, *Anal. Chem.*, 62 (1990) 615–622.
[18] D. R. Gere, *Science*, 222 (1983) 253–259.
[19] P. R. Sassiat, P. Mourier, M. H. Caude, R. H. Rosset, *Anal. Chem.*, 59 (1987) 1164–1170.
[20] T. A. Berger, *J. High Resolut. Chromatogr.*, 14 (1991) 312– 316.
[21] T. A. Berger, W. H. Wilson, *J. Chromatogr. Sci.*, 31 (1993) 127–132.
[22] T. A. Berger and J. F. Deye, *J. Chromatogr. Sci.*, 29 (1991) 54–59.
[23] D. R. Gere, R. Board, D. McManigill, *Anal. Chem.*, 54 (1982) 736–740.
[24] T. A. Berger, W. H. Wilson, *Anal. Chem.*, 65 (1993) 1451– 1455.

[25] P. Mourier, P. Sassiat, M. Caude, R. Rosset, *J. Chromatogr.*, 353 (1986) 61–75.
[26] K. G. Lynam, E. C. Nicolas, *J. Pharm. Biomed. Anal.*, 11 (1993) 1197–1206.
[27] T. A. Berger, *J. Chromatogr. A*, 785 (1997) 3–33.
[28] P. A. Mourier, E. Eliot, M. H. Caude, R. H. Rosset, A. G. Tambuté, *Anal. Chem.*, 57 (1985) 2819–2823.
[29] P. R. Massey, M. J. Tandy, *Chirality*, 6 (1994) 63–71.
[30] L. Siret, P. Macaudière, N. Bargmann-Leyder, A. Tambuté, M. Caude, E. Gougeon, *Chirality*, 6 (1994) 440–445.
[31] K. L. Williams, L. C. Sander, S. A. Wise, *Chirality*, 8 (1996) 325–331.
[32] K. W. Phinney, unpublished results.
[33] Y. Okamoto, R. Aburatani, K. Hatano, K. Hatada, *J. Liq. Chromatogr.*, 11 (1988) 2147–2163.
[34] K. Anton, J. Eppinger, L. Frederiksen, E. Francotte, T. A. Berger, W. H. Wilson, *J. Chromatogr. A*, 666 (1994) 395–401.
[35] R. W. Stringham, K. G. Lynam, C. C. Grasso, *Anal. Chem.*, 66 (1994) 1949–1954.
[36] M. S. Villeneuve and R. J. Anderegg, *J. Chromatogr. A*, 826 (1998) 217–225.
[37] G. J. Terfloth, W. H. Pirkle, K. G. Lynam, E. C. Nicolas, *J. Chromatogr. A*, 705 (1995) 185–194.
[38] D. J. Mathre, W. A. Schafer, B. D. Johnson, presented at the 6th International Symposium on SFC and SFE, Uppsala, Sweden, 1995.
[39] A. Kot, P. Sandra, A. Venema, *J. Chromatogr. Sci.*, 32 (1994) 439–448.
[40] W. H. Pirkle, C. J. Welch, *J. Chromatogr. A*, 731 (1996) 322–326.
[41] K. W. Phinney, L. C. Sander, S. A. Wise, *Anal. Chem.*, 70 (1998) 2331–2335.
[42] G. Fuchs, L. Doguet, D. Barth, M. Perrut, *J. Chromatogr.*, 623 (1992) 329–336.
[43] M. Perrut, *J. Chromatogr. A*, 658 (1994) 293–313.
[44] J. Whatley, *J. Chromatogr. A*, 697 (1995) 251–255.
[45] C. J. Welch, *J. Chromatogr.*, 666 (1994) 3–26.
[46] P. Macaudière, A. Tambuté, M. Caude, R. Rosset, M. C. Alembik, I. W. Wainer, *J. Chromatogr.*, 371 (1986) 177–193.
[47] L. Siret, N. Bargmann, A. Tambuté, M. Caude, *Chirality*, 4 (1992) 252–262.
[48] N. Bargmann-Leyder, C. Sella, D. Bauer, A. Tambuté, M. Caude, *Anal. Chem.*, 67 (1995) 952–958.
[49] N. Bargmann-Leyder, D. Thiébaut, F. Vergne, A. Bégos, A. Tambuté, M. Caude, *Chromatographia*, 39 (1994) 673–681.
[50] A. M. Blum, K. G. Lynam, E. C. Nicolas, *Chirality*, 6 (1994) 302–313.
[51] D. W. Armstrong, W. DeMond, B. Czech, *Anal. Chem.*, 57 (1985) 481–484.
[52] S. C. Chang, G. L. Reid, III, S. Chen, C. D. Chang, D. W. Armstrong, *Trends Anal. Chem.*, 12 (1993) 144–153.
[53] P. Macaudière, M. Caude, R. Rosset, A. Tambuté, *J. Chromatogr.*, 405 (1987) 135–143.
[54] K. L. Williams, L. C. Sander, S. A. Wise, *J. Chromatogr. A*, 746 (1996) 91–101.
[55] Y. Okamoto, Y. Kaida, *J. High. Resolut. Chromatogr.*, 13 (1990) 708–712.
[56] P. Biermanns, C. Miller, V. Lyon, W. Wilson, *LC-GC*, 11 (1993) 744–747.
[57] C. R. Lee, J.-P. Porziemsky, M.-C. Aubert, A. M. Krstulovic, *J. Chromatogr.*, 539 (1991) 55–69.
[58] N. Bargmann-Leyder, A. Tambuté, M. Caude, *Chirality*, 7 (1995) 311–325.
[59] M. Z. Wang, M. S. Klee, S. K. Yang, *J. Chromatogr. B*, 665 (1995) 139–146.
[60] A. V. Overbeke, P. Sandra, A. Medvedovici, W. Baeyens, H. Y. Aboul-Enein, *Chirality*, 9 (1997) 126–132.
[61] W. H. Wilson, *Chirality*, 6 (1994) 216–219.
[62] D.W. Armstrong, Y. Tang, S. Chen, Y. Zhou, C. Bagwill, J.-R. Chen, *Anal. Chem.*, 66 (1994) 1473–1484.
[63] A. Medvedovici, P. Sandra, L. Toribio, F. David, *J. Chromatogr. A*, 785 (1997) 159–171.
[64] P. Macaudière, M. Caude, R. Rosset, A. Tambuté, *J. Chromatogr. Sci.*, 27 (1989) 583–591.
[65] S. M. Wilkins, D. R. Taylor, R. J. Smith, *J. Chromatogr. A*, 697 (1995) 587–590.
[66] R. E. Majors, *LC-GC*, 15 (1997) 412–422.
[67] R. W. Stringham, *Chirality*, 8 (1996) 249–257.
[68] S. Hara, A. Dobashi, K. Kinoshita, T. Hondo, M. Saito, M. Senda, *J. Chromatogr.*, 371 (1986) 153–158.
[69] J. A. Blackwell, R. W. Stringham, D. Xiang, R.E. Waltermire, *J. Chromatogr. A*, 852 (1999) 383–394.
[70] J. A. Blackwell, R.W. Stringham, *Chirality*, 11 (1999) 98–102.

[71] G. O. Cantrell, R. W. Stringham, J. A. Blackwell, J. D. Weckwerth, P. W. Carr, *Anal. Chem.*, 68 (1996) 3645–3650.

[72] J. A. Blackwell, *Chirality*, 10 (1998) 338–342.

[73] J. A. Blackwell, R. W. Stringham, J. D. Weckwerth, *Anal. Chem.*, 69 (1997) 409–415.

[74] J. A. Blackwell, R. W. Stringham, *Chirality*, 9 (1997) 693– 698.

[75] C. Wolf, W. H. Pirkle, *J. Chromatogr. A*, 785 (1997) 173–178.

[76] R. W. Stringham, J. A. Blackwell, *Anal. Chem.*, 68 (1996) 2179–2185.

13 International Regulation of Chiral Drugs[1]

Sarah K. Branch

13.1 Introduction

The regulation of chiral drugs provides a good demonstration of the mutual relationship between progress in scientific methodology and the development of regulatory guidelines. It is also an example of international debate between regulatory authorities and the pharmaceutical industry leading to a consensus in recognition of the global nature of pharmaceutical development. The significance of optical isomerism in drug action is well-established; for example, it was known in 1926 that the biological activity of atropine resided in only one stereoisomer [1]. However, the absence of suitable methods for either the large-scale preparation of pure enantiomers or for stereoselective analysis meant that the majority of stereochemically pure drugs on the market were of natural origin. That drug metabolism could be stereoselective was already acknowledged by the 1970s [2], but it was only in the following decade that the wider implications of chirality in clinical pharmacokinetics and drug safety began to be recognized and then only after some vigorous campaigning, for example, by Ariëns [3, 4]. The "enantiomer-versus-racemate" debate will not be reiterated here as the published literature is quite extensive and has been previously reviewed [e.g., 5–8]. A summary, however, is given in Table 13-1 of the different scenarios, illustrated by single examples, associated with the administration of a racemate which might need be taken into account during development of new drug substances.

Historically, synthetic chiral drugs were mainly presented as the racemate due to the technical difficulties of either synthesizing the pure enantiomers or separating them to yield the individual isomers. The 1980s saw the introduction of new methods for the preparation of single enantiomers accompanied by advances in chiral analytical procedures. The field of asymmetric synthesis in organic chemistry burgeoned and a wide range of chiral precursors and reagents are now commercially available as a result. There has also been exploitation of biosynthetic pathways in micro-organisms to produce drug substances, either employing chiral starting mate-

---

[1] The views expressed in this chapter are those of the author and do not necessarily represent the views or the opinions of the Medicines Control Agency, other regulatory authorities or any of their advisory committees.

Table 13-1. Enantiospecific drug action and pharmacokinetics.

Type of effect	Examples[1]
Pharmacodynamic effects Quantitative difference in pharmacological effect *One isomer may be less active or inert*	Isoprenaline *β-agonist* *(−)-isomer 800 × more potent than* *(+)-isomer*
Qualitative difference in pharmacological effect *Opposite enantiomer may cause unwanted effects*	Ketamine *parenteral anesthetic* *S(+)-isomer has fewer post-* *operative side-effects*
Opposite enantiomer may cause different effects	(+)-Propoxyphene (2S,3R) *predominantly analgesic* (−)-Propoxyphene (2R,3S) *predominantly antitussive*
Enantiomer may have opposing actions	Picenadol *opioid analgesic* *(+)-isomer (3S,4R)* *μ-receptor agonist* *(−)-isomer (3R,4S)* *weak μ-receptor antagonist*
Pharmacokinetic differences Differential metabolism	Prilocaine *local anesthetic* *S(+)-isomer slowly hydrolyzed* *by amidase* *R(−)-isomer rapidly hydrolyzed to* *toluidine (cause of methemoglobinemia)*

Type of effect	Examples[1]
Differential protein binding	Disopyramide *anti-arrhythmic* S(+)-isomer more strongly protein bound
Drug interactions *Stereoselective inhibition of metabolism*	Fluvastatin *anti-lipidemic* (+)-isomer (3R,5S) more strongly inhibits enzyme CYP 2C9 responsible for metabolism of phenytoin and oral anticoagulants
Disease states	Ibuprofen *anti-inflammatory* Enantiomeric ratio changes in cirrhosis of the liver
Age	Hexobarbitone *sedative* (−)-isomer is cleared 2 × faster in young compared to elderly (+)-isomer shows no difference

Type of effect	Examples[1]
Gender	Methylphenobarbitone *sedative* R-isomer is cleared more rapidly in young males than young females or elderly of either sex
Genetic factors	Metoprolol *antihypertensive* poor metabolizers deficient in enzyme CYP 2D6 have lower proportion of the active S-isomer when racemate is administered

[1]These examples have been taken from the reviews cited in the text.

rials for subsequent derivatization (achiral or otherwise) or by using well-characterized enantioselective microbial reactions at strategic points in a chemical synthesis to introduce the desired chiral center. Where the option of synthesizing the required stereoisomer is not feasible, the enantiomers of the drug must be separated. The traditional method is fractional crystallization using a chiral counter-ion to produce diastereoisomeric salts. This technique is still commonly used, but finding a suitable counter-ion is mainly based on trial and error. Chiral high-performance liquid chromatography (HPLC) may provide a suitable preparative method for small amounts of enantiomers, but is less likely to be commercially viable. New developments such as simulated moving bed (SMB) chromatography provide exciting possibilities for the future production of large-scale quantities of pure enantiomers.

Alongside the improvement in procedures for obtaining enantiomerically pure compounds have come new methods of analysis which are necessary to demonstrate the effectiveness of manufacturing procedures and to control the quality of the active ingredient and corresponding medicinal product. The greatest area of evolution has been in chiral chromatography, particularly HPLC, but methods based on nuclear magnetic resonance (NMR) spectroscopy are also used and latterly capillary electrophoresis (CE) techniques have become available. Bioanalytical methods are also required for monitoring the fate of individual isomers after administration [9].

As the new synthetic, preparative and analytical methods arrived, they allowed growth of the body of evidence demonstrating the pharmacological, pharmacokinetic and clinical significance of drug chirality until the stage was reached where the issue needed to be addressed by both the pharmaceutical industry and the regulatory authorities responsible for licensing medicinal products. A consultative approach was adopted by the regulators. In 1992, a Drug Information Association (DIA) workshop on chirality was held with a concurrent independent discussion on regulatory requirements for chiral drugs [10]. Regulators and representatives from the pharmaceutical industry in the European Union (EU), Japan and the US debated chiral issues relating to quality, safety, and efficacy – the three principles which form the basis for approval of a medicinal product. The workshop recommended a pragmatic approach to the regulation of chiral drugs whereby the choice of the stereochemical form should be based on scientific data relating to quality, safety, efficacy and risk–benefit. The decision as to whether a racemate or enantiomer ought to be developed should reside with the applicant for the marketing authorization or sponsor of the product. The workshop recommended that the regulation of chiral drugs should be consistent with these principles.

Regulatory guidance on the development of chiral drugs was published subsequently, first in the US in 1992 and then shortly afterwards by the EU, and indeed has adopted the pragmatic approach advocated by the DIA workshop. A similar attitude prevails in Japan, although formal guidance has not been produced. The Canadian guidance is also based on the same principles [11]. There is a wider international effect of these and other regulatory policies through exchange schemes with other countries. The same dossier of information may be accepted to register a medicinal product elsewhere and in some cases an authorization is granted on the basis of the original assessment report. An example of the latter is the Pharmaceutical Evaluation Report (PER) scheme which was initiated by the EFTA countries but now has world-wide membership. The publication of guidance by the various regulatory authorities has itself in turn further stimulated the development of single enantiomers by the pharmaceutical industry and thus ensured the continued introduction of new enantioselective methods of synthesis and analysis.

The progress made in this area has been such that, in retrospect, it is difficult to see why the biological significance of chirality was not initially more widely appreciated and why there was so much controversy when the concept of regulating chiral drugs was first aired. Nowadays, it is widely accepted that enantiomers should be treated as separate compounds from the point of view of their pharmacological action, although there are still lessons to be learned with respect to drug safety [12]. The presentation of a drug as a racemate now requires full justification before a marketing authorization can be granted, which means that studies with individual enantiomers using chiral analytical procedures are a necessity during product development even if it is the racemate that is finally marketed. The issue of chirality has been reviewed with respect to classes of drugs such as nonsteroidal anti-inflammatory agents [13] and antimicrobials [14].

The remainder of this chapter outlines the regulatory requirements for investigation of chiral drugs in the three regions of Europe, the US and Japan. These countries provide the sponsoring bodies for the International Conference on Harmonization which aims to unify the process of drug registration globally (see Section 13.5). The requirements are discussed with particular emphasis on their pharmaceutical and chemical (quality) aspects: enantiopurity is a significant factor to be considered when seeking approval of a chiral drug.

13.2 Requirements in the European Union

13.2.1 Introduction

Common legislation governs the criteria for the approval of human medicines throughout the countries of the EU whether applications for marketing authorizations are made through the centralized or national routes. There are several volumes of the *Rules Governing Medicinal Products in the European Community* [15] which include the relevant Directives and also provide interpretation of the pharmaceutical legislation. The *Rules* are available on the Internet web-site for Directorate General III of the European Commission which is responsible for pharmaceuticals (http://dg3.eudra.org). Recommendations on the studies to be conducted in support of an application for a marketing authorization are made in *Guidelines on Quality, Safety and Efficacy of Medicinal Products for Human Use* contained in Volume III of the *Rules*. These notes for guidance begin life under the auspices of the various working parties of the Committee of Proprietary Medicinal Products (CPMP) and include guidelines which are the result of international harmonization (see Section

13.5). Such guidelines are not legally binding, but the applicant would be expected to provide a satisfactory (scientifically based) justification in cases where the recommendations had not been followed. European guidelines are available at their draft consultation stages and in their final form on the Eudranet web-site of the European Medicines Evaluation Agency (EMEA) at http://www.eudra.org/emea.html. The note for guidance of primary interest for drugs which may exist as optical isomers is called *Investigation of Chiral Active Substances* (CPMP/III/3501/91). Discussion on this guideline started in 1991, it was adopted in October 1993 and came into force in April 1994. Its requirements are discussed below together with relevant aspects of other notes for guidance concerning quality issues.

13.2.2   Note for Guidance on Investigation of Chiral Active Substances

The contents of this guideline are additional to other guidelines relating to the quality, safety and efficacy of medicinal products licensed in the EU. Manufacturers must decide whether to market a drug as a single enantiomer or a racemate. They should provide sufficient justification of their decision so that the licensing authority can assess the risk:benefit ratio. The key to a successful outcome for an application for a marketing authorization is proper justification of the decisions made concerning a product during development. The guideline sets out requirements for studies to justify the chosen strategy in the areas corresponding to the three technical parts of the dossier accompanying the application.

13.2.3   Chemistry and pharmacy aspects

Chemical aspects of the information required to support an application for a medicinal product containing a new drug were set out in the original guideline *Chemistry of Active Ingredients* (Eudra/Q/87/011) published in Volume III of the *Rules*. This guidance has since been updated in the draft note for guidance on the *Chemistry of new active substances* (CPMP/QWP/130/96). The guideline on chiral active substances supplements these texts.

13.2.3.1 Synthesis of the Active Substance

In addition to providing the usual information concerning the manufacturing procedure, the step where the chiral center is formed must be described in detail, and the measures taken to maintain the desired configuration during subsequent stages of synthesis must be shown. The ability of the process to provide adequate stereochemical control must be validated and thus analytical methods for determination of chiral compounds are of fundamental importance in the control and regulation of medicines containing such drugs. The synthetic product must be fully characterized with respect to identity, related substances and other impurities as for any other drug substance, but with the additional requirement of establishing stereochemical purity.

Several synthetic strategies are possible and demand that different types of information be provided. In cases where the starting material, whether a racemate or enantiomer, already contains the required chiral center, full characterization of that substance is required, including stereochemical purity and validation of chiral analytical procedures. Where a racemate or other intended enantiomeric mixture is required, evidence should be provided that these are the result unless obvious from the synthetic route employed. Where the preferred enantiomer is obtained by isolation, the resolution step is considered part of the overall manufacturing process and the usual details of the procedure should be given together with the number of cycles used. If a nonequimolar mixture of enantiomers is needed, then the manufacturing process must be validated to ensure consistent composition of the active ingredient.

There are circumstances in which it is not possible to obtain the required enantiomer at manufacturing scale either by synthesis or isolation, e.g. because of difficulties with scale-up or failure to obtain material in a suitable physical form for pharmaceutical manufacture. In such cases, all the experimental results available should be described and the reason for the failure given. Likewise, if enantiomeric material could not be obtained for preclinical and clinical studies (see below), this should also be discussed. Advances in preparative techniques should eventually make this scenario less common.

13.2.3.2 Quality of the Active Substance

The quality of a drug substance is controlled by its specification. An internationally harmonized guideline on specifications and tests for chemical substances as active ingredients and in drug products makes reference to chiral compounds. This has recently been finalized and is discussed in Section 13.5.2.

The guideline on chiral active substances states that particular attention should be paid to identity and stereochemical purity. It states that specifications for a racemate should include a test to show that the substance is indeed a racemate and this is a position supported by the requirements of the *European Pharmacopoeia* for drug substance monographs [16].

The chiral drugs guideline lists examples of methods that may be used for the control of drug substances, ranging from the simpler ones such as optical rotation, melting point, chiral HPLC to the more sophisticated techniques including optical rotatory dispersion, circular dichroism, or NMR with chiral shift reagents. It is not expected that this list would preclude the adoption of other methods or those that may be introduced in the future. It is the responsibility of applicants to decide on the techniques that are appropriate for the satisfactory control of each drug substance and to ensure that they are fully validated. The guidelines on analytical validation have been internationally harmonized and are discussed below (Section 13.5.4). Stereoisomeric reference substances may be required for test procedures for chiral drugs. The stereochemical purity of reference materials must be stated by giving a value for their assay determination. Care should be taken in the characterization of such materials when they are required to support the stereochemical identification of the drug substance. It is all too easy to fall into a circular argument when trying to establish the absolute configuration of a compound based on mechanistic arguments and/or the chirality of starting materials. Single crystal X-ray diffraction studies of the final drug substance with methods appropriate for the determination of absolute configuration provide the greatest confidence.

Stereoisomers may arise during synthesis of a drug substance, or they may arise as degradation products on storage. The guideline on chiral active substances states that when a chiral drug substance is presented as a single enantiomer, the unwanted enantiomer is considered to be an impurity. The internationally harmonized guideline on impurities (see Section 13.5.3) applies in principle to substances containing enantiomeric or diastereoisomeric impurities as it would to active ingredients containing any other organic impurities. However, the limits normally expected do not apply to chiral impurities. The limits for the control of enantiomers in drug substances are usually relaxed compared to tests using achiral methods because it is recognized that the chiral separation methods may not be able to achieve the same sensitivity.

The use to which development and commercial scale batches were put must also be detailed by the applicant so that each can be linked to a particular safety or clinical study. This information assists in the *qualification* of impurities which is the process by which the biological safety of an individual impurity, or an impurity profile, is established at a specified level. If the new drug substance containing a particular level of impurity has been adequately tested in safety and/or clinical studies then that level is considered to be qualified. Metabolism studies with chiral drugs should demonstrate whether or not chiral inversion occurs. Impurities that are also significant human metabolites do not need further qualification, as exposure to them would be automatic on administration of the drug in clinical trials. Together with the batch analysis data, the qualification studies should be used to justify the specification limits for individual known, unknown and total impurities. The applicant should demonstrate that unacceptable changes in stereochemical purity or enantiomeric ratio do not occur on storage of the active ingredient.

Chemical development: Proof of structure and configuration are required as part of the information on chemical development. The methods used at batch release should be validated to guarantee the identity and purity of the substance. It should be established whether a drug produced as a racemate is a true racemate or a conglomerate by investigating physical parameters such as melting point, solubility and crystal properties. The physicochemical properties of the drug substance should be characterized, e.g. crystallinity, polymorphism and rate of dissolution.

Finished product: The applicant should show that the manufacturing process produces no unacceptable changes in the stereochemical purity of the active ingredient and that such changes do not occur on storage for the proposed shelf-life.

13.2.4   Preclinical and Clinical Studies

13.2.4.1 Single Enantiomer

The development of a single enantiomer as a new active substance should be described in the same manner as for any other new chemical entity. Studies should be carried out with the single enantiomer, but if development began with the racemate then these studies may also be taken into account. Chiral conversion should be considered early on so that enantiospecific bioanalytical methods may be developed. These methods should be described in chemistry and pharmacy part of the dossier. If the opposite enantiomer is formed in vivo, then it should be evaluated in the same way as other metabolites. For endogenous human chiral compounds, enantiospecific analysis may not be necessary. The enantiomeric purity of the active ingredient used in preclinical and clinical studies should be stated.

13.2.4.2 Racemate

The applicant should provide justification for using the racemate. Where the interconversion of the enantiomers in vivo is more rapid than the distribution and elimination rates, then use of the racemate is justified. In cases where there is no such interconversion or it is slow, then differential pharmacological effects and fate of the enantiomers may be apparent. Use of the racemate may also be justified if any toxicity is associated with the pharmacological action and the therapeutic index is the same for both isomers. For preclinical assessment, pharmacodynamic, pharmacokinetic (using enantiospecific analytical methods) and appropriate toxicological studies of the individual enantiomers and the racemate will be needed. Clinical studies on human pharmacodynamics and tolerance, human pharmacokinetics and pharmacotherapeutics will be required for the racemate and for the enantiomers as appropriate.

13.2.4.3 New Single Enantiomer from Approved Racemate or New Racemate from Approved Single Enantiomer

These situations are treated as completely new applications which should include an explanation of the decision to develop the enantiomer or racemate. Data on the existing racemate or enantiomer may be included where appropriate with bridging studies as necessary.

13.2.4.4 Nonracemic Mixture from Approved Racemate or Single Enantiomer

This can be viewed as optimization of the pharmacotherapeutic profile and therefore is treated as a fixed combination product, for which a separate note for guidance applies. Data on stereoisomers not previously approved should be provided together with justification for the fixed combination.

13.2.4.5 Abridged Applications

Generic applications for chiral medicinal products should be supported by bioequivalence studies using enantiospecific bioanalytical methods unless both products contain the same, stable, single enantiomer or both products contain a racemate where both enantiomers show linear pharmacokinetics.

The guideline concludes with a note that there is no intention to require further data on established medicinal products which contain a racemic drug unless new evidence emerges concerning the safety or efficacy of one enantiomer. If new claims related to the chiral nature of the active substance are made, then supporting studies on the individual enantiomers will be required.

13.3   Requirements in the United States

13.3.1   Introduction

The Food and Drug Administration (FDA) is responsible for authorizing human medicinal products in the USA through its Center for Drug Evaluation and Research (CDER). Policy and guidance relating to drug registration for chemical substances is published in the Federal Register and is available on the FDA Internet web-site at http://www.fda.gov/cder/guidance/index.htm. The relevant guidance for chiral drugs is the policy statement for the development of new stereoisomeric drugs which is described below.

13.3.2   Policy Statement for the Development of New Stereoisomeric Drugs

The FDA has taken essentially the same view as the EU with respect to the development of chiral drugs but emphasizes different aspects in its guidance. The FDA's *Policy statement for the development of new stereoisomeric drugs* was first published in January 1992, with corrections made in January 1997. The statement was produced in response to the technological advances which permitted production of many single enantiomers on a commercial scale. The policy relates only to enantiomers and not to geometric isomers or to diastereoisomers which have chemically and pharmacologically distinct actions. Except in rare cases where biotransformation occurs, such compounds are treated as separate drugs, and mixtures are not developed unless fortuitously as a fixed-dose combination. The guideline acknowledges that the development of racemates may continue to be appropriate, but identifies two areas which should be considered in product development.

The first is the manufacture and control of a product to assure its stereoisomeric composition with respect to identity, strength, quality and purity. The quantitative composition of the material used in the pharmacological, toxicological and clinical studies conducted during development must be known.

The second point of consideration is the pharmacokinetic evaluation of a chiral drug. Results from such studies will be misleading if the disposition of the enantiomers is different, unless a chiral assay is used. Such an assay would have to be established for in vivo use early in the drug development process as results from initial pharmacokinetic measurements, including information on interconversion of enantiomer in vivo, will inform the decision as to whether the individual enantiomer or racemate should be developed. If the drug product is to contain a racemate and the pharmacokinetic profiles of the individual isomers are different, appropriate studies should be conducted to measure characteristics such as the dose linearity, the effects of altered metabolism and excretion and drug–drug interactions for the individual enantiomers. An achiral assay or monitoring of only one enantiomer is acceptable if the pharmacokinetics of the optical isomers is the same or in a fixed ratio in the target population. The in vivo measurement of individual enantiomers would be of assistance in assessing the results of toxicological studies, but if this is not possible then human pharmacokinetic studies would be sufficient.

The pharmacological activities of the isomers should be compared in vitro and in vivo in both animals and humans. Separate toxicological evaluation of the enantiomers would not usually be required when the profile of the racemate was relatively benign but unexpected effects – especially if unusual or near-effective doses in animals or near planned human exposure – would warrant further studies with the individual isomers.

The guideline notes that the FDA invites discussion with sponsors on whether to pursue development of the racemate or single enantiomer. This reflects the somewhat different regulatory approach in the US where there is greater interaction between the FDA and sponsor during the drug development process than occurs in Europe. All information obtained by the sponsor or available in published literature relating to the chemistry, pharmacology, toxicology or clinical actions of the stereoisomers should be included in the investigational new drug (IND) or new drug (NDA) submissions.

13.3.3 Chemistry, manufacturing and controls

The policy gives further recommendations on the information which should be provided on chemistry, manufacturing and controls (CMC) in addition to that found in other guidance (see Section 13.3.6).

13.3.3.1 Methods and Specifications

For drug substances and drug products, applications for enantiomers and racemates should include a stereochemically specific identity test and/or a stereochemically selective assay. The choice of control tests should be based on the method of manufacture and stability characteristics and, in the case of the finished product, its composition.

13.3.3.2 Stability

Methods of assessing the stereochemical integrity of enantiomeric drug substances and drug products should be included in the stability protocols for both, but stereoselective tests may not be required once it has been shown that racemization does not occur.

13.3.3.3 Impurity Limits

It is essential to determine the concentration of each isomer and define limits for all isomeric components, impurities, and contaminants of the compound tested preclinically that is intended for use in clinical trials. The maximum level of impurities in a stereoisomeric product used in clinical studies should not exceed that in the material evaluated in nonclinical toxicity studies. This point is expanded in the ICH impurities guideline (Section 13.5.3).

13.3.4 Pharmacology/Toxicology

The activity of the individual enantiomers should be characterized according to the principal and any other important pharmacological effects with respect to the usual parameters including potency, specificity and maximum effect. The pharmacokinetic profile of each isomer should be established in animals and later compared to the human pharmacokinetic profile found in Phase I studies which are conducted in healthy volunteers. It is normally sufficient to carry out toxicity studies on the racemate. If the drug causes toxic effects other than that predicted from its pharmacology at relatively low exposure in comparison with planned clinical trials, then the studies should be repeated with the individual isomers to ascertain whether a single enantiomer is responsible for the effect. If this is the case, then it would be desirable to eliminate the toxicity by developing the appropriate single enantiomer with only the desired effect.

13.3.4.1 Developing a Single Enantiomer after a Racemate is Studied

Once a mixture of stereoisomers has been investigated nonclinically, an abbreviated evaluation of pharmacology and toxicity could be conducted to allow the existing knowledge of the racemate available to the applicant to be applied to the pure enantiomer. No further studies would be needed if the single enantiomer and the racemate had the same toxicological profile. However, if the single enantiomer should appear more toxic, then further investigations would be required to produce an explanation and the implications for dosing in humans would have to be considered.

13.3.5 Clinical and Biopharmaceutical Studies

Where the individual enantiomers and the racemate show little difference in activity and pharmacokinetics, the development of the racemate is justifiable. In other instances, the development of the single enantiomer is especially desirable, for example where one isomer is toxic and the other is not. Cases where unexpected toxicity or pharmacological effects occur at clinical doses of the racemate should be further investigated with respect to the properties of the individual enantiomers and their active metabolites. Such investigations might take place in animals, but human studies may be essential. The unexpected effects may not relate to the parent enantiomer but may be associated with an isomer-specific metabolite. Generally, it is not as important to consider developing only one enantiomer if the opposite isomer is pharmacologically inert. Clinical evaluation of both enantiomers and potential development of a single enantiomer is more important when both enantiomers are pharmacologically active but differ significantly in their potency, specificity or maximum effect. If both enantiomers carry desirable but different properties, then development of a mixture of the two – not necessarily as a racemate – as a fixed combination might be reasonable.

Where the drug studied is a racemate, the pharmacokinetics, including potential interconversion, of the individual enantiomers should be investigated in Phase I clinical studies. Phase I or II data in the target population should indicate whether an achiral assay, or monitoring of only one optical isomer where a fixed ratio is confirmed, will be adequate for pharmacokinetic evaluation. If the racemate has already been marketed and the sponsor wishes to develop the single enantiomer, additional studies should include determination of any conversion to the other isomer and whether there is any difference in pharmacokinetics between the single enantiomer administered alone or as part of the racemate.

13.3.6 Other Relevant FDA Guidance

The FDA's *Guideline for submitting supporting documentation in drug applications for the manufacture of drug substances* makes some specific references to chiral drug substances. The requirements are similar to those in the EU. Elucidation of the structure of a chiral drug molecule should include determination of its configuration. This analytical information should be supplemented by an knowledge of the synthesis and the way in which the chiral center is produced (e.g. from a starting material, stereospecific reaction or by resolution of intermediates). Optically active starting materials may require additional testing compared to those without asymmetric centers if their chirality is of significance in the manufacture of the drug substance, e.g. they are used in a resolution step. The policy notes that enantiomers may be considered as impurities (even in racemates) and as such require proper control during manufacture and in the final drug substance. The guidance specifically addresses the issue of key intermediates, those compounds in which the essential molecular characteristics necessary for the desired pharmacological activity are first introduced into the structure. Key intermediates will often be those where a chiral center of the correct stereochemistry is introduced, and as such they should be subjected to quantitative tests to limit the content of undesired isomers. The control of drug substances is discussed in the policy, but specifications and tests are now addressed by the internationally harmonized guideline discussed in Section 13.5.2. The need for stereochemical characterization of reference materials used during analytical procedures is noted.

The FDA's *Reviewer guidance on the validation of chromatographic methods* issued in November 1994 also refers to chiral methods. The guidance incorporates the ICH analytical terms (see Section 13.5.4). It is noted that separation of enantiomers can be achieved by HPLC with chiral stationary phases or with achiral stationary phases by formation of diastereoisomers using derivatizing agents or by the use of mobile phase additives. When the chromatographic method is used in an impurity test, the sensitivity is enhanced if the enantiomeric impurity elutes before the enantiomeric drug (to avoid the tail of the main peak).

13.4 Requirements in Japan

The Japanese regulatory authority is the Ministry of Health and Welfare (MHW) and the Pharmaceutical and Medical Safety Bureau (PSMB) is responsible for the promulgation of national and international guidelines in the form of Notifications. Guidelines are available on the Internet web-site of the National Institute of Health and Science (http://www.nihs.go.jp). The MHW has not issued specific guidance on the development of chiral drugs, but has nonetheless responded to the "enantiomer-versus-racemate" scientific debate. The attitude of the MHW and its advisory body, the Central Pharmaceutical Affairs Council (CPAC) is discussed in two articles by Shindo and Caldwell published in 1991 and 1995 [17, 18]. The latter paper analyzes the results of a survey of the Japanese pharmaceutical industry which sought responses on chirality issues.

Shindo and Caldwell reported that specific reference was made to chiral drugs in only two places in the Japanese Requirements for Drug Manufacturing Approval.

The first was an amendment to the original document on *Points to consider when preparing data* published in 1989 in the section on 'Data concerning physicochemical properties and standards and test methods'. It stated that, for mixtures of optical isomers, it was recommended that chromatographic tests were performed in addition to optical rotatory tests and indicated the MHW's response to the growing development of chiral stationary phases in HPLC. The second reference was added in 1985 and appeared in the section on 'Test data concerning absorption, distribution, metabolism and excretion'. It stated that when the drug concerned was a racemate, investigation of the absorption, distribution, metabolism and excretion (ADME) of each optical isomer was recommended.

Further references to chiral drugs have been found in the Japanese guidelines on establishing specifications for new active substances issued in 1994. These indicate firstly that consideration should be given to the solvent used in a test for optical rotation and its effect on the result explained, and secondly that where the active ingredient is an optical isomer, a method of discriminating between enantiomers should be investigated and the ratio of enantiomers determined. The ICH guideline on specifications and tests (Section 13.5.2) now applies in Japan.

While the official requirements offer limited guidance on investigation of chiral drugs, there is considerable correspondence on individual cases either with individual applicants or pharmaceutical industry associations. Shindo and Caldwell [17] report that in 1986 this led to a distinction by the MHW between racemates, where the ADME patterns of each enantiomer should be investigated together with the possibility of interconversion in vivo, and mixtures of diastereoisomers, where the kinetics and contribution of each isomer to the efficacy of the drug should be established. The CPAC also publishes an annual review of answers to industry questions. These comments allow an insight into the drug approval process in Japan and offer interpretations of the official guidelines. Shindo and Caldwell [18] provide examples of some responses from the CPAC which show that the approval of a racemate is not precluded, but that the selection of the optical form for marketing should be based on a consideration of the efficacy and toxicity of each isomer. Investigations should be performed on enantiomeric composition, pharmacological effect, metabolism, toxicity, in vivo interconversion, etc. for the racemate and different isomers. In this respect, the Japanese authorities operate on the same principles as other regulatory authorities. However, it is also true that the requirements for drug registration in Japan are more stringent than elsewhere, for example, in the battery of preclinical tests required preapproval.

Thus, although there is a lack of formal guidance in Japan, it is apparent that there is a considerable degree of concordance with the regulatory principles established elsewhere. Approval is not proscriptive, is based on the data for individual cases and the applicant is required to justify their reasons for developing a racemate if that is the case.

13.5 Guidelines from the International Conference on Harmonization

13.5.1 Introduction

The International Conference on Harmonization of Technical Requirements for the Registration of Pharmaceuticals for Human Use (ICH) is a tripartite body sponsored by regulatory authorities and research-based industry representatives from the three regions discussed above: the European Union, Japan and the United States. In addition, the ICH Steering Committee includes observers from the World Health Organization (WHO), the Canadian Drugs Directorate and the European Free Trade Association (EFTA). It is the aim of ICH to promote international harmonization of regulatory requirements. Such harmonization avoids the duplication of the development work required for registering new medicinal products and is of importance to the pharmaceutical industry which is becoming increasingly globalized.

The Steering Committee is responsible for identifying topics for which harmonized guidelines are then developed. These guidelines can be obtained from the ICH Internet web-site which is maintained by the International Federation of Pharmaceutical Manufacturers Associations (IFPMA) who provide the secretariat support for ICH (http://www.ifpma.org). ICH also holds biennial conferences and workshops. There are five stages in the ICH process for developing a guideline, represented in Fig. 13-1, which starts with consideration of the topic and development of a consensus by the relevant Expert Working Group (EWG). The EWG members are nominated from the regulatory and industrial bodies in the three regions. The draft consensus resulting from the EWG is then released by the ICH Steering Committee for wider consultation in the three sponsor regions. Comments from other geographical areas are received through IFPMA and WHO contacts. The comments received are consolidated and the final guideline is issued for adoption and implementation in the three regions.

Some internationally harmonized guidelines regarding specifications and tests, impurities and validation of analytical methods have particular relevance to the development of chiral drugs and are discussed below. In addition, the impact of work on the common technical document is considered.

13.5.2 Specifications and Tests

ICH Topic Q6A *Specifications: Test Procedures and Acceptance Criteria for new drug substances and new drug products: Chemical Substances* reached Step 4 in October 1999 and was approved by the CPMP in November 1999 with a date of May 2000 (Step 5) for coming into operation in the EU [CPMP/ICH/367/96]. It provides guidance on the selection of test procedures and the setting and justification of acceptance criteria for new drug substances of synthetic chemical origin, and drug products produced from them, that have not been previously registered in the EU,

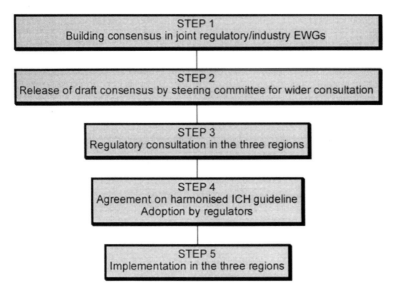

Fig. 13-1. International Conference on Harmonization (ICH) process for developing harmonized guidelines.

Japan or US. Detailed recommendations are made regarding the specifications for active ingredients and different types of dosage forms, and reference is made to chiral drugs. Thus, this ICH guideline may supersede, or at least provide additional guidance to, the recommendations in the regional guidelines described above. It should be noted that this ICH guideline does not apply to drugs of natural origin.

Guidance on specifications is divided into universal tests/criteria which are considered generally applicable to all new substances/products and specific tests/criteria which may need to be addressed on a case-by-case basis when they have an impact on the quality for batch control. Tests are expected to follow the ICH guideline on analytical validation (Section 13.5.4). Identification of the drug substance is included in the universal category, and such a test must be able discriminate between compounds of closely related structure which are likely to be present. It is acknowledged here that optically active substances may need specific identification testing or performance of a chiral assay in addition to this requirement.

Tests for chiral drug substances are included in the category of specific tests/criteria. A decision tree (Fig. 13-2) summarizes when and if chiral identity tests, impurity tests and assays may be needed in drug substance and finished product specifications. For a drug substance, an identity test should be capable of distinguishing between the enantiomers and the racemate for a drug substance developed as a single enantiomer. A chiral assay or enantiomeric impurity procedure may serve to provide a chiral identity test. When the active ingredient is a racemate, a stereospecific test is appropriate where there is a significant possibility that substitution of an enantiomer for a racemate may occur or when preferential crystallization may lead to unintentional production of a nonracemic mixture. Such a test is generally not

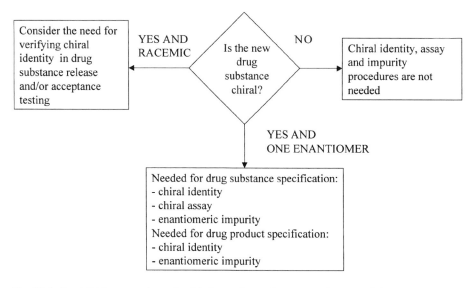

Fig. 13-2. Establishing procedures for chiral new drug substances and new medicinal products containing new chiral drug substances.

needed in the finished product specification if there is insignificant racemization during manufacture of the dosage form or on storage and a test is included in the drug substance specification. If the opposite enantiomer is formed on storage, then a chiral assay or enantiomeric impurity testing will serve to identify the substance as well.

With respect to impurities, it is acknowledged that, where the substance is predominantly one enantiomer, the opposite isomer is excluded from the qualification and identification thresholds given in the ICH guideline on impurities (Section 13.5.3) because of practical difficulties in quantification at the recommended levels. Otherwise, it is expected that the principles of that guidance apply. The guideline allows that appropriate testing of a starting material or intermediate, with suitable justification from studies conducted during development, could give assurance of control. This approach may necessary, for example, when there are multiple chiral centers present in the drug molecule. Control of the other enantiomer in the finished product is needed unless racemization during manufacture of the dosage form or on storage is insignificant. The procedure used may be the same as the assay, or it may be separate.

Determination of the drug substance is expected to be enantioselective, and this may be achieved by including a chiral assay in the specification or an achiral assay together with appropriate methods of controlling the enantiomeric impurity. For a drug product where racemization does not occur during manufacture or storage, an achiral assay may suffice. If racemization does happen, then a chiral assay should be used or an achiral method combined with a validated procedure to control the presence of the other enantiomer.

13.5.3 Impurities

There are two ICH guidelines on impurities: Topics Q3A makes recommendations on *Impurities in new drug substances* and Topic Q3B on *Impurities in new medicinal products*. In Europe, the guidelines are published respectively as CPMP/ICH/142/95 and its Annex CPMP/ICH/282/95. Revisions to these to guidelines are at Step 2 of the ICH process (regulatory consultation in the three regions) and affect the way rounding of analytical results is related to the limits for impurities denoted in the texts. As previously stated, enantiomeric impurities are excluded from the guideline, but the principles expressed are expected to apply. There are two aspects of control of impurities: firstly, their chemical classification and identification; and secondly, assessment of their safety at the level imposed by the drug substance specification. The latter is the process of qualification already mentioned in Section 13.2.3.2, Quality of the active substance.

The guidance on impurities in new drug substances state that the sources of actual and potential impurities, whether arising from synthesis, purification or degradation, should be discussed. Analytical data are required that show the level of individual and total impurities in development and commercial scale batches. The impurity profiles, e.g. chromatograms, must be available if requested. Samples should be intentionally degraded so that potential impurities arising from storage can be identified. Such studies would reveal whether racemization of single enantiomers was likely to occur. In normal application of the guideline, identification of organic impurities is required above certain specified thresholds, usually by isolation and spectroscopic characterization, or if this has not been possible, the unsuccessful laboratory studies described. Below these thresholds, identification is not required but it is useful to present this data if available and identification should be attempted in any case for compounds expected to be unusually potent or toxic. Chiral analysis would enable the identification of enantiomeric impurities.

The guideline gives thresholds depending on the maximum daily dose of the drug above which qualification studies are required. Lower or higher qualification thresholds may be appropriate for certain classes of drugs. Where the qualification threshold is exceeded, additional safety studies may be required according to a decision tree provided in the guideline. Similar qualification of enantiomeric impurities by their presence in batches of drug substance used in safety and/or clinical studies would be expected, although they are not strictly covered by the guideline. The revisions to the guideline currently under consultation include the designation of reporting thresholds for impurities.

The guideline on impurities in new medicinal products parallels the drug substance text, but the designated thresholds concern only degradation products. The thresholds should be applied to the product at the end of its shelf-life, as that is when the greatest level of degradation is expected to have occurred.

13.5.4 Analytical Validation

There are two ICH guidelines on analytical validation. The first provides a glossary of terms and the second addresses methodology. The first guideline, ICH Topic Q2A *Validation of analytical procedures: Definitions and terminology*, reached Step 4 in October 1994. It sought only to present a collection of terms and definitions and not to provide direction on how to accomplish validation. The guideline was intended to bridge the differences which could exist between the various compendia and regulators in the three regions of the ICH. In the EU, the guideline was approved by the CPMP in November 1994 (CPMP/III/5626/93) and came into operation in June 1995. As mentioned above, the FDA incorporated the ICH definitions of analytical terms into its guidance on validation of chromatographic methods in November 1994.

The guideline states that the objective of validation is to demonstrate that an analytical method is fit for its purpose and summarizes the characteristics required of tests for identification, control of impurities and assay procedures (Table 13-2). As such, it applies to chiral drug substances as to any other active ingredients. Requirements for other analytical procedures may be added in due course.

Table 13-2. Characteristics of analytical procedures requiring validation (indicated by a tick).

	Identity	Control of impurities		Assay
		Quantification	Limit test	
Accuracy and precision		✓		✓
Specificity	✓	✓	✓	✓
Limit of detection		(✓)	✓	
Limit of quantitation		✓		
Linearity and range		✓		✓

Assays may be applied to the active moiety in the drug substance or drug product or to other selected components of the product. They are used for content/potency determinations and for measurement of dissolution. Precision includes repeatability (intra-assay precision) and intermediate precision (within laboratory) except the latter is not required where reproducibility (inter-laboratory) has been performed. If there is lack of specificity in one analytical procedure, compensation by other supporting methods is allowed. The characteristics listed in Table 13-2 are considered typical, but allowance is made for dealing with exceptions on a case-by-case basis. Robustness is not listed, but should be considered at an appropriate stage in development. Revalidation of analytical procedures is required following changes in the synthesis of a drug substance, composition of the finished product or in the analytical procedure.

The second guideline, ICH Topic Q2B, *Validation of analytical procedures: Methodology*, reached Step 4 in November 1996, was approved by the CPMP in Europe in December 1996 (CPMP/ICH/281/95) and came into operation in June 1997. It is complementary to the first guideline and provides some guidance and recommendations on acceptable methods for validating the characteristics of an analytical procedure. An indication of the data which should be provided in an application for a marketing authorization is given. It discusses the following characteristics separately: specificity; linearity; range; accuracy; precision; detection limit; quantitation limit; robustness and system suitability testing.

13.5.5   Common Technical Document

This ICH Topic (M4) is mentioned here as an indication of a future trend in the global authorization of new medicinal products. It is probably the ICH's most ambitious undertaking, aiming as it does to provide the basis for a single set of registration documents to support an application for marketing authorization in any of the three ICH regions. The EWG for the Common Technical Document (CTD) has been extended to include the observers to ICH and representatives of the generics industry and manufacturers of products for self-medication. There is also liaison with the ICH topic (M2) on electronic submission of documents supporting a drug registration. The magnitude of the task is such that the progress to date on reaching a consensus on the table of contents for the CTD has only been achieved after considerable debate. There are differences in regulatory practice in the three regions, in particular in the way that authorities interact with the drug development process, which will provide considerable hurdles to be overcome by the harmonization process. The CTD has been broken down into modules and the tables of contents (TOC) for the quality, safety and efficacy sections have been released for Step 2 consultation. The TOC give cross-references to appropriate harmonized guidelines. The CTD will provide the focus of the Fifth ICH Conference in November 2000.

13.6   The Effect of Regulatory Guidelines

The regulatory guidelines and attitudes of the regulatory authorities described above have affected the number of submissions for authorization of medicinal products containing drugs which are single enantiomers. Table 13-3 summarizes the data available from various surveys [17–20] using the categories first established by Ariëns in 1984 [19]. The figures for the Medicines Control Agency (MCA) result from an informal analysis of cases assessed between July 1996 and June 1999. Table 13-3 shows that the proportion of synthetic chiral drugs developed as single enantiomers appears to have genuinely risen between 1982 and 1999, even though the figures have been obtained from different surveys. This increase reflects both the regulatory requirements introduced in the early 1990s and the availability of the necessary scientific techniques to synthesize and control the enantiopurity of chiral drugs. The proportion of synthetic drugs which are presented in a nonchiral form is somewhat variable over the period of time represented by the figures in Table 13-1. In the surveys conducted between 1982 and 1985, the proportion was about 60 %, and between 1986 and 1991 it was about 25 %, whereas from 1992 to 1999 it appeared to increase from 30 % to 40 %, thus making any conclusions about trends rather difficult.

Table 13-3. New chemical entities (NCEs)[1] categorized according to their origin and chirality.

	Drugs in use 1982 [19]	NCEs approved 1983–85 [19]	NCEs approved in Japan 1986–89 [17]	Drugs in use 1991 [20]	NCEs approved in Japan 1992–93 [18]	NCEs assessed by MCA 1996–99
Natural/Semi-synthetic	**475**	**39**	**53**	**147**	**–**	**17**
Racemate	8	0	5	8	–	1
Single enantiomer	461	39	47	119	–	14
Achiral	6	0	1	8	–	2
Synthetic	**1200**	**91**	**47**	**521**	**47**	**61**
Racemate	422 (88 %)	36 (95 %)	29 (80 %)	140 (56 %)	22 (67 %)	13 (35 %)
Single enantiomer	58 (12 %)	2 (5 %)	7 (20 %)	110 (44 %)	11 (33 %)	24 (65 %)
Achiral	720	53	11	140	14	24
Total	**1675**	**130**	**100**	**668**		**78**

[1] In the US, NCEs are referred to as new molecular entities (NMEs).

The area of "racemic switches" where a single enantiomer is developed subsequently to a corresponding racemate which is already on the market has attracted much interest [7, 8]. A description of the preclinical and clinical development of dexketoprofen provides a detailed example of one of these racemic switches [21]. The regulations in Europe and the US both allow for the development of a single enantiomer from a racemate by the use of bridging studies between the old and new applications. One problem to be considered is how a company which was not responsible for the original development can provide equivalent data.

Apart from any intrinsically beneficial effects to patients from the administration of pure enantiomers, it has been speculated that such switches may provide a mechanism for extending the patentable period of a new drug. While this may have been attempted, the theoretical advantages of single enantiomers have not always been realized in practice. Examples of drugs which were first marketed as racemates where the single enantiomer is now available are dexfenfluramine, levofloxacin, levobupivacaine, dexketoprofen and dexibuprofen (in a limited number of countries). Others are either in development or in the process of registration. Inspection of the indications and precautions granted for such compounds reveals that the claimed advantages of the single enantiomer are not necessarily borne out by the clinical studies. Some of the pitfalls in developing chiral drugs from the clinical point of view have been outlined previously [12]. One problem is interconversion of stereoisomers which can offset any differences in pharmacological effect. For example, the inactive form of ibuprofen, the $R(-)$-isomer, is incompletely converted to the active $S(+)$-isomer; thus the reduction in dose of the single enantiomer required to achieve the same clinical effect is not half that of the racemate [13]. Interconversion is also the reason why it is unlikely that development of the single $R(+)$-enantiomer of thalidomide would prevent the well-known teratogenic side-effects of this drug which are probably associated with the $S(-)$-enantiomer. The R-isomer is converted following administration to the S-isomer in man, but not in some other species. This example illustrates the need for studies on a case-by-case basis to establish the patterns of activity and pharmacokinetics for each pair of enantiomers and for careful assessment of the future development of either the racemate or pure enantiomer.

13.7 Concluding Remarks

Regulation of chiral drugs is now well established, and has had the effect of producing a higher ratio of single enantiomer to racemic compounds for new synthetic drugs on the market. New analytical and preparative techniques will make it easier in the future to develop single enantiomers. There is no real evidence that the number of achiral drugs is increasing to avoid the problems associated with enantiopurity. Therefore it seems safe to assume that the technical methods for controlling chiral drugs have developed to a stage where many of the challenges presented to the analyst can be solved.

There will be a continued need for enantiospecific methods of preparation and analysis, not only to ensure the quality of the final drug substance and reference materials, but also to control starting materials used for their manufacture, and key intermediates during synthesis. Likewise, specific and sensitive bioanalytical methods will be required to follow the fate of individual enantiomers after their administration.

References

[1] A. R. Cushny, *Biological relations of optically isomeric substances,* Balliere, Tindall and Cox, London, 1926.
[2] P. Jenner, B. Testa, Influence of stereochemical factors on drug disposition, *Drug Metab. Rev.* 1973, 2, 117–184.
[3] E. J. Ariëns, Stereochemistry, a basis for sophisticated non-sense in pharmacokinetics and clinical pharmacology, *Eur. J. Clin. Pharmacol.* 1984, 26, 663–668.
[4] E. J. Ariëns, Stereochemistry: a source of problems in medicinal chemistry, *Med. Res. Rev.* 1986, 6, 451–466.
[5] A. R. Fassihi, Racemates and enantiomers in drug development, *Int. J. Pharmaceutics* 1993, 92, 1–14.
[6] R. R. Shah, Clinical pharmacokinetics: current requirements and future perspectives from a regulatory point of view, *Xenobiotica*, 1993, 23, 1159–1193.

[7] A. J. Hutt, S. C. Tan, Drug chirality and its clinical significance, *Drugs,* 1996, 52 (Suppl 5), 1–12.

[8] D. J. Triggle, *Drug Discovery Today,* 1997, 2, 138–147.

[9] J. Caldwell, Importance of stereospecific bioanalytical monitoring in drug development, *J. Chromatogr. A* 1996, 719, 3–13.

[10] M. Gross, A. Cartwright, B. Campbell, R. Bolton, K. Holmes, K. Kirkland, T. Salmonson, J.-L. Robert. Regulatory Requirements for chiral drugs, *Drug Information J.* 1993, 27, 453–457.

[11] A. G. Rauws, K. Groen, Current regulatory (draft) guidance on chiral medicinal products: Canada, EEC, Japan, United States, *Chirality,* 1994, 6, 72–75.

[12] R. R. Shah, J. M. Midgley, S. K. Branch, Stereochemical origin of some clinically significant drug safety concerns: lessons for future drug development, *Adverse Drug React. Toxicol. Rev.* 1998, 17, 145–190.

[13] P. J. Hayball, Chirality and nonsteroidal anti-inflammatory drugs, *Drugs,* 1996, 52 (Suppl. 5), 47–58.

[14] A. J. Hutt, J. O'Grady, Drug chirality: a consideration of the significance of the stereochemistry of antimicrobial drugs, *J. Antimicrobial Chemother.,* 1996, 37, 7–32.

[15] Eudralex: *The rules governing medicinal products in the European Union,* Vols. 1–9, Office for Official Publications of the European Communities, Luxembourg, 1998.

[16] *European Pharmacopoeia technical guide for elaboration of monographs 2nd* ed. *Pharmeuropa,* Special Issue, November 1996.

[17] H. Shindo, J. Caldwell, Regulatory aspects of the development of chiral drugs in Japan: a status report, *Chirality,* 1991, 3, 91–93.

[18] H. Shindo, J. Caldwell, Development of chiral drugs in Japan: an update on regulatory and industrial opinion, *Chirality,* 1995, 7, 349–352.

[19] E. J. Ariëns, E. W. Wuis, E. J. Verings, Stereoselectivity of bioactive xenobiotics, *Biochem. Pharmacol.* 1988, 37, 9–15.

[20] J. S. Millership, A. Fitzpatrick, Commonly used chiral drugs: a survey, *Chirality,* 1993, 5, 573–576.

[21] D. Mauleon, R. Artigas, M. L. Garcia, G. Carganico, Pre-clinical and clinical development of dexketoprofen, *Drugs,* 1996, 52 (Suppl. 5) 24–46.

Index